Advances in Molecular Modeling in Chemistry

Advances in Molecular Modeling in Chemistry

Heng Zhang
Shiling Yuan

Basel • Beijing • Wuhan • Barcelona • Belgrade • Novi Sad • Cluj • Manchester

Editors

Heng Zhang
School of chemistry and
chemical engineering
Shandong University
Jinan
China

Shiling Yuan
School of chemistry and
chemical engineering
Shandong University
Jinan
China

Editorial Office
MDPI AG
Grosspeteranlage 5
4052 Basel, Switzerland

This is a reprint of articles from the Special Issue published online in the open access journal *Molecules* (ISSN 1420-3049) (available at: www.mdpi.com/journal/molecules/special_issues/W83C91Y936).

For citation purposes, cite each article independently as indicated on the article page online and as indicated below:

Lastname, A.A.; Lastname, B.B. Article Title. *Journal Name* **Year**, *Volume Number*, Page Range.

ISBN 978-3-7258-2478-6 (Hbk)
ISBN 978-3-7258-2477-9 (PDF)
doi.org/10.3390/books978-3-7258-2477-9

© 2024 by the authors. Articles in this book are Open Access and distributed under the Creative Commons Attribution (CC BY) license. The book as a whole is distributed by MDPI under the terms and conditions of the Creative Commons Attribution-NonCommercial-NoDerivs (CC BY-NC-ND) license.

Contents

Preface . . . . **vii**

He Zhou, Heng Zhang and Shiling Yuan
Comparison of H_2O Adsorption and Dissociation Behaviors on Rutile (110) and Anatase (101) Surfaces Based on ReaxFF Molecular Dynamics Simulation
Reprinted from: *Molecules* **2023**, *28*, 6823, doi:10.3390/molecules28196823 . . . . **1**

Tong Meng, Zhiguo Wang, Hao Zhang, Zhen Zhao, Wanlin Huang and Liucheng Xu et al.
In Silico Investigations on the Synergistic Binding Mechanism of Functional Compounds with Beta-Lactoglobulin
Reprinted from: *Molecules* **2024**, *29*, 956, doi:10.3390/molecules29050956 . . . . **15**

Dian Song, Jie Li, Kun Liu, Junnan Guo, Hui Li and Artem Okulov
Size- and Voltage-Dependent Electron Transport of C_2N-Rings-Based Molecular Chains
Reprinted from: *Molecules* **2023**, *28*, 7994, doi:10.3390/molecules28247994 . . . . **31**

Xiaowei Xu, Hao Li, Andleeb Mehmood, Kebin Chi, Dejun Shi and Zhuozheng Wang et al.
Mechanistic Studies on Aluminum-Catalyzed Ring-Opening Alternating Copolymerization of Maleic Anhydride with Epoxides: Ligand Effects and Quantitative Structure-Activity Relationship Model
Reprinted from: *Molecules* **2023**, *28*, 7279, doi:10.3390/molecules28217279 . . . . **48**

Jie Li, Yuchen Zhou, Kun Liu, Yifan Wang, Hui Li and Artem Okulov
Tunable Electronic Transport of New-Type 2D Iodine Materials Affected by the Doping of Metal Elements
Reprinted from: *Molecules* **2023**, *28*, 7159, doi:10.3390/molecules28207159 . . . . **60**

Junnan Guo, Xinyue Dai, Lishu Zhang and Hui Li
Electron Transport Properties of Graphene/WS_2 Van Der Waals Heterojunctions
Reprinted from: *Molecules* **2023**, *28*, 6866, doi:10.3390/molecules28196866 . . . . **74**

Silvana Ceauranu, Alecu Ciorsac, Vasile Ostafe and Adriana Isvoran
Evaluation of the Toxicity Potential of the Metabolites of Di-Isononyl Phthalate and of Their Interactions with Members of Family 1 of Sulfotransferases—A Computational Study
Reprinted from: *Molecules* **2023**, *28*, 6748, doi:10.3390/molecules28186748 . . . . **86**

Chengfeng Zhang, Lulu Cao, Yongkang Jiang, Zhiyao Huang, Guokui Liu and Yaoyao Wei et al.
Molecular Dynamics Simulations on the Adsorbed Monolayers of N-Dodecyl Betaine at the Air–Water Interface
Reprinted from: *Molecules* **2023**, *28*, 5580, doi:10.3390/molecules28145580 . . . . **104**

Binbin Jiang, Huan Hou, Qian Liu, Hongyuan Wang, Yang Li and Boyu Yang et al.
Detachment of Dodecane from Silica Surfaces with Variable Surface Chemistry Studied Using Molecular Dynamics Simulation
Reprinted from: *Molecules* **2023**, *28*, 4765, doi:10.3390/molecules28124765 . . . . **116**

Linjie Wang, Pengtu Zhang, Yali Geng, Zaisheng Zhu and Shiling Yuan
Harmonic Vibrational Frequency Simulation of Pharmaceutical Molecules via a Novel Multi-Molecular Fragment Interception Method
Reprinted from: *Molecules* **2023**, *28*, 4638, doi:10.3390/molecules28124638 . . . . **130**

Yuanping Zhang, Boyu Ma, Xinlei Jia and Conghua Hou
Prediction of Ethanol-Mediated Growth Morphology of Ammonium Dinitramide/Pyrazine-1,4-Dioxide Cocrystal at Different Temperatures
Reprinted from: *Molecules* **2023**, *28*, 4534, doi:10.3390/molecules28114534 . . . . **148**

Yang Zhao, Liping Fang, Pei Guo, Ying Fang and Jianhua Wu
A MD Simulation Prediction for Regulation of *N*-Terminal Modification on Binding of CD47 to CD172a in a Force-Dependent Manner
Reprinted from: *Molecules* **2023**, *28*, 4224, doi:10.3390/molecules28104224 . . . . **162**

Jin Peng, Xiaoju Song, Xin Li, Yongkang Jiang, Guokui Liu and Yaoyao Wei et al.
Molecular Dynamics Study on the Aggregation Behavior of Triton X Micelles with Different PEO Chain Lengths in Aqueous Solution
Reprinted from: *Molecules* **2023**, *28*, 3557, doi:10.3390/molecules28083557 . . . . **176**

Preface

Molecular modeling is playing a crucial role in chemistry investigations. With the current developments in computing power, it is possible to achieve large-scale simulations. Molecular modeling has been applied successfully in many areas of chemistry, e.g., the behavior of liquid solutions, proteins, DNA, polysaccharides, lipid membranes, crystals, amorphous solids, or any combination of them; the process of adsorption or desorption at interfaces; protein folding; self-assembly; etc.

Aside from the widely spread application of molecular modeling, the techniques of simulation also developed rapidly. Many simulation techniques have emerged, including ab initio molecular dynamics, polarizable force field, reactive molecular dynamics, machine learning accelerated simulation, metadynamics, etc.

This reprint includes 13 original papers reporting on molecular simulation works, including quantum chemistry calculation, molecular dynamic simulation, etc., or combined experiments and simulation studies. We hope that this collection of research studies can bring inspiration to readers regarding the application of molecular modeling methods.

Heng Zhang and Shiling Yuan
Editors

Article

Comparison of H₂O Adsorption and Dissociation Behaviors on Rutile (110) and Anatase (101) Surfaces Based on ReaxFF Molecular Dynamics Simulation

He Zhou, Heng Zhang and Shiling Yuan *

Key Lab of Colloid and Interface Chemistry, Shandong University, Jinan 250100, China
* Correspondence: shilingyuan@sdu.edu.cn

Abstract: The relationship between structure and reactivity plays a dominant role in water dissociation on the various TiO_2 crystallines. To observe the adsorption and dissociation behavior of H_2O, the reaction force field (ReaxFF) is used to investigate the dynamic behavior of H_2O on rutile (110) and anatase (101) surfaces in an aqueous environment. Simulation results show that there is a direct proton transfer between the adsorbed H_2O (H_2O_{ad}) and the bridging oxygen (O_{br}) on the rutile (110) surface. Compared with that on the rutile (110) surface, an indirect proton transfer occurs on the anatase (101) surface along the H-bond network from the second layer of water. This different mechanism of water dissociation is determined by the distance between the 5-fold coordinated Ti (Ti_{5c}) and O_{br} of the rutile and anatase TiO_2 surfaces, resulting in the direct or indirect proton transfer. Additionally, the hydrogen bond (H-bond) network plays a crucial role in the adsorption and dissociation of H_2O on the TiO_2 surface. To describe interfacial water structures between TiO_2 and bulk water, the double-layer model is proposed. The first layer is the dissociated H_2O on the rutile (110) and anatase (101) surfaces. The second layer forms an ordered water structure adsorbed to the surface O_{br} or terminal OH group through strong hydrogen bonding (H-bonding). Affected by the H-bond network, the H_2O dissociation on the rutile (110) surface is inhibited but that on the anatase (101) surface is promoted.

Keywords: water dissociation; hydrogen bond network; TiO_2; ReaxFF; molecular dynamics simulation

Citation: Zhou, H.; Zhang, H.; Yuan, S. Comparison of H₂O Adsorption and Dissociation Behaviors on Rutile (110) and Anatase (101) Surfaces Based on ReaxFF Molecular Dynamics Simulation. *Molecules* **2023**, *28*, 6823. https://doi.org/10.3390/molecules28196823

Academic Editor: Erich A. Müller

Received: 29 August 2023
Revised: 21 September 2023
Accepted: 25 September 2023
Published: 27 September 2023

Copyright: © 2023 by the authors. Licensee MDPI, Basel, Switzerland. This article is an open access article distributed under the terms and conditions of the Creative Commons Attribution (CC BY) license (https://creativecommons.org/licenses/by/4.0/).

1. Introduction

TiO_2 is an important photocatalyst in photocatalytic hydrogen production and photooxidation of organic pollutants [1–3]. The interaction of H_2O with TiO_2 in an aqueous environment plays an important role in many practical applications [4]. As a fundamental process, water splitting is one of the most important chemical reactions [5]. It affects the reactivity, surface chemistry, and overall performance of the material. For the dissociation of H_2O by the reaction $H_2O \rightarrow OH^- + H^+$, reactive energy is required and gained from the catalyst or charge transfer [6,7].

Among surfaces relevant to heterogeneous catalysis, rutile and anatase have been extensively investigated for their interaction with water, owing to their photocatalytic activity. To study the adsorption and dissociation characteristics of H_2O on rutile (110) and anatase (101) surfaces, extensive research using experiments and theoretical calculations has been carried out in the past decades [8]. Many scholars confirmed that H_2O dissociated at the Ti_{5c} site on the surface of defect-free rutile (110) [9,10]. Using scanning tunneling microscopy (STM), Tan et al. indicated that the dissociative state of H_2O was more stable than the molecular state of H_2O at the Ti_{5c} site on the rutile (110) surface [11]. Additionally, the adsorption of H_2O on the anatase (101) surface was in the molecular form rather than the dissociated form under the vacuum condition [12–14].

At the atomic level, it is still a great challenge to characterize the adsorption and dissociation behavior of H_2O on the TiO_2 surface under an aqueous environment in experiments. There are still some unsettled fundamental problems and controversies. For example, the nature of adsorbed water, the extent of H_2O dissociation, and the generation of surface hydroxyl species on the perfect TiO_2 surface remain controversial. Some experimental measurements and theoretical investigations suggested that molecular adsorption mechanism or mixed adsorption mechanisms may occur simultaneously at room temperature [12,15–19]. Additionally, Ángel et al. directly determined that the terminal hydroxyl group and the bridge hydroxyl group were easily formed on the anatase TiO_2 surface in an aqueous environment [20,21]. Under this circumstance, even in the well-defined crystal plane, the adsorption and dissociation mechanism of H_2O on the TiO_2 surface remains incompletely elucidated [16]. Therefore, it is still necessary to clarify the adsorption and dissociation behavior of H_2O on the TiO_2 surface in the real environment.

In the real environment, there is an intermolecular interaction between adsorbed and non-adsorbed H_2O in bulk water. By the first-principle calculation, double or triple layers of water can be found on TiO_2 surfaces [22,23]. Additionally, some in situ experiments were also carried out. Using STM measurement, Tan et al. confirmed that the interfacial H-bond facilitated the dehydrogenation of H_2O on the rutile (110) surface [11]. Yang et al. found that the dissociation of H_2O on the rutile (110) surface was suppressed at high coverage [24,25]. However, the probability of H_2O dissociation on the anatase (101) surface remained constant with the increase in H_2O coverage [26]. An STM experiment showed that the H-bond network on the anatase (001) surface promoted hydrolysis dissociation [27]. However, the role of the H-bond network in H_2O dissociation on rutile (110) and anatase (101) surfaces has not been completely revealed yet. Hence, it is essential to explore the role of the H-bond network on H_2O adsorption and dissociation on rutile (110) and anatase (101) surfaces.

As mentioned above, there are still two problems that need to be solved. One is the adsorption and dissociation mechanism of H_2O on H_2O–TiO_2 interfaces in an aqueous environment. The other is that the water structure formed on TiO_2 surfaces may be affected by the strength of H-bonding. For this purpose, reactive molecular dynamics (RMD) with ReaxFF is employed to study the atomic and molecular behavior of H_2O on H_2O–TiO_2 heterogeneous interaction in an aqueous environment. Here, we demonstrate the significance of subtle surface structures in water dissociation on rutile and anatase TiO_2 in aqueous environments. This study will offer strategies to achieve efficient catalysis via matching proper surface structures with targeted reaction characteristics. Additionally, the double-layer model is proposed to describe the water structure on rutile (110) and anatase (101) surfaces theoretically. The result shows that the two problems mentioned above are illustrated clearly.

The rest of the paper is organized as follows. Section 3 introduces the main methods such as the ReaxFF force field. The results and discussion are given in Section 2. The conclusions are summarized in Section 4.

2. Results and Discussion

2.1. Adsorption and Dissociation Mechanism of H_2O on Rutile (110)

Among various rutile surfaces, rutile (110) is the most stable one and it is widely discussed [28]. Figure 1 shows the RMD simulation snapshot of the water distribution with the coverage of 2.0 ML on the rutile (110) surface. Here, the adsorption and dissociation of H_2O at the Ti_{5c} site are observed. In Figure 1, the blue ball represents the dissociated H_2O at the Ti_{5c} site. As shown in Figure 2a,c, the result shows that the mixed state containing the molecular and dissociative adsorption of H_2O on the rutile (110) surface is favorable. For molecular adsorption, the O atom of H_2O_{ad} forms a coordination bond with the surface Ti_{5c} atom, and the length of the Ti_{5c}-O_{ad} bond is 2.29 Å in Figure 2a. The terminal H_2O_{ad} forms an intermolecular H-bond with the surface O_{br}, in which the distance of the H-bond is 1.35 Å. The interfacial H-bond can effectively assist proton transfer and exchange across the surface. Figure 2b shows that the generated OH group (OH_{ad}) is stably adsorbed at the Ti_{5c} site, where the length of the Ti_{5c}-O bond is 1.80 Å. This result is consistent with other research using STM experiments and DFT calculations [11,29]. For dissociative adsorption, the terminal free H_2O (H_2O_f) can easily combine with the surface O_{br}, resulting in a free OH group (OH_f) and an $O_{br}H$, as shown in Figure 2c,d. The formation of $O_{br}H$ promotes the dissociation of H_2O_f. Subsequently, the generated OH_f group is stably adsorbed at the adjacent Ti_{5c} site with the length of the Ti_{5c}-O bond being 2.04 Å. Alternatively, the OH_f group recombines with the $O_{br}H$. Additionally, the indirect dissociation mechanism is also observed on the rutile (110) surface. As shown in Figure 2e,f, the H_2O in the second layer donates an H-bond to an adjacent O_{br} and transfers its proton to the O_{br}. Simultaneously, the H_2O in the second layer receives a proton from the H_2O_{ad} at the Ti_{5c} site. This is consistent with other results of DFT studies [30]. However, this indirect proton transfer involves at least two proton transfers, whose sequential occurrence is less likely than the direct proton transfer between the H_2O_{ad} and the surface O_{br}.

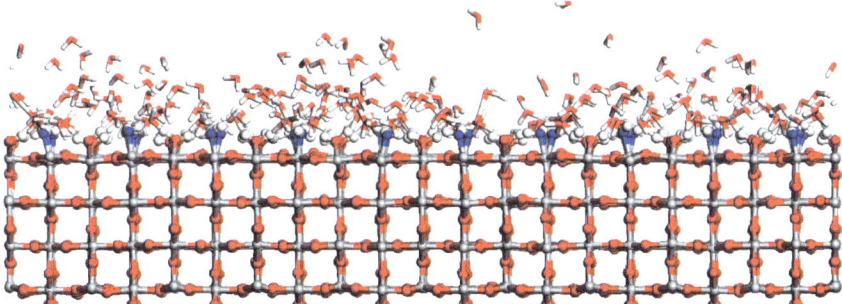

Figure 1. The RMD simulation snapshot of water on rutile (110) at the coverage of 2.0 ML. The blue ball represents the O atom from the adsorbed and dissociated H_2O.

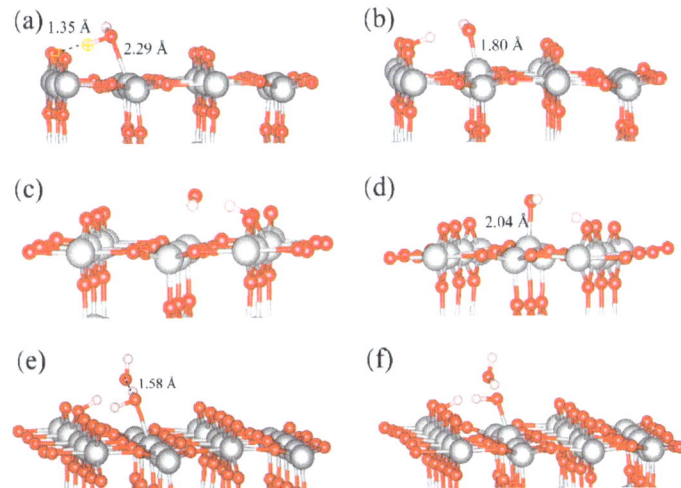

Figure 2. Snapshots of H_2O adsorption and dissociation on the rutile (110) surface: (**a**,**b**) represent direct dissociation of the H_2O_{ad} molecule at the Ti_{5c} site; (**c**,**d**) represent the dissociation and adsorption process of H_2O_f molecule at the Ti_{5c} site; (**e**,**f**) represent the indirect dissociation of H_2O_{ad} at the Ti_{5c} site. Red, grey and pink balls represent O, Ti and H atoms, respectively.

To manifest the interfacial character more clearly, Figure 3 shows the radial distribution function (RDF) of Ti_{5c}-O_w (O_w represents the O from H_2O) at the coverage of 3.0 ML. The first peak of $r(Ti_{5c}$-$O_w)$ is about ~1.85 Å, which corresponds to terminal hydroxyl groups

(OH$_{ad}$) from dissociative H$_2$O molecules on the rutile (110) surface. The second peaks of r(Ti$_{5c}$-O$_w$) appear at about ~3.55 Å and ~3.75 Å. They correspond to the H$_2$O in the second layer connected with the surface O$_{br}$ and terminal OH through the H-bond network. The water above the second layer can be regarded as bulk water. These results are consistent with the simulation data obtained by Předota et al. [31]. They observed the first layer of oxygen from the terminal OH group at the top of Ti$_{5c}$ sites with a distance of ~1.9–2.4 Å, and the second layer of water appears at about ~3.8 Å. The double-layered structure of water is also obtained by Mamontov and co-workers [32,33]. As illustrated on the right of Figure 3, the dashed red line represents the H-bond network between the second layer of water and surface O$_{br}$ or OH groups, which affects water dissociation.

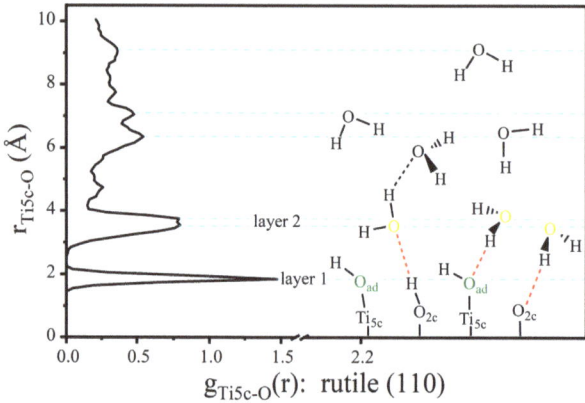

Figure 3. On the left is the RDF of Ti$_{5c}$-O. On the right is the cartoon illustration of the H-bond network on the rutile TiO$_2$ (110) surface. The red dotted line represents the enhanced H-bond between the first layer and the second layer of water, and the black dotted line represents the H-bond of ordinary H$_2$O-H$_2$O.

To investigate the effects of the H-bond network for H$_2$O dissociation on the rutile (110) surface, the RMD simulation under different initial coverage is shown in Figure 4a,b. The result shows that the amount of water dissociation (AWD) reaches a maximum value at 1.5 ML coverage with the increase in water coverage. The result is consistent with the previous ab initio MD simulation by Bandura et al. [34]. The possible reason is that the H-bond network formed by the second layer of water inhibits the direct dissociation of H$_2$O to a certain extent. At low coverage (<1.5 ML), there is a direct proton transfer between the terminal H$_2$O$_{ad}$ molecule and a nearby surface O$_{br}$. At higher coverage (>1.5 ML), H$_2$O in the second layer shares the H-bonding with the surface O$_{br}$ and OH group. The direct dissociation of H$_2$O$_{ad}$ is inhibited because the surface O$_{br}$ is occupied by the H$_2$O in the second layer. However, the possibility of H$_2$O$_{ad}$ dissociation via the indirect proton transfer is relatively low. This demonstrates that the dissociation of H$_2$O will be suppressed at high coverage on the rutile (110) surface. These results confirm the observation results of the STM experiment by Yang et al. [24]. They suggested that the reaction of H$_2$O dissociation was strongly suppressed as the coverage of water increases on the rutile (110) surface. Through the statistics and ensemble averaging of these microscopic processes, our RMD results predict macroscopic properties well.

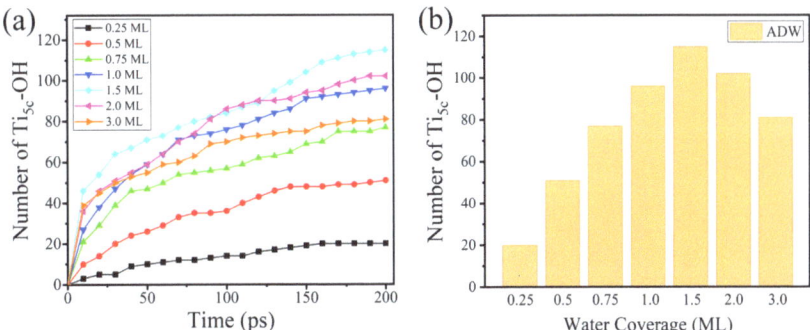

Figure 4. (a) The amount of water dissociation varies with time under the different coverage. (b) The AWD changes with the coverage of water on the rutile (110) surface.

2.2. Adsorption and Dissociation Mechanism of H_2O on Anatase (101)

Anatase (101) is the lowest-energy surface of the anatase TiO_2 polymorph [35]. Figure 5 shows the equilibrium trajectory snapshot of the water distribution over anatase (101) at the coverage of 2.0 ML, in which the blue ball represents the dissociated H_2O at the Ti_{5c} site. As shown in Figure 6a, the simulation result shows that molecular adsorption is favorable on the anatase (101) surface, which matches with the observation in the experimental result [12]. For molecular adsorption, the O atom of H_2O_{ad} forms a coordination bond with the surface Ti_{5c} atom, and the length of Ti_{5c}-O_{ad} is 2.03 Å in Figure 6a. The H_2O in the second layer shares the H-bond with both the H_2O_{ad} and surface O_{br}, and forms a water layer in close contact with bulk water. The simulated snapshot in Figure 6b describes the dissociation progress of H_2O_{ad} on the anatase (101) surface. The H_2O in the second layer provides a cascaded channel for the proton transfer from the H_2O_{ad} to the surface O_{br}. The RMD simulation result is consistent with the DFT calculation by Selloni et al. [18]. However, during the reaction on the anatase (101) surface, no direct proton transfer is observed between the terminal H_2O_{ad} molecule and a nearby surface O_{br}. This behavior is contrary to that of the rutile (110) surface investigated in this paper. As shown in Figure 6a and Figure S3, the H-bond between the H_2O_{ad} and the H_2O in the second layer is 1.52 Å, and that between H_2O_{ad} and the nearby O_{br} is 2.50 Å. The larger distance between H_2O_{ad} and the O_{br} site makes the direct proton transfer unfavorable on the anatase (101) surface, which is consistent with the calculation result by others [18,36]. The varied behavior of the water dissociation is found to be related to the subtle structure difference of surface Ti_{5c} and O_{br} sites on rutile (110) and anatase (101). As shown in Figure 7a,b, the rutile (110) surface is flat with O_{br} or O_{br}^{2-} bound to 6-fold-coordinated Ti cations and is projected out of the surface plane, whereas the anatase (101) surface has a terraced structure and exposes O_{br} or O_{br}^{2-} at the step edge. The distance between Ti_{5c} and O_{br} is about 3.55–3.56 Å on the rutile (110) surface, while that distance is about 3.85 Å on the anatase (101) surface. The larger distance between Ti_{5c} and O_{br} makes the transfer of H atom from H_2O_{ad} to O_{br} difficult on the anatase (101) surface. Therefore, for the dissociation of H_2O_{ad} on the anatase (101) surface, it is necessary to assist proton transfer through the H-bond network.

In order to illustrate interfacial characteristics more clearly, Figure 8 shows the RDF of Ti_{5c}-O_w on the anatase (101) surface at the coverage of 3.0 ML. The first peak of r(Ti_{5c}-O_w) is estimated at ~1.85 Å, corresponding to the terminal OH_{ad} group. The second peak of r(Ti_{5c}-O_w) appears at about ~3.85 Å, which corresponds to the second layer of H_2O. The H_2O in the second layer is connected to the surface O_{br} and terminal OH group through the H-bond network. As shown in Figure 8, the dashed red line represents the H-bond network between the first layer and the second layer of water. The water above the second layer can be regarded as bulk water. The double-layer model explains the experimental and theoretical results well [37–39]. These results are consistent with the simulation data obtained by Sumita and co-workers [39]. They showed that the O atom in the first layer was consistent with dissociated H_2O_{ad} at the Ti_{5c} site, and the RDF of the first peak was at ~1.82 Å by DFT.

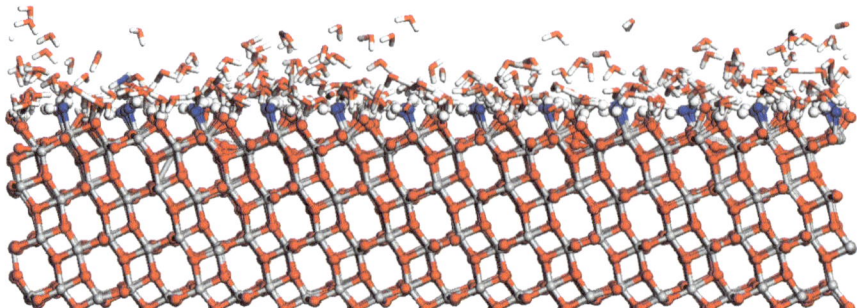

Figure 5. The RMD simulation snapshot of water distribution on the anatase (101) surface at the coverage of 2.0 ML. The blue ball represents the O atom from the adsorbed H_2O.

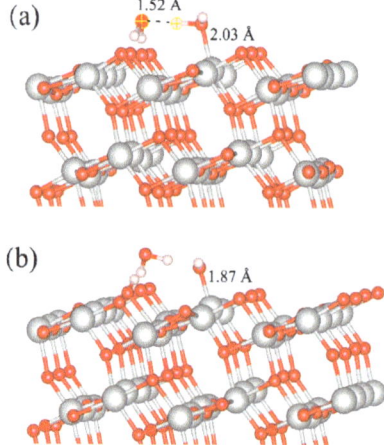

Figure 6. Snapshots of H_2O adsorption and dissociation on the anatase (101) surface: (**a**) the molecular adsorption of H_2O_{ad} at the Ti_{5c} site, and (**b**) the indirect dissociation of H_2O_{ad} at the Ti_{5c} site.

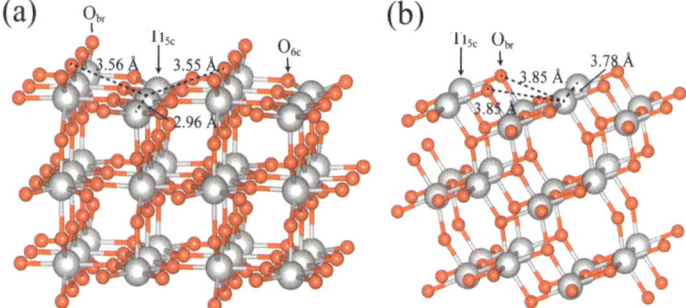

Figure 7. The surface structures of (**a**) rutile (110) and (**b**) anatase (101).

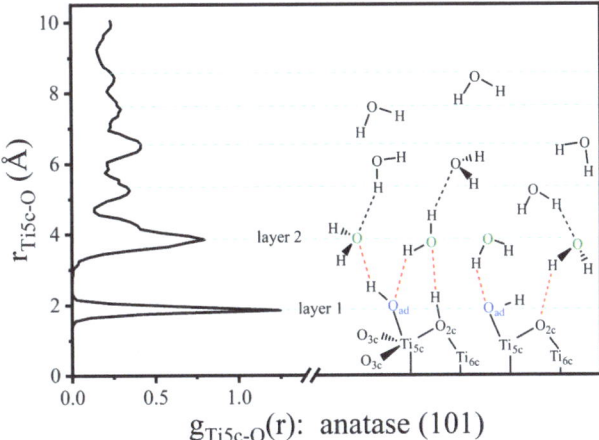

Figure 8. On the left is the RDF of Ti_{5c}-O. On the right is the cartoon illustration of the H-bond network on the anatase TiO_2 (101) surface. The red dotted line represents the enhanced H-bond between the first layer and the second layer of water, and the black dotted line represents the H-bond of ordinary H_2O-H_2O.

To investigate the effect of the H-bond network for H_2O dissociation on the anatase (101) surface, Figure 9a,b exhibit the change of the AWD with different initial coverage in the RMD simulation. However, it is worth noting that the AWD monotonically increases with the increase in coverage. The result suggests that the H_2O in the second layer participates in and assists the dissociation of H_2O_{ad}. This phenomenon is contrary to the results of the rutile (110) surface studied in this paper. The main reason is that the dissociation of H_2O happens in different ways on rutile (110) and anatase (101) surfaces. The dissociation of H_2O on rutile (110) is mainly driven by the proton transfer directly to the surface O_{br}, and the H-bond network between the first layer of water and the second layer of water may greatly reduce the dissociation of H_2O, while the indirect proton transfer on the anatase (101) surface needs to be assisted by the H_2O in the second layer.

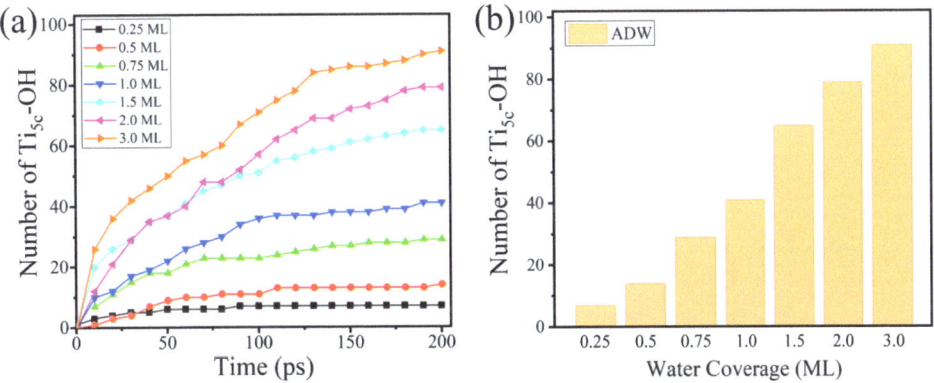

Figure 9. (a) The amount of water dissociation varies with time under the different coverage. (b) The AWD changes with the coverage of water on the anatase (101) surface.

2.3. The Roles of the H-Bond Network in Water Dissociation

As discussed above, the H-bond network, which is ubiquitous in a practical aqueous environment, plays a crucial role in the dissociation of H_2O on rutile (110) and anatase

(101) surfaces. This paper suggests the double-layer model on rutile (110) and anatase (101) surfaces: the first layer is the dissociated H_2O_{ad} at the Ti_{5c} site, and the second layer is defined as the H_2O adsorbed onto the O_{br} or terminal OH group through H-bonding. Simulated snapshots in Figure 10a,b show the ordered H-bond network geometry of the second layer of water on the rutile (110) surface at 2.6 ps and the anatase (101) surface at 3.5 ps, respectively. It is observed that the H_2O in the second layer adsorbed on O_{br} through strong H-bonding interaction. To further understand the effect of strong H-bonding in the dissociation of H_2O on TiO_2 surfaces, this paper goes on to investigate the property of the H-bond network in the second layer of water. Figure 10c,d and Figure 11a show the RDFs of O-H on the rutile (110) surface, anatase (101) surface, and bulk water, respectively. The first peak represents the distance of the intramolecular O-H bond. The second peak of r(O-H) corresponds to the strong H-bonding between the H_2O in the second layer and the surface O_{br} or terminal OH group. For convenience, the second peak of r(O-H) is labeled as $r_2(O_{ad}-H_w)$. As shown in Figure 10c,d, the $r_2(O_{ad}-H_w)$ of rutile (110) and anatase (101) is about ~1.63 Å. For comparison, the $r_2(O_{ad}-H_w)$ in bulk water is estimated at ~1.78 Å in Figure 11a. The shorter $r_2(O_{ad}-H_w)$ on rutile (110) and anatase (101) surfaces suggests a stronger H-bonding interaction between the H_2O in the second layer and the surface O_{br} or terminal OH group.

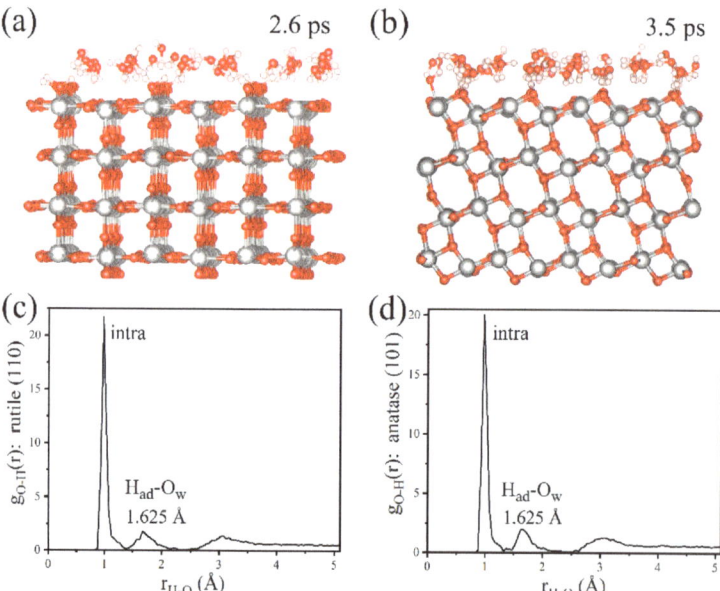

Figure 10. (**a**) A simulated snapshot of local water distribution at 2.6 ps on the rutile (110) surface with the coverage of 3.0 ML; (**b**) a simulated snapshot of local water distribution at 3.5 ps on the anatase (101) surface with the coverage of 3.0 ML; (**c**,**d**) RDFs of O-H on rutile (110) and anatase (101) surfaces, respectively.

Figure 11b–d, respectively, calculate the RDFs of O-O on the rutile (110) surface, anatase (101) surface, and bulk water. The first peak of $r(O-O_w)$ represents the distance from the surface O_{br} or O_{ad} to the H_2O in the second layer on rutile (110) and anatase (101) surfaces. As shown in Figure 11b, the first peak of $r(O_w-O_w)$ is located at about ~2.78 Å in the bulk water, which is in accord with the experimental measurement and DFT calculations [40,41]. Compared with that in the bulk water, the first peak of $r(O-O_w)$ on rutile (110) is estimated at ~2.68 Å in Figure 11c, and that on anatase (101) is about ~2.63 Å in Figure 11d. The shorter $r(O-O_w)$ indicates the strong interaction between the

H_2O in the second layer and the surface O_{br} or OH group on rutile (110) and anatase (101) surfaces. Therefore, the H-bonding between the surface O_{br} or terminal OH group and the second layer of water appears to be stronger than the H-bonding of ordinary H_2O-H_2O in bulk water.

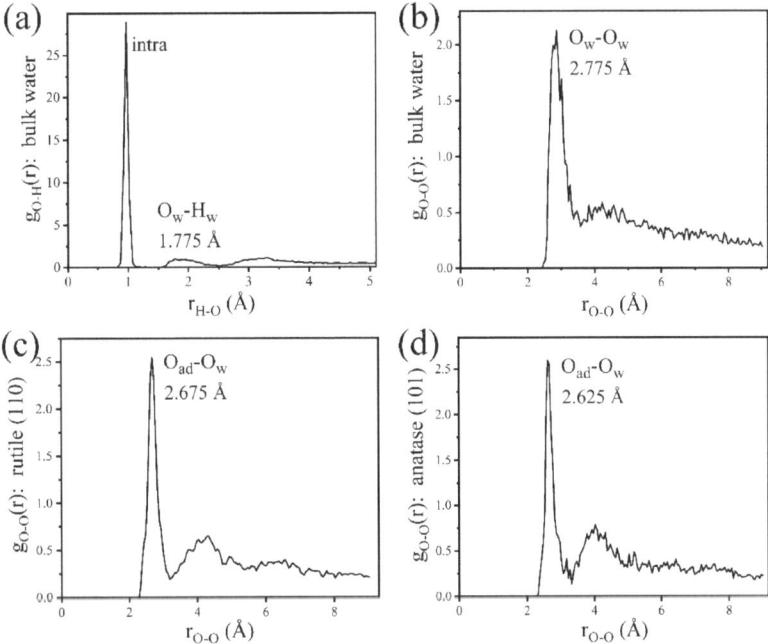

Figure 11. RDFs of (**a**) O_w-H_w in bulk water; (**b**) O_w-O_w in bulk water; (**c**) O-O_w on the rutile (110) surface; and (**d**) O-O_w on the anatase (101) surface.

Figure 12 shows the change of surface O_{br} and O_{ad} atomic charge during the reaction process. As illustrated in Figure 12a,d, the mean charge of surface O_{br} atoms is close to -0.65 e on the rutile (110) surface, and that on the anatase (101) surface is about -0.70 e before the dissociation reaction of H_2O. As shown in Figure 12b,c,e,f, the mean charge of O_{br} and O_{ad} atoms dramatically decreases until the reaction is completed. The charge of surface O_{br} and O_{ad} atoms is roughly stable after the reaction. The mean charge of O_{br} and O_{ad} on rutile (110) is -0.77 e and that on anatase (101) is about -0.80 e. In Figure 12c,f, surface O_{br} and O_{ad} atoms are more electron-rich than O_w in bulk water (-0.71 e). Correspondingly, the polarization of surface O_{br} and O_{ad} is enhanced. It is indicated that the strong H-bonding between the surface O_{br} or O_{ad} group and the second layer of water is formed. These results provide insights to reveal the role of the H-bond network for water dissociation on rutile (110) and anatase (101) surfaces. More specifically, the H-bond network inhibits the dissociation of H_2O at high coverage on the rutile (110) surface. But the H-bond network may facilitate the dissociation of H_2O by linking the proton transfer channel on the anatase (101) surface.

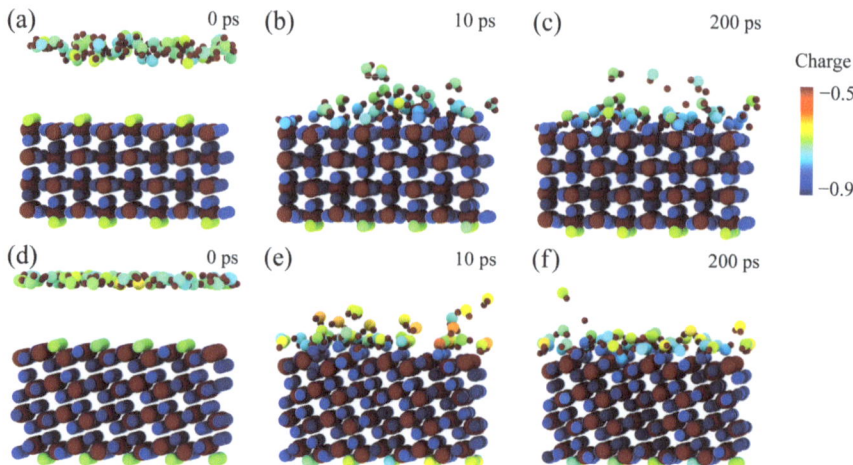

Figure 12. Snapshot views of the local structures of the rutile (110) surface: (**a**) before the interaction at 0 ps, (**b**) after the reaction occurs at 10 ps, and (**c**) after the reaction occurs at 200 ps with the coverage of 1.5 ML. Snapshot views of the local structures of the anatase (101) surface: (**d**) before the interaction at 0 ps, (**e**) after the reaction occurs at 10 ps, and (**f**) after the reaction occurs at 200 ps with the coverage of 1.5 ML. The atom is colored by charge. Green, yellow and blue balls represent O atoms. Smaller and bigger brown balls represent H and Ti atoms, respectively.

3. Methods

In this paper, molecular dynamic simulation with ReaxFF force field [42,43] is employed to investigate the behavior of H_2O on rutile (110) and anatase (101) surfaces. The force field parameters are determined from quantum mechanics (QM) based on training sets and experimental results, which can ensure the accuracy of the RMD simulations. The ReaxFF method developed by van Duin uses the bond order relation obtained from the interatomic distance [43,44], which is updated at each RMD or energy minimization step. It allows continuous bond dissociation for all orders at the same time. Therefore, the ReaxFF can be used to describe chemical reactions, including bond formation and bond breaking [42].

In the ReaxFF reactive force field, the total (system) energy is given by [45]

$$E_{system} = E_{bond} + E_{over} + E_{under} + E_{lp} + E_{val} + E_{vdWaals} + E_{coulomb} \quad (1)$$

The terms in Equation (1) include bond energies (E_{bond}), the energy to penalize over-coordination of atoms (E_{over}), the energy to stabilize under-coordination of atoms (E_{under}), lone-pair energies (E_{lp}), valence-angle energies (E_{val}), van der Waals interactions ($E_{vdwaals}$) and terms to handle nonbonded Coulomb ($E_{coulomb}$), respectively.

This paper employs the Ti/O/H ReaxFF interatomic potential, which is developed by Kim et al. [45]. The force field is carefully used and validated in previous research on the H_2O–TiO_2 interface [46–49]. Using the large-scale atomic/molecular massively parallel simulator (LAMMPS) package [50,51], the RMD calculation with ReaxFF is capable of simulating systems larger than 10^6 atoms in nanosecond time scales.

The general flow chart of the molecular dynamics simulation is shown in Figure 13. Rutile (110) and anatase (101) surfaces are carved from bulk rutile and anatase TiO_2 crystals, respectively. The dimensions of simulation cells are 59.18 Å (x) × 64.97 Å (y) and 67.96 Å (x) × 61.26 Å (y) for rutile (110) and anatase (101) surfaces in Figure S1a,b, respectively. Cleaved rutile (110) and anatase (101) surfaces are composed of four-layer TiO_2 slabs. Their bottom two layers are fixed in the bulk configuration to simulate the bulk-like environment. There are 200 and 216 Ti_{5c} reactive sites on rutile (110) and anatase (101) surfaces, respectively.

In all discussions that follow, one monolayer (ML) is defined as the number of Ti_{5c} sites on rutile (110) and anatase (101) surfaces. H_2O molecules are placed over rutile (110) and anatase (101) surfaces with a coverage of 0.25, 0.50, 0.75, 1.0, 1.5, 2.0, and 3.0 ML. The simulation box is constructed with periodic boundary conditions in both the X and Y directions. And the fixed boundary condition is applied along the Z direction. To avoid interaction between H_2O and the bottom of TiO_2, a reflecting wall is used at the top of the box along the Z direction. The RMD simulation is performed in the canonical ensemble (NVT) with the time step of 0.25 fs. The conjugate gradient (CG) approach is used to minimize energy. During the simulation, the ambient temperature of 300 K is constantly controlled by the Nosé–Hoover thermostat with a 50 fs damping constant [52]. The velocity Verlet algorithm is employed to calculate Newton's equation of motion. The atomic charge is equilibrated at every time step using the QEq (charge equilibration) model. Ovito is employed to generate snapshots of the simulation. In this paper, all systems reach equilibrium after 200 ps, which can be monitored by the convergence of potential energy in Figure S2.

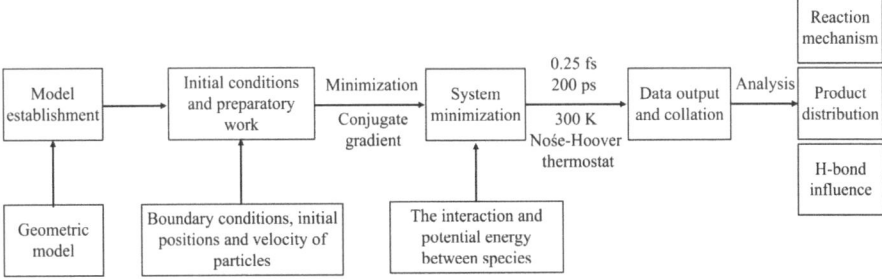

Figure 13. Schematic diagram of the calculation step.

4. Conclusions

In this paper, the ReaxFF RMD simulation is employed to investigate the adsorption and dissociation mechanisms of H_2O on rutile (110) and anatase (101) surfaces in an aqueous environment. Furthermore, this paper explores the vital role of the H-bond network in understanding the underlying mechanisms for water dissociation at a deeper level. Here are several significant findings and conclusions from this paper:

(1) There is a mixed adsorption trend with both molecular and dissociative adsorption on the rutile (110) surface. Compared with that on the rutile (110) surface, molecular adsorption is dominant on the anatase (101) surface.

(2) The dissociation of H_2O is mainly the direct dissociation on the rutile (110) surface. The interfacial H-bond between the adsorbed H_2O_{ad} molecule and the surface O_{br} promotes proton transfer for H_2O dissociation on the rutile (110) surface. Compared with that on the rutile (110) surface, the dissociation of H_2O is dominated by indirect proton transfer on the anatase (101) surface. This different catalytic function is solely determined by the distance between Ti_{5c} and O_{br} on the surface, which determines the behavior of water dissociation.

(3) The H-bond network plays a crucial role in the dissociation of H_2O on rutile (110) and anatase (101) surfaces. At high coverage (>1.5 ML), the H-bond network structure of the second layer of water on the rutile (110) surface inhibits the dissociation of H_2O to some extent. Compared with that on the rutile (110) surface, the RMD simulation shows that H-bond could assist the proton transfer on the anatase (101) surface. In an aqueous environment, the dissociation of H_2O_{ad} is promoted by the enhanced H-bond network structure of the second layer of water on the anatase (101) surface.

Overall, this paper provides a meaningful insight to understand the behavior of H_2O adsorption and dissociation on TiO_2 surfaces in an aqueous environment. It is hoped that the findings reported here will motivate further experimental and theoretical work to achieve a complete understanding of this technologically relevant interface.

Supplementary Materials: The following supporting information can be downloaded at: https://www.mdpi.com/article/10.3390/molecules28196823/s1. Figure S1. Initial models of simulation are established. (a) H_2O molecules on the rutile (110) surface at the coverage of 2.0 ML. (b) H_2O molecules on the anatase (101) surface at the coverage of 2.0 ML. Grey, red, and white balls represent Ti, O, and H atoms, respectively. The upper H_2O is represented by a stick model. Figure S2. The time evolution of potential energy on (a) the rutile (110) surface and (b) the anatase (101) surface during the NVT RMD simulation of water dissociation. Figure S3. The molecular absorption of H_2O on the anatase (101) surface.

Author Contributions: Conceptualization, H.Z. (He Zhou) and S.Y.; Methodology, H.Z. (Heng Zhang); Software, H.Z. (Heng Zhang); Formal analysis, S.Y.; Data curation, H.Z. (He Zhou); Writing—original draft, H.Z. (He Zhou); Writing—review & editing, H.Z. (He Zhou). All authors have read and agreed to the published version of the manuscript.

Funding: This research was funded by [Natural Science Foundation of Shandong Province] grant number [ZR2021MB040].

Institutional Review Board Statement: Not applicable.

Informed Consent Statement: Not applicable.

Data Availability Statement: Not applicable.

Conflicts of Interest: The authors declare no conflict of interest.

Sample Availability: Not applicable.

References

1. Wagstaffe, M.; Dominguez-Castro, A.; Wenthaus, L.; Palutke, S.; Kutnyakhov, D.; Heber, M.; Pressacco, F.; Dziarzhytski, S.; Gleißner, H.; Gupta, V.K.; et al. Photoinduced Dynamics at the Water/TiO_2(101) Interface. *Phys. Rev. Lett.* **2023**, *130*, 108001. [CrossRef] [PubMed]
2. Mannaa, M.A.; Qasim, K.F.; Alshorifi, F.T.; El-Bahy, S.M.; Salama, R.S. Role of NiO Nanoparticles in Enhancing Structure Properties of TiO_2 and Its Applications in Photodegradation and Hydrogen Evolution. *ACS Omega* **2021**, *6*, 30386–30400. [CrossRef] [PubMed]
3. Alasri, T.M.; Ali, S.L.; Salama, R.S.; Alshorifi, F.T. Band-Structure Engineering of TiO_2 Photocatalyst by Ause Quantum Dots for Efficient Degradation of Malachite Green and Phenol. *J. Inorg. Organomet. Polym. Mater.* **2023**, *33*, 1729–1740. [CrossRef]
4. Li, F.; Chen, J.-F.; Gong, X.-Q.; Hu, P.; Wang, D. Subtle Structure Matters: The Vicinity of Surface Ti_{5c} Cations Alters the Photooxidation Behaviors of Anatase and Rutile TiO_2 under Aqueous Environments. *ACS Catal.* **2022**, *12*, 8242–8251. [CrossRef]
5. Kudo, A.; Miseki, Y. Heterogeneous Photocatalyst Materials for Water Splitting. *Chem. Soc. Rev.* **2009**, *38*, 253–278. [CrossRef]
6. Bikondoa, O.; Pang, C.L.; Ithnin, R.; Muryn, C.A.; Onishi, H.; Thornton, G. Direct Visualization of Defect-Mediated Dissociation of Water on TiO_2 (110). *Nat. Mater.* **2006**, *5*, 189–192. [CrossRef]
7. Migani, A.; Blancafort, L. What Controls Photocatalytic Water Oxidation on Rutile TiO_2 (110) under Ultra-High-Vacuum Conditions? *J. Am. Chem. Soc.* **2017**, *139*, 11845–11856. [CrossRef]
8. Guo, Q.; Ma, Z.; Zhou, C.; Ren, Z.; Yang, X. Single Molecule Photocatalysis on TiO_2 Surfaces. *Chem. Rev.* **2019**, *119*, 11020–11041. [CrossRef]
9. Tan, S.; Feng, H.; Ji, Y.; Wang, Y.; Zhao, J.; Zhao, A.; Wang, B.; Luo, Y.; Yang, J.; Hou, J.G. Observation of Photocatalytic Dissociation of Water on Terminal Ti Sites of TiO_2(110)-1x1 Surface. *J. Am. Chem. Soc.* **2012**, *134*, 9978–9985. [CrossRef]
10. Guo, Q.; Xu, C.; Ren, Z.; Yang, W.; Ma, Z.; Dai, D.; Fan, H.; Minton, T.K.; Yang, X. Stepwise Photocatalytic Dissociation of Methanol and Water on TiO_2(110). *J. Am. Chem. Soc.* **2012**, *134*, 13366–13673. [CrossRef]
11. Tan, S.; Feng, H.; Zheng, Q.; Cui, X.; Zhao, J.; Luo, Y.; Yang, J.; Wang, B.; Hou, J.G. Interfacial Hydrogen-Bonding Dynamics in Surface-Facilitated Dehydrogenation of Water on TiO_2(110). *J. Am. Chem. Soc.* **2020**, *142*, 826–834. [CrossRef] [PubMed]
12. He, Y.; Tilocca, A.; Dulub, O.; Selloni, A.; Diebold, U. Local Ordering and Electronic Signatures of Submonolayer Water on Anatase TiO_2(101). *Nat. Mater.* **2009**, *8*, 585–589. [CrossRef] [PubMed]
13. Herman, G.S.; Dohnálek, Z.; Ruzycki, N.; Diebold, U. Experimental Investigation of the Interaction of Water and Methanol with Anatase-TiO_2 (101). *J. Phys. Chem. B* **2003**, *107*, 2788–2795. [CrossRef]
14. Sun, C.; Liu, L.-M.; Selloni, A.; Lu, G.Q.; Smith, S.C. Titania-Water Interactions: A Review of Theoretical Studies. *J. Mater. Chem.* **2010**, *20*, 10319–10334. [CrossRef]
15. Schaefer, A.; Lanzilotto, V.; Cappel, U.; Uvdal, P.; Borg, A.; Sandell, A. First Layer Water Phases on Anatase TiO_2 (101). *Surf. Sci.* **2018**, *674*, 25–31. [CrossRef]

16. Walle, L.E.; Borg, A.; Johansson, E.M.J.; Plogmaker, S.; Rensmo, H.; Uvdal, P.; Sandell, A. Mixed Dissociative and Molecular Water Adsorption on Anatase TiO_2 (101). *J. Phys. Chem. C* **2011**, *115*, 9545–9550. [CrossRef]
17. Bourikas, K.; Kordulis, C.; Lycourghiotis, A. Titanium Dioxide (Anatase and Rutile): Surface Chemistry, Liquid–Solid Interface Chemistry, and Scientific Synthesis of Supported Catalysts. *Chem. Rev.* **2014**, *114*, 9754–9823. [CrossRef]
18. Calegari Andrade, M.F.; Ko, H.Y.; Zhang, L.; Car, R.; Selloni, A. Free Energy of Proton Transfer at the Water-TiO_2 Interface from Ab Initio Deep Potential Molecular Dynamics. *Chem. Sci.* **2020**, *11*, 2335–2341. [CrossRef]
19. Martinez-Casado, R.; Mallia, G.; Harrison, N.M.; Pérez, R. First-Principles Study of the Water Adsorption on Anatase (101) as a Function of the Coverage. *J. Phys. Chem. C* **2018**, *122*, 20736–20744. [CrossRef]
20. Mino, L.; Morales-García, Á.; Bromley, S.T.; Illas, F. Understanding the Nature and Location of Hydroxyl Groups on Hydrated Titania Nanoparticles. *Nanoscale* **2021**, *13*, 6577–6585. [CrossRef]
21. Recio-Poo, M.; Morales-García, Á.; Illas, F.; Bromley, S.T. Crystal Properties without Crystallinity? Influence of Surface Hydroxylation on the Structure and Properties of Small TiO_2 Nanoparticles. *Nanoscale* **2023**, *15*, 4809–4820. [CrossRef] [PubMed]
22. Mattioli, G.; Filippone, F.; Caminiti, R.; Bonapasta, A.A. Short Hydrogen Bonds at the Water/TiO_2 (Anatase) Interface. *J. Phys. Chem. C* **2008**, *112*, 13579–13586. [CrossRef]
23. Kimmel, G.A.; Baer, M.; Petrik, N.G.; VandeVondele, J.; Rousseau, R.; Mundy, C.J. Polarization- and Azimuth-Resolved Infrared Spectroscopy of Water on TiO_2 (110): Anisotropy and the Hydrogen-Bonding Network. *J. Phys. Chem. Lett.* **2012**, *3*, 778–784. [CrossRef] [PubMed]
24. Yang, W.; Wei, D.; Jin, X.; Xu, C.; Geng, Z.; Guo, Q.; Ma, Z.; Dai, D.; Fan, H.; Yang, X. Effect of the Hydrogen Bond in Photoinduced Water Dissociation: A Double-Edged Sword. *J. Phys. Chem. Lett.* **2016**, *7*, 603–608. [CrossRef]
25. Xu, C.; Xu, F.; Chen, X.; Li, Z.; Luan, Z.; Wang, X.; Guo, Q.; Yang, X. Wavelength-Dependent Water Oxidation on Rutile TiO_2 (110). *J. Phys. Chem. Lett.* **2021**, *12*, 1066–1072. [CrossRef]
26. Geng, Z.; Chen, X.; Yang, W.; Guo, Q.; Xu, C.; Dai, D.; Yang, X. Highly Efficient Water Dissociation on Anatase TiO_2 (101). *J. Phys. Chem. C* **2016**, *120*, 26807–26813. [CrossRef]
27. Ma, X.; Shi, Y.; Liu, J.; Li, X.; Cui, X.; Tan, S.; Zhao, J.; Wang, B. Hydrogen-Bond Network Promotes Water Splitting on the TiO_2 Surface. *J. Am. Chem. Soc.* **2022**, *144*, 13565–13573. [CrossRef]
28. Sun, H.; Mowbray, D.J.; Migani, A.; Zhao, J.; Petek, H.; Rubio, A. Comparing Quasiparticle H_2O Level Alignment on Anatase and Rutile TiO_2. *ACS Catal.* **2015**, *5*, 4242–4254. [CrossRef]
29. Zhou, H.; Zhang, X.; Zhang, J.; Ma, H.; Jin, F.; Ma, Y. A New Deep Hole-Trapping Site for Water Splitting on the Rutile TiO_2 (110) Surface. *J. Mater. Chem. A* **2021**, *9*, 7650–7655. [CrossRef]
30. Wen, B.; Calegari Andrade, M.F.; Liu, L.-M.; Selloni, A. Water Dissociation at the Water–Rutile TiO_2 (110) Interface from Ab Initio-Based Deep Neural Network Simulations. *Proc. Natl. Acad. Sci. USA* **2023**, *120*, e2212250120. [CrossRef]
31. Předota, M.; Bandura, A.V.; Cummings, P.T.; Kubicki, J.D.; Wesolowski, D.J.; Chialvo, A.A.; Machesky, M.L. Electric Double Layer at the Rutile (110) Surface. 1. Structure of Surfaces and Interfacial Water from Molecular Dynamics by Use of Ab Initio Potentials. *J. Phys. Chem. B* **2004**, *108*, 12049–12060. [CrossRef]
32. Mamontov, E.; Vlcek, L.; Wesolowski, D.J.; Cummings, P.T.; Wang, W.; Anovitz, L.M.; Rosenqvist, J.; Brown, C.M.; Sakai, V.G. Dynamics and Structure of Hydration Water on Rutile and Cassiterite Nanopowders Studied by Quasielastic Neutron Scattering and Molecular Dynamics Simulations. *J. Phys. Chem. C* **2007**, *111*, 4328–4341. [CrossRef]
33. Koparde, V.N.; Cummings, P.T. Molecular Dynamics Study of Water Adsorption on TiO_2 Nanoparticles. *J. Phys. Chem. C* **2007**, *111*, 6920–6926. [CrossRef]
34. Bandura, A.V.; Kubicki, J.D.; Sofo, J.O. Comparisons of Multilayer H_2O Adsorption onto the (110) Surfaces of A-TiO_2 and SnO_2 as Calculated with Density Functional Theory. *J. Phys. Chem. B* **2008**, *112*, 11616–11624. [CrossRef]
35. Yates, J.T. Photochemistry on TiO_2: Mechanisms Behind the Surface Chemistry. *Surf. Sci.* **2009**, *603*, 1605–1612. [CrossRef]
36. Li, J.-Q.; Sun, Y.; Cheng, J. Theoretical Investigation on Water Adsorption Conformations at Aqueous Anatase TiO_2/Water Interfaces. *J. Mater. Chem. A* **2023**, *11*, 943–952. [CrossRef]
37. Nosaka, A.Y.; Fujiwara, T.; Yagi, H.; Akutsu, H.; Nosaka, Y. Characteristics of Water Adsorbed on TiO_2 Photocatalytic Systems with Increasing Temperature as Studied by Solid-State ^1H NMR Spectroscopy. *J. Phys. Chem. B* **2004**, *108*, 9121–9125. [CrossRef]
38. Atsuko, Y.N.; Yoshio, N. Characteristics of Water Adsorbed on TiO_2 Photocatalytic Surfaces as Studied by ^1H NMR Spectroscopy. *Bull. Chem. Soc. Jpn.* **2005**, *78*, 1595–1607.
39. Sumita, M.; Hu, C.; Tateyama, Y. Interface Water on TiO_2 Anatase (101) and (001) Surfaces: First-Principles Study with TiO_2 Slabs Dipped in Bulk Water. *J. Phys. Chem. C* **2010**, *114*, 18529–18537. [CrossRef]
40. Sorenson, J.M.; Hura, G.; Glaeser, R.M.; Head-Gordon, T. What Can X-Ray Scattering Tell Us About the Radial Distribution Functions of Water? *J. Chem. Phys.* **2000**, *113*, 9149–9161. [CrossRef]
41. Grossman, J.C.; Schwegler, E.; Draeger, E.W.; Gygi, F.; Galli, G. Towards an Assessment of the Accuracy of Density Functional Theory for First Principles Simulations of Water. *J. Chem. Phys.* **2004**, *120*, 300–311. [CrossRef] [PubMed]
42. Senftle, T.P.; Hong, S.; Islam, M.M.; Kylasa, S.B.; Zheng, Y.; Shin, Y.K.; Junkermeier, C.; Engel-Herbert, R.; Janik, M.J.; Aktulga, H.M.; et al. The Reaxff Reactive Force-Field: Development, Applications and Future Directions. *NPJ Comput. Mater.* **2016**, *2*, 15011. [CrossRef]
43. Van Duin, A.C.T.; Dasgupta, S.; Lorant, F.; Goddard, W.A. Reaxff: A Reactive Force Field for Hydrocarbons. *J. Phys. Chem. A* **2001**, *105*, 9396–9409. [CrossRef]

44. Chenoweth, K.; van Duin, A.C.; Goddard, W.A. Reaxff Reactive Force Field for Molecular Dynamics Simulations of Hydrocarbon Oxidation. *J. Phys. Chem. A* **2008**, *112*, 1040–1053. [CrossRef]
45. Kim, S.Y.; Kumar, N.; Persson, P.; Sofo, J.; van Duin, A.C.; Kubicki, J.D. Development of a Reaxff Reactive Force Field for Titanium Dioxide/Water Systems. *Langmuir* **2013**, *29*, 7838–7846. [CrossRef]
46. Yuan, S.; Liu, S.; Wang, X.; Zhang, H.; Yuan, S. Atomistic Insights into Uptake of Hydrogen Peroxide by TiO_2 Particles as a Function of Humidity. *J. Mol. Liq.* **2022**, *346*, 117097. [CrossRef]
47. Raju, M.; Kim, S.-Y.; van Duin, A.C.T.; Fichthorn, K.A. Reaxff Reactive Force Field Study of the Dissociation of Water on Titania Surfaces. *J. Phys. Chem. C* **2013**, *117*, 10558–10572. [CrossRef]
48. Huang, L.; Gubbins, K.E.; Li, L.; Lu, X. Water on Titanium Dioxide Surface: A Revisiting by Reactive Molecular Dynamics Simulations. *Langmuir* **2014**, *30*, 14832–14840. [CrossRef]
49. Groh, S.; Saßnick, H.; Ruiz, V.G.; Dzubiella, J. How the Hydroxylation State of the (110)-Rutile TiO_2 Surface Governs Its Electric Double Layer Properties. *Phys. Chem. Chem. Phys.* **2021**, *23*, 14770–14782. [CrossRef]
50. Plimpton, S. Fast Parallel Algorithms for Short-Range Molecular Dynamics. *J. Comput. Phys.* **1995**, *117*, 1–19. [CrossRef]
51. Aktulga, H.M.; Fogarty, J.C.; Pandit, S.A.; Grama, A.Y. Parallel Reactive Molecular Dynamics: Numerical Methods and Algorithmic Techniques. *Parallel Comput.* **2012**, *38*, 245–259. [CrossRef]
52. Evans, D.J.; Holian, B.L. The Nose–Hoover Thermostat. *J. Chem. Phys.* **1985**, *83*, 4069–4074. [CrossRef]

Disclaimer/Publisher's Note: The statements, opinions and data contained in all publications are solely those of the individual author(s) and contributor(s) and not of MDPI and/or the editor(s). MDPI and/or the editor(s) disclaim responsibility for any injury to people or property resulting from any ideas, methods, instructions or products referred to in the content.

Article

In Silico Investigations on the Synergistic Binding Mechanism of Functional Compounds with Beta-Lactoglobulin

Tong Meng [1], Zhiguo Wang [2,*], Hao Zhang [1], Zhen Zhao [1], Wanlin Huang [3], Liucheng Xu [3], Min Liu [4], Jun Li [1] and Hui Yan [1,*]

1. School of Pharmaceutical Sciences, Liaocheng University, Liaocheng 252059, China; mengtong19950613@sina.com (T.M.); 15163592080@163.com (H.Z.); 18354060812@163.com (Z.Z.); lijun1982@lcu.edu.cn (J.L.)
2. Institute of Ageing Research, School of Basic Medical Sciences, Hangzhou Normal University, Hangzhou 311121, China
3. School of Basic Medical Sciences, Hangzhou Normal University, Hangzhou 311121, China; 2020211301217@stu.hznu.edu.cn (W.H.); 2021211301175@stu.hznu.edu.cn (L.X.)
4. School of Chemistry and Chemical Engineering, Liaocheng University, Liaocheng 252059, China; panpanliumin@163.com
* Correspondence: zhgwang@hznu.edu.cn (Z.W.); yanhui@lcu.edu.cn (H.Y.)

Citation: Meng, T.; Wang, Z.; Zhang, H.; Zhao, Z.; Huang, W.; Xu, L.; Liu, M.; Li, J.; Yan, H. In Silico Investigations on the Synergistic Binding Mechanism of Functional Compounds with Beta-Lactoglobulin. *Molecules* **2024**, *29*, 956. https://doi.org/10.3390/molecules29050956

Academic Editor: Bernard Maigret

Received: 30 December 2023
Revised: 3 February 2024
Accepted: 20 February 2024
Published: 22 February 2024

Copyright: © 2024 by the authors. Licensee MDPI, Basel, Switzerland. This article is an open access article distributed under the terms and conditions of the Creative Commons Attribution (CC BY) license (https://creativecommons.org/licenses/by/4.0/).

Abstract: Piceatannol (PIC) and epigallocatechin gallate (EGCG) are polyphenolic compounds with applications in the treatment of various diseases such as cancer, but their stability is poor. β-lactoglobulin (β-LG) is a natural carrier that provides a protective effect to small molecule compounds and thus improves their stability. To elucidate the mechanism of action of EGCG, PIC, and palmitate (PLM) in binding to β-LG individually and jointly, this study applied molecular docking and molecular dynamics simulations combined with in-depth analyses including noncovalent interaction (NCI) and binding free energy to investigate the binding characteristics between β-LG and compounds of PIC, EGCG, and PLM. Simulations on the binary complexes of β-LG + PIC, β-LG + EGCG, and β-LG + PLM and ternary complexes of (β-LG + PLM) + PIC, (β-LG + PLM) + EGCG, β-LG + PIC) + EGCG, and (β-LG + EGCG) + PIC were performed for comparison and characterizing the interactions between binding compounds. The results demonstrated that the co-bound PIC and EGCG showed non-beneficial effects on each other. However, the centrally located PLM was revealed to be able to adjust the binding conformation of PIC, which led to the increase in binding affinity with β-LG, thus showing a synergistic effect on the co-bound PIC. The current study of β-LG co-encapsulated PLM and PIC provides a theoretical basis and research suggestions for improving the stability of polyphenols.

Keywords: polyphenols; palmitate; β-lactoglobulin; molecular docking; molecular dynamics; MM/GBSA

1. Introduction

Polyphenols are widely found in plants and have many biological functions and health benefits. Epigallocatechin gallate (Figure 1a), which is one of the most common polyphenols, can effectively inhibit the growth and induce apoptosis in human breast cancer cells [1–4], lung cancer cells [5], prostate cancer cells [6], etc. Piceatannol (Figure 1b) is a hydroxylated analog of resveratrol. Many studies have shown that PIC is a potent inhibitor of apoptosis in lymphoma cell lines [7] and primary leukemia cells [8]. However, polyphenolic compounds contain multiple phenolic hydroxyl groups in their structures, which leads to their reduced pH/light stability and photosensitivity [9]. Protection of polyphenolic compounds can be achieved by co-coating them with proteins, in which lactoglobulin is a natural delivery carrier that binds to small molecules to form protein–ligand complexes, thereby protecting polyphenolic compounds [10–12]. Palmitate (PLM,

Figure 1c) is a medium- to long-chain fatty acid found in saturated fat-containing foods with potential benefits in skin health, anti-inflammation, and metabolism enhancement [13].

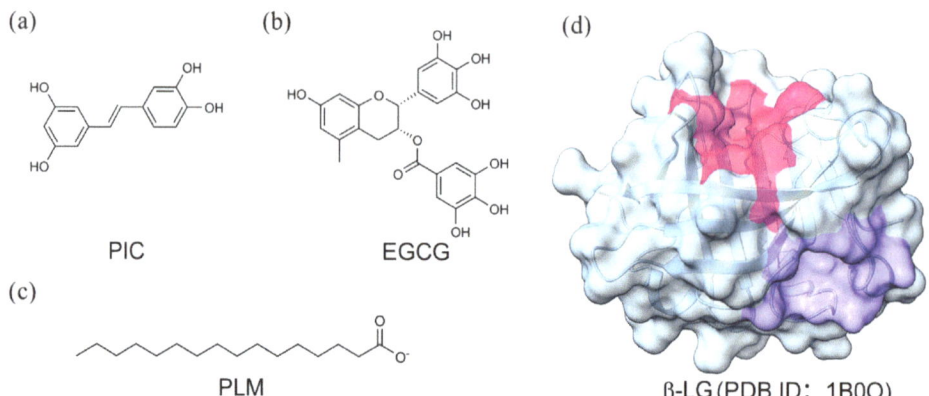

Figure 1. Schematic diagram of PIC (**a**), EGCG (**b**), and PLM (**c**) together with the structure of β-lactoglobulin (**d**). The central and peripheral binding pockets are outlined by magenta and slate blue surfaces, respectively (**d**).

In addition, there are synergistic effects among the compounds when proteins encapsulate two or more polyphenols. Studies have shown that epigallocatechin gallate is more biologically active and stable when administered in combination with other biologically active ingredients than when administered as a single substance [14]. The antioxidant properties of polyphenols and palmitate are well studied, but using polyphenols and polyphenol/palmitate combinations in conjunction with proteins is relatively understudied and warrants further research.

Direct observation of these microscopic interactions is challenging due to limitations in experimental techniques. Therefore, investigating these interactions between polyphenols and proteins through computational methods can provide theoretical insights into the microscopic mechanism for protein drug carriers, which is essential for further developing polyphenol-based anticancer drugs. Over the past decades, molecular dynamics (MD) simulations have proven to be a powerful technique that can provide complementary and microscopic insights into experimental observations. Many computational studies have attempted to gain insight into the microscopic behavior of polyphenol molecules interacting with proteins. Kinetic studies related to the binding of proteins by a single EGCG small molecule have been reported [15]. However, studies on the molecular dynamics of EGCG and PIC co-binding to proteins have not been reported, especially the cooperative interactions among drug molecules bound to protein carriers, which deserve further investigation.

In our recent study [16], we used beta-lactoglobulin (β-LG, Figure 1d) as a carrier to encapsulate EGCG, PIC, and oxidized resveratrol (OXY) and we investigated the effects of ligand–protein binding on these three active ingredients' antioxidant activity, stability, solubility, and cytotoxicity. Stability and solubility experiments showed that the addition of β-LG significantly improved the stability and solubility of the three polyphenols. β-LG, one of the most widely studied food proteins, plays a vital role in the milk of mammals. It is rich in nutrients and has multiple functional properties, making it an ideal vehicle to load a variety of natural active ingredients, which can achieve the protection of polyphenol function [17] and provide multiple health benefits. Therefore, in-depth studies on the interaction mechanism between proteins and polyphenols and the effect of palmitic acid (PLM) on the complexes between proteins and polyphenols are essential for developing effective polyphenolic anticancer drugs.

In this work, by performing molecular docking and explicit molecular dynamics simulations, we revealed the mechanism of action between polyphenol compounds and β-LG through the characterization of root-mean-square deviation (RMSD), root-mean-square fluctuation (RMSF), noncovalent interactions (NCI), principal component analysis (PCA), dynamic cross-correlation (DCC), and binding free energies, which will provide theoretical bases for improving the further research and development of polyphenol compounds.

2. Results and Discussion
2.1. Binding Modes of PLM, EGCG, and PIC with β-LG

The β-LG crystal structure retrieved from the protein data bank is a binding complex of β-LG + PLM (PDB ID: 1B0O). We first redocked PLM into β-LG to verify the accuracy of the docking method. The PLM conformation obtained by molecular docking was well superimposed with the crystal form (Figure S1), validating the accuracy of the current docking method.

We docked EGCG and PIC to β-LG separately and exported twenty possible binding poses for each compound (Figure 2). Though three to four potential binding sites were predicted for both EGCG and PIC, the peripheral pocket defined by Y20, E44, Q59, E157, Q159, and adjacent residues should be most probable since docked poses were most populated here and showed advantages in binding affinity (Figure 2a,b). The identified binding mode is consistent with previous reports. Specifically, Kanakis et al. found that EGCG binds to the sidewalls of the β-barrel structure of β-LG by spectroscopic and molecular docking studies [18]. Liu Min's team revealed that the PLM located at the central pocket of β-LG showed neglectable effects on the binding of EGCG and the PIC [16], further strengthening the reliability of our docking calculations. For compound PLM, all the docked poses were located at the center of β-LG and showed a good superimposition with the crystal configuration (Figrues 2c and S1).

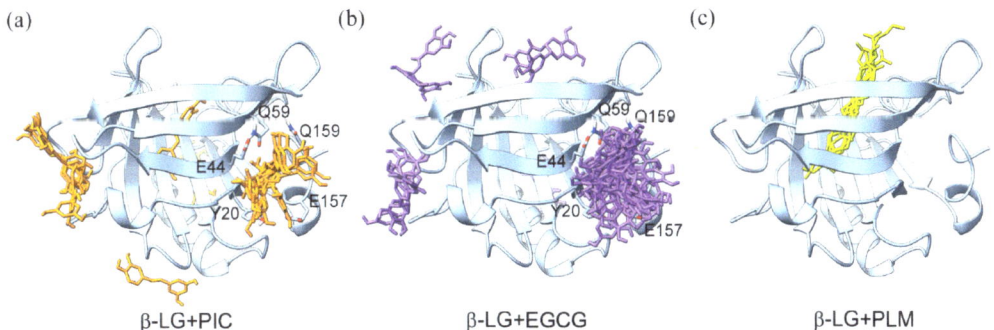

Figure 2. Docking predicted binding conformations of PIC (**a**), EGCG (**b**), and PLM (**c**) with β-LG. The molecules of PIC, EGCG, and PLM are colored in orange, purple, and yellow, respectively.

The docked poses of PIC, EGCG, and PLM molecules were ranked by the binding energies, and the root-mean-square deviation (RMSD) values relative to the first pose that showed the highest binding affinity were listed to show the conformational differences among all poses (Tables S1–S3). For PIC and EGCG, the poses that showed the most negative binding energies were recognized as potent bioactive binding conformations and were subjected to subsequent MD simulations to characterize their dynamic binding features with β-LG. For PLM, the crystal configuration was directly used (Figure S2). Moreover, Amber score [19], a physics-based scoring function embedded in Dock6 [20], was applied to re-rank the top ten poses from the Vina calculation. Two additional poses of PIC and EGCG that performed best under the Amber score were submitted to MD simulations as well, in order to investigate the effect of initial structures on the convergence of the equilibrated states. For PLM, the additional MD simulations were still based on the crystal structure.

To investigate the interactions between the bound polyphenol compounds, a molecule of EGCG/PIC was further docked to the β-LG + PIC/EGCG binary complex (Figure S3) to form the corresponding ternary binding complexes, which were denoted as (β-LG + PIC) + EGCG and (β-LG + EGCG) + PIC, respectively. Interestingly, both EGCG and PIC were found to prefer binding to the β-barrel cavity (Figure S3). The ternary binding complexes of (β-LG + PLM) + PIC and (β-LG + PLM) + EGCG were constructed by incorporating the crystal configuration of PLM and the best-scored poses of EGCG and PIC (Vina score), respectively (Figure S3). Similarly, two additional MD simulations based on different starting structures were performed for each ternary binding complex.

2.2. Binding Characteristics of the β-LG and Compounds of PIC, EGCG, and PLM

The RMSD results of the binary complexes indicated that β-LG and the binding compounds of PIC, EGCG, and PLM achieved equilibrium within 600 ns (Figure 3a–c). The RMSF profiles of β-LG that were calculated based on the MD trajectories showed a consistent fluctuation pattern with the one that was converted from the PDB B-factors (Figure 3d–f), enhancing the credibility of our MD simulations. Observations on the MD-equilibrated conformations of the binary complexes found that, in spite of some orientational alternations, the bound PIC, EGCG, and PLM remained located at their initial binding sites (Figures 3g–i and S2), indicating their high binding stability with β-LG.

PCA calculation showed that over 20% of essential motions can be characterized by the first two eigenvectors of the PIC/EGCG/PLM-bound β-LG (Figure S4), which were, therefore, represented with a porcupine plot (Figure 4a–c). These two eigenvectors mainly corresponded to the motions of peripheral loop structures, and the central β-barrels were of high stability, as indicated by the small arrows. Notably, the regions presenting large motions are the ones with high RMSF values (Figure 3d–f), mutually verifying the reliability of the two results.

Dynamics cross-correlations between the Cα atom pairs are shown in Figure 4d–f. The pairwise cross-correlation coefficients indicate the extent to which the fluctuation of an atom is correlated or anticorrelated with another atom. For the PIC-bound β-LG, strong positive correlations such as amino acids (aa) 22–40 and aa 100–120, and strong negative correlations such as aa 2–20 and aa 22–40, were identified (Figure 4d). The EGCG and PLM-bound β-LG showed a similar correlation pattern, with the latter being weaker in correlation strength (Figure 4e,f).

In order to accurately identify the close interactions between β-LG and the bound compounds, NCIplot analysis was performed to present the interactions as isosurfaces (Figure 4g–i). All three compounds interacted with β-LG through vdW interactions, as indicated by the extensive green isosurfaces. In addition, both PIC and EGCG formed hydrogen bonds with the main chain of V43, and PLM tended to form hydrogen bonds and electrostatic interactions with K60 and K69.

MD simulations on the additional replicas of β-LG + PIC/EGCG/PLM showed that the equilibrated PIC and EGCG in replica 2 and PLM in both replica 2 and replica 3 superimposed well with the ones in replica 1, indicating the convergence of our main (replica 1) MD simulations (Figure S5). The locational difference of compounds PIC and EGCG in replica 3 is probably due to their over-deviated initial structures relative to those in replica 1 (Figure S2).

2.3. Characteristics of the (β-LG + PLM) + PIC/EGCG Ternary Complexes

In order to explore the effect of PLM presence in the β-LG central cavity on the binding of PIC and EGCG, molecular dynamics simulations of the (β-LG + PLM) + PIC and (β-LG + PLM) + EGCG ternary complexes were performed. The converged RMSD profiles in Figures 5a and 6a indicate that both binding complexes achieved MD equilibrium within a 600 ns simulation. Affected by the binding compounds, the RMSF profiles showed obvious deviations at the C-terminal and loop of aa 125–129 for the PLM + PIC and PLM + EGCG-bound β-LG, respectively (Figure S6a,b). PCA revealed that over 20% of essential motions can be characterized by the first two eigenvectors for the PLM + PIC- and PLM + EGCG-bound β-LG (Figure S4d,e). For the PLM + PIC-bound β-LG, major motions happened at the C-terminal and the loop of aa 122–129 (Figure 5b). For the PLM + EGCG-bound β-LG, major motions happened at the loops of aa 47–53, aa 83–89, and aa 102–107 with relatively lower scales (Figure 6b). Correspondingly, for the PLM + PIC and PLM + EGCG-bound β-LG, the DCC maps presented relatively strong (both positive and negative) and weak (mainly positive) correlations within and between the Cα pairs of loop residues, respectively (Figures 5c and 6c).

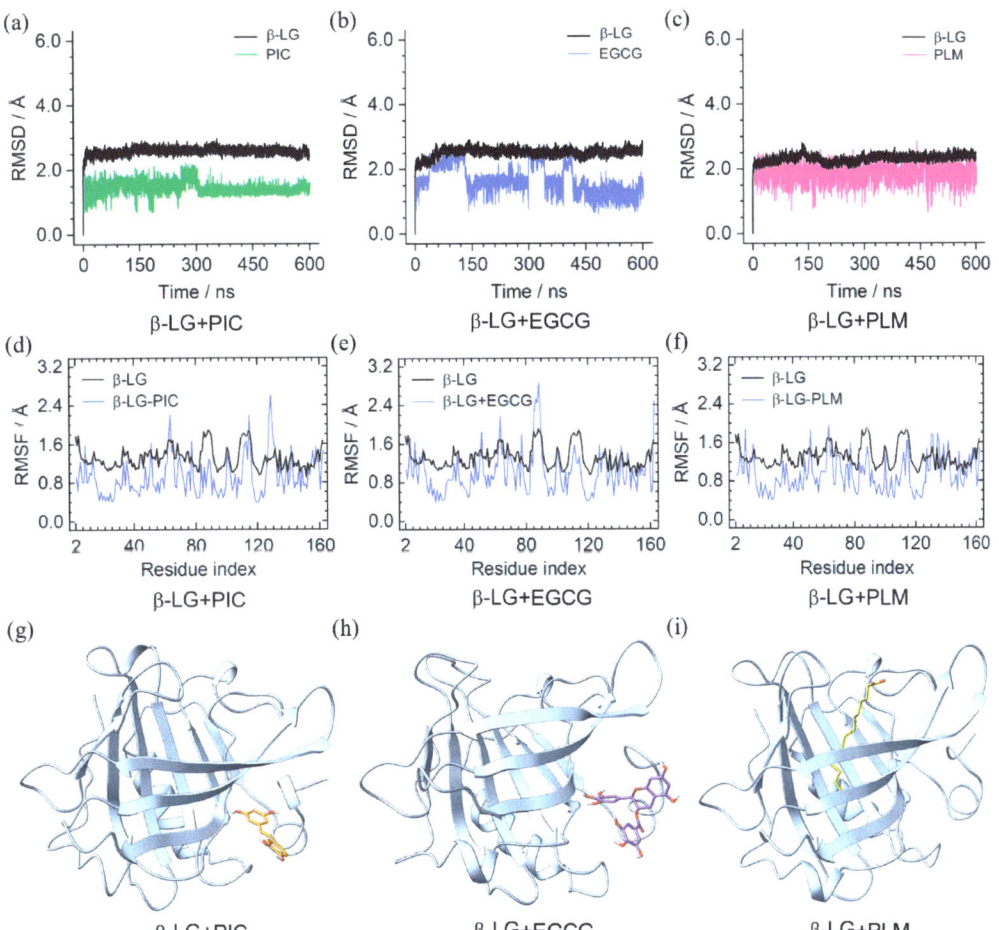

Figure 3. MD simulations of the binary complexes of β-LG + PIC/EGCG/PLM. (**a–c**) RMSD profiles of β-LG and the bound compounds; (**d–f**) root-mean-square fluctuations (RMSFs) of the compound-bound β-LG; (**g–i**) equilibrated binding conformations of the binary complexes. The molecules of PIC, EGCG, and PLM are colored in orange, purple, and yellow, respectively.

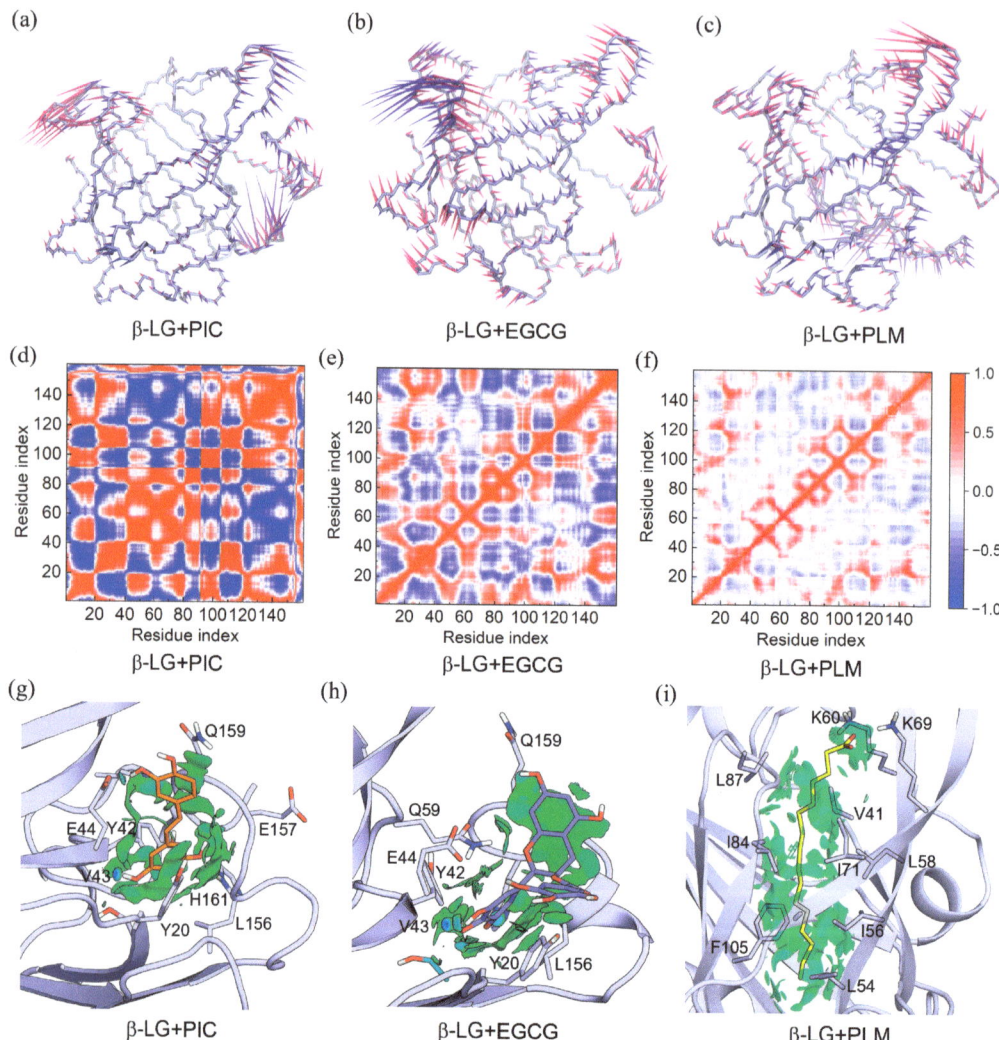

Figure 4. The dynamics feature of the PIC/EGCG/PLM-bound β-LG and their intermolecular interactions. (**a–c**) Porcupine plots of the PIC/EGCG/PLM-bound β-LG, with the first and second eigenvectors colored in violet and magenta, respectively; (**d–f**) dynamic cross-correlation map for the Cα atom pairs within the PIC/EGCG/PLM-bound β-LG. Correlation coefficients are shown as different colors, with values from 0 to 1 representing positive correlations, whereas values from −1 to 0 represent negative correlations; (**g–i**) noncovalent interactions between β-LG and the binding compounds (isovalue of 0.3 au). The green and blue isosurfaces indicate the vdW and hydrogen bond interactions, respectively.

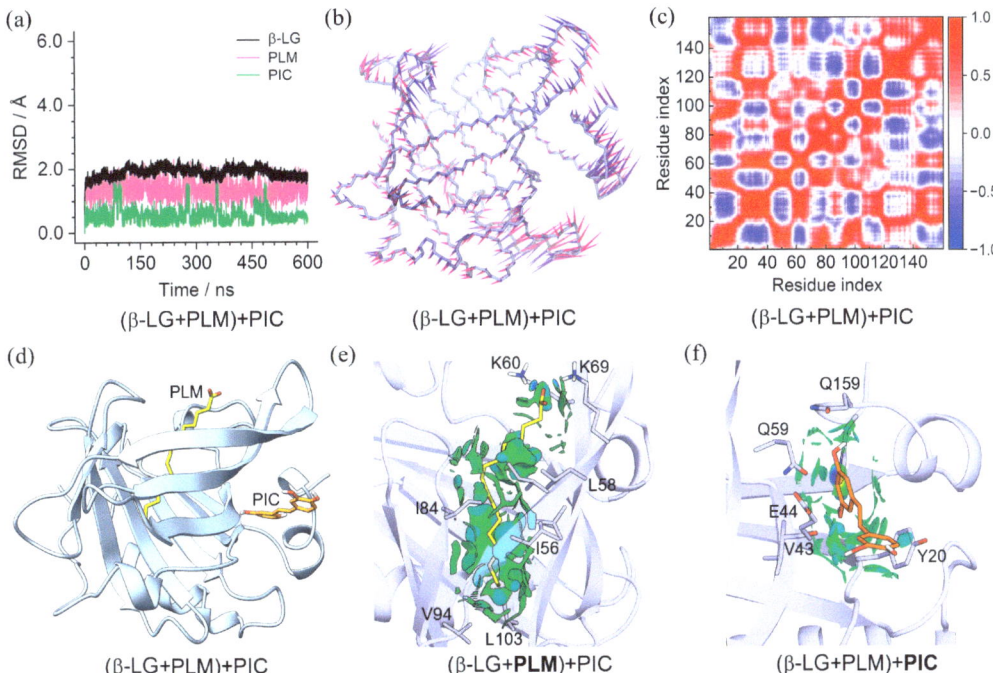

Figure 5. Characteristics of the (β-LG + PLM) + PIC binding complex. (**a**) RMSDs of β-LG and the binding PLM and PIC; (**b**) porcupine plot of the first (violet) and the second (magenta) eigenvectors of the PLM and PIC-bound β-LG; (**c**) dynamic cross-correlation map for the Cα atom pairs within the PLM and PIC-bound β-LG. Correlation coefficients are shown as different colors, with values from 0 to 1 representing positive correlations, whereas values from −1 to 0 represent negative correlations; (**d**) MD-equilibrated binding conformation of (β-LG + PLM) + PIC; (**e**,**f**) NCI surface around PLM and PIC in the binding site of β-LG (isovalue of 0.3 au). The green and blue isosurfaces indicate the vdW and hydrogen bond interactions, respectively.

The equilibrated binding conformations shown in Figures 5d and 6d demonstrate that PLM and PIC/EGCG located at the same binding site as their initial binding and similar cases were found for replica 2 and replica 3 (Figure S7c,f), indicating their converged binding states. A closer observation of the intermolecular interactions through NICplot revealed that the hydrophobic residues of the b-barrel played a key role in the binding of PLM through vdW interactions, with K60 and K69 providing essential contributions as well through hydrogen bond and electrostatic interactions (Figures 5e and 6e). For PIC, vdW interactions with hydrophobic residues such as Y20 and hydrogen bond interaction with H161 were discovered (Figure 5f). For EGCG, more extensive vdW interactions with surrounding hydrophobic residues and hydrogen bond interactions with the main chains of V43 and L156 and the side chains of E44 and Q59 were identified (Figure 6f).

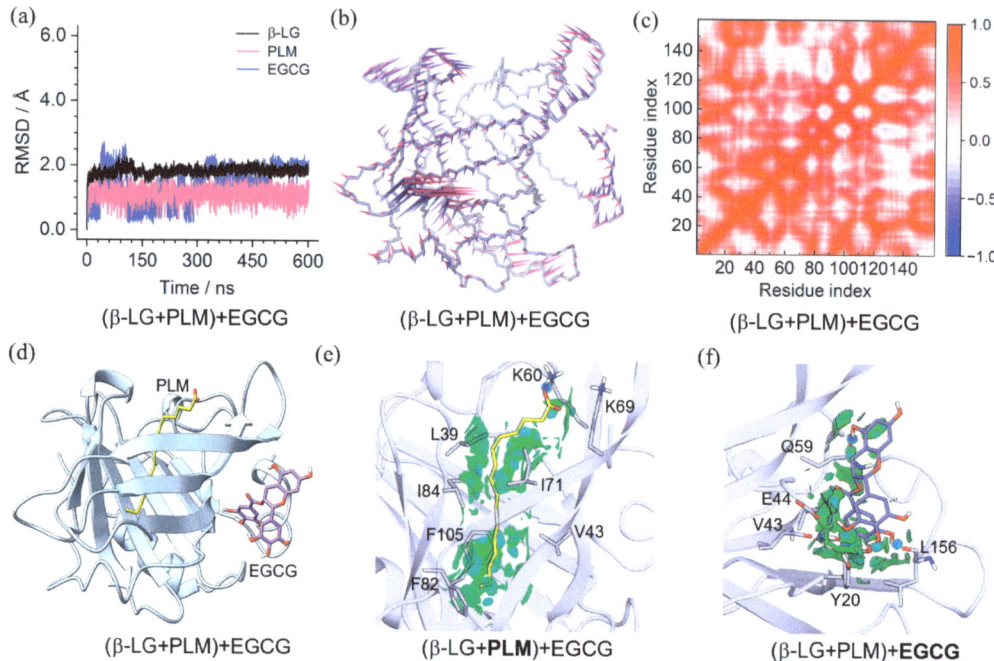

Figure 6. Characteristics of the (β-LG + PLM) + EGCG binding complex. (**a**) RMSDs of β-LG and the binding PLM and EGCG; (**b**) porcupine plot of the first (violet) and the second (magenta) eigenvectors of the PLM and EGCG-bound β-LG; (**c**) dynamic cross-correlation map for the Cα atom pairs within the PLM and EGCG-bound β-LG. Correlation coefficients are shown as different colors, with values from 0 to 1 representing positive correlations, whereas values from −1 to 0 represent negative correlations; (**d**) MD-equilibrated binding conformation of (β-LG + PLM) + EGCG; (**e**,**f**), NCI surface around PLM and EGCG in the binding site of β-LG (isovalue of 0.3 au). The green and blue isosurfaces indicate the vdW and hydrogen bond interactions, respectively.

2.4. Binding Characteristics of the (β-LG + PIC/EGCG) + EGCG/PIC Ternary Complexes

To investigate the co-binding feature of PIC and EGCG with β-LG, MD simulations were performed on the ternary complexes of (β-LG + PIC) + EGCG and (β-LG + EGCG) + PIC. The results shown in Figures 7 and 8 indicate that both complexes reached an equilibrium state within 600 ns. PCA showed that 56.36% and 29.27% of essential motions of the PIC + EGCG- and EGCG + PIC-bound β-LG can be represented by the first two eigenvectors, with the corresponding motions mainly located at the loop of aa 83–90 ((β-LG + PIC) + EGCG) and at the C-terminal and the loops of aa 83–90 and aa 107–112 ((β-LG + EGCG) + PIC), respectively (Figure S4f,g). The DCC maps shared a similar correlation pattern of Cα pairs, with the ones in (β-LG + PIC) + EGCG presenting slightly higher correlation strength (Figures 7f and 8f). Moreover, the highest RMSF values were discovered at the aa 83–90 in both RMSF profiles (Figure S6c,d), consistent with the PCA result.

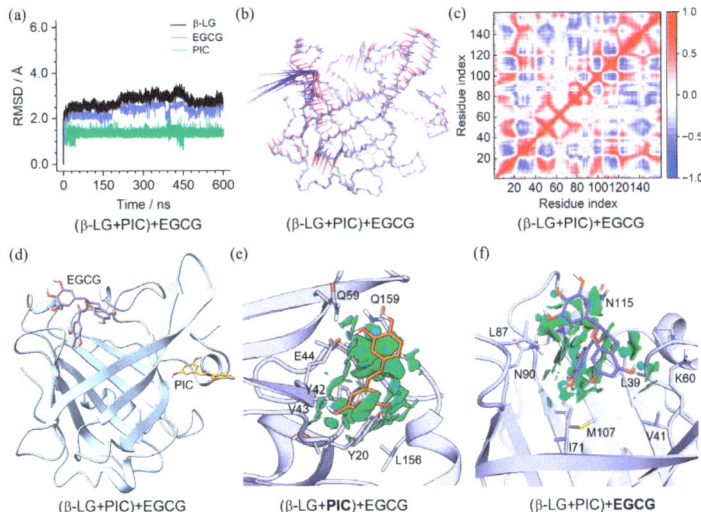

Figure 7. Characteristics of the (β-LG + PIC) + EGCG binding complex. (**a**) RMSDs of β-LG and the binding PIC and EGCG; (**b**) porcupine plot of the first (violet) and the second (magenta) eigenvectors of the PIC and EGCG-bound β-LG; (**c**) dynamic cross-correlation map for the Cα atom pairs within the PIC and EGCG-bound β-LG. Correlation coefficients are shown as different colors, with values from 0 to 1 representing positive correlations, whereas values from −1 to 0 represent negative correlations; (**d**) MD-equilibrated binding conformation of (β-LG + PIC) + EGCG; (**e,f**) NCI surface around PIC and EGCG in the binding site of β-LG (isovalue of 0.3 au). The green and blue isosurfaces indicate the vdW and hydrogen bond interactions, respectively.

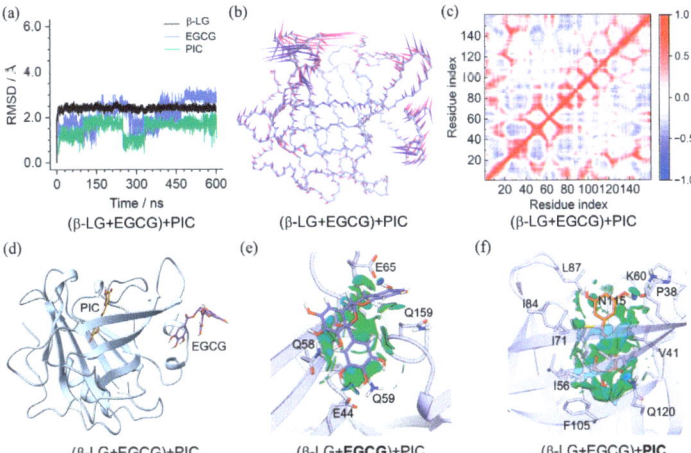

Figure 8. Characteristics of the (β-LG + EGCG) + PIC binding complex. (**a**) RMSDs of β-LG and the binding EGCG and PIC; (**b**) porcupine plot of the first (violet) and the second (magenta) eigenvectors of the EGCG and PIC-bound β-LG; (**c**) dynamic cross-correlation map for the Cα atom pairs within the EGCG and PIC-bound β-LG. Correlation coefficients are shown as different colors, with values from 0 to 1 representing positive correlations, whereas values from −1 to 0 represent negative correlations; (**d**) MD-equilibrated binding conformation of (β-LG + EGCG) + PIC; (**e,f**) NCI surface around EGCG and PIC in the binding site of β-LG (isovalue of 0.3 au). The green and blue isosurfaces indicate the vdW and hydrogen bond interactions, respectively.

The equilibrated binding conformation of (β-LG + PIC) + EGCG shown in Figures 7d and S7i demonstrated a converged binding mode, although EGCG that bound to the central cavity of β-LG exhibited relatively larger conformational variations. PIC in the peripheral pocket showed the same binding mode as in the β-LG + PIC complex (Figures 4g and 7e). However, EGCG in the central site adopted a different binding mode compared to PLM; it can only be located at the upper part of the pocket due to its much larger molecular size (Figure 7f), whereas the equilibrated conformations of (β-LG + EGCG) + PIC shown in Figures 8d and S7l indicate a converged binding mode of PIC and a loose binding of EGCG. EGCG in all replicas is apparently more outward compared to the one in the β-LG + EGCG (Figures 3h and 8d), resulting in more fragmented NCI isosurfaces (Figure 8e). PIC in the central site showed a similar binding mode to PLM, though the binding strength may be probably lowered due to its disadvantage in molecular length (Figure 8f).

2.5. Binding Free Energies between β-LG and the Binding Compounds

To evaluate the affinities between β-LG and binding compounds, binding free energies were obtained through molecular mechanics/generalized Born surface (MM/GBSA) calculations. The results in Table 1 indicate that the vdW interaction (ΔE_{vdW}) contributed most to the binding in all cases. The contributions from electrostatic interactions (ΔE_{ele}) and the polar solvation effects (ΔG_{GB}) are always opposite, and the net contributions from ΔE_{ele} and solvation effects (including the polar part ΔG_{GB} and the nonpolar part ΔG_{SA}) are small relative to ΔE_{vdW}. According to the equation of $\Delta G_{bind} = \Delta H - T\Delta S$, the entropy item always makes unfavorable contributions since the binding led to the decrease in degrees of freedom.

The binding affinities of compounds followed an order of PIC (-10.32 kcal·mol^{-1}) < EGCG (-17.20 kcal·mol^{-1}) < PLM (-19.30 kcal·mol^{-1}) in the binary complexes. Interestingly, PLM binding in the central cavity was found to enhance PIC's binding affinity to -12.83 kcal·mol^{-1}, and in the meantime, the PLM's binding affinity remained basically unchanged, presenting a synergistic effect. Such an effect was not observed between the co-bound PLM and EGCG; PLM binding led to the decrease in EGCG's binding affinity (-15.71 kcal·mol^{-1}), but, conversely, the bound EGCG increased the binding strength of PLM (-21.88 kcal·mol^{-1}). With PIC binding at the peripheral site, EGCG showed a comparable affinity (-17.03 kcal·mol^{-1}) by binding at the central site, indicating that PIC and EGCG may concomitantly bind to β-LG ((β-LG + PIC) + EGCG) without affecting each other's binding strength. However, PIC binding at the central site would significantly decrease the binding affinity of EGCG (-6.37 kcal·mol^{-1}), indicative of the unfavorable binding mode of (β-LG + EGCG) + PIC.

Table 1. Binding free energies between β-LG and the binding compounds [a].

Receptor	Ligand	Energy Component					
		ΔE_{ele}	ΔE_{vdW}	ΔG_{GB}	ΔG_{SA}	$-T\Delta S$	ΔG_{bind}
β-LG	PIC	-13.66 ± 1.47	-24.28 ± 3.28	13.81 ± 0.94	-3.68 ± 0.12	17.50 ± 11.76	-10.32
β-LG	EGCG	-13.74 ± 3.23	-35.50 ± 3.03	15.76 ± 2.75	-4.91 ± 0.23	21.18 ± 10.27	-17.20
β-LG	PLM	12.36 ± 7.19	-36.38 ± 2.76	-12.82 ± 5.77	-6.23 ± 0.31	23.78 ± 11.07	-19.30
β-LG	PLM	19.47 ± 7.58	-37.04 ± 0.05	-18.87 ± 6.15	-6.16 ± 0.17	23.43 ± 10.49	-19.17
	PIC	-12.84 ± 1.54	-23.33 ± 3.10	13.01 ± 0.94	-3.75 ± 0.09	14.09 ± 10.99	-12.83
β-LG	PLM	15.4 ± 8.84	-35.85 ± 2.84	-15.54 ± 7.15	-6.26 ± 0.21	20.38 ± 11.04	-21.88
	EGCG	-17.07 ± 5.37	-36.93 ± 4.78	19.1 ± 4.25	-5.22 ± 0.20	24.41 ± 11.44	-15.71
β-LG	PIC	-12.85 ± 1.58	-22.91 ± 3.23	12.85 ± 0.99	-3.63 ± 0.11	16.72 ± 10.55	-10.00
	EGCG	-3.95 ± 2.52	-37.32 ± 5.03	7.83 ± 2.14	-5.40 ± 0.40	21.81 ± 11.98	-17.03
β-LG	EGCG	-9.45 ± 4.74	-26.87 ± 6.59	11.73 ± 3.97	-4.09 ± 0.48	22.31 ± 10.25	-6.37
	PIC	-3.71 ± 1.22	-32.83 ± 2.45	6.78 ± 0.81	-4.73 ± 0.12	17.31 ± 9.25	-17.18

[a] Energies are in kcal·mol^{-1}.

The binding free energies were further decomposed to evaluate per-residue contributions, under conditions in which entropic contributions were excluded. Intriguingly, the per-residue free energy contributions calculated based on the trajectories of the last

100 ns of MD simulations were consistent with the information shown in the NCIplot (Figures 4–9), which was based on the MD-equilibrated structures. Moreover, except for the charged residues (such as E44, K60, and K69) that contributed to the binding mainly through electrostatic interactions, the other residues preferred to bind with the compounds mainly through vdW interactions, as indicated by the green columns in Figure 9. The residues that made significant contributions to the binding in the binary and ternary complexes were clearly identified, providing valuable information for the understanding of the protection mechanism of β-LG to functional compounds.

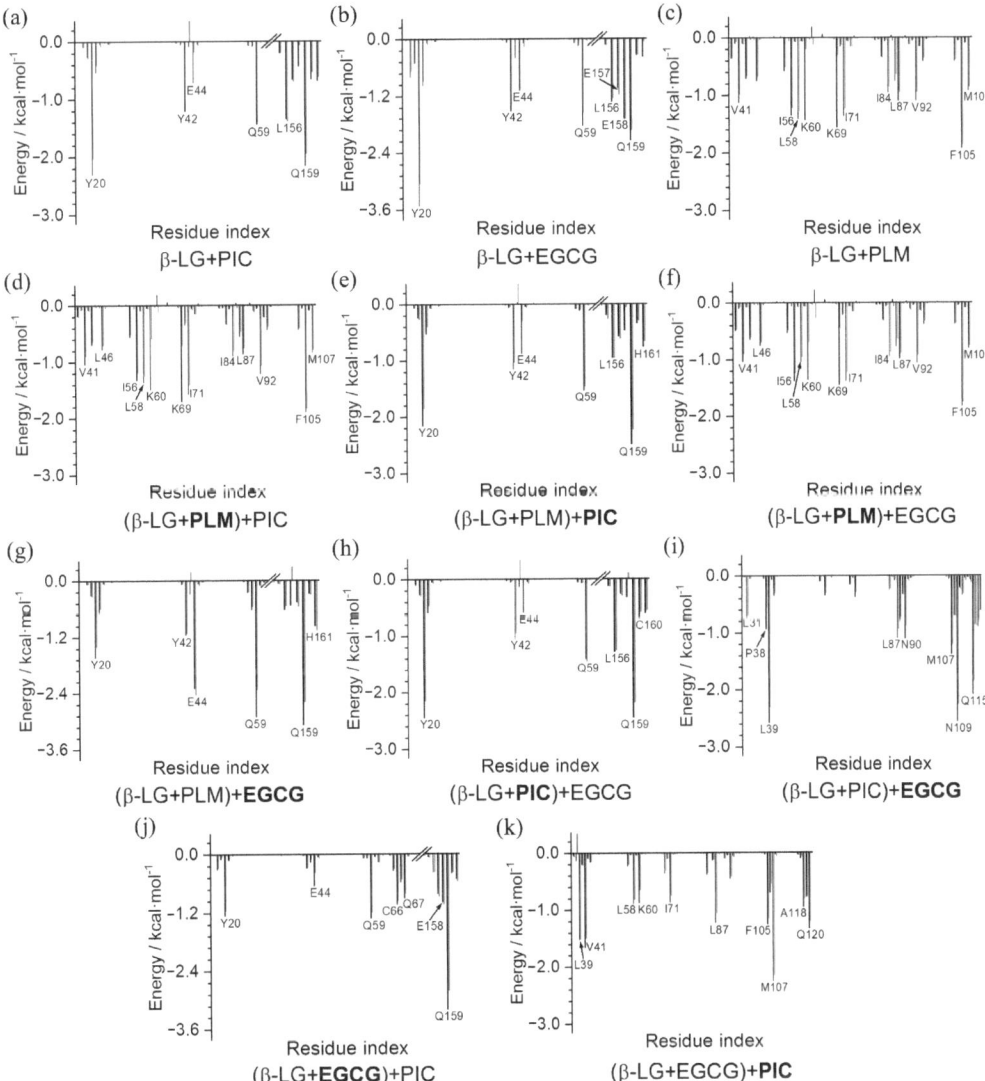

Figure 9. Per-residue decomposition of the binding free energy. (**a**–**k**) The contributions from the vdW interactions and the combined overall contributions are represented by the green and black columns, respectively.

3. Methods

3.1. Data

The initial structure of β-LG was retrieved from the PDB data bank with the ID of 1B0O [21]. The structures of PIC/EGCG were generated using the GaussView software version 6.0.16 (Gaussian, Wallingford, CT, USA) and were optimized at the DFTB3LYP/6-31G(d) level [22,23]. The atomic partial charges of PIC, EGCG, and PLM were calculated using the restricted electrostatic potential (RESP) method with a basis set of HF/6-31G(d) [24]; the force field parameters of these compounds were then generated using AmberTools (AMBER, San Francisco, CA, USA) [25].

3.2. Molecular Docking

Molecular docking calculations were performed using the AutoDock Vina 1.2.5 software [26]. The protonation states of the titratable residues of β-LG were determined with PropKa 3.0 [27] and the compounds of PIC, EGCG, and PLM were prepared with the Auto-DockTools software (version 1.5.6) [28]. The Gasteiger charges were computed for both β-LG and compounds, with the nonpolar hydrogen atoms merged. All the rotatable bonds of compounds were set as flexible, while β-LG was set as the rigid receptor. In each docking calculation, a cubic box centered at the geometric center of β-LG comprising 50 × 50 × 50 grids with a grid spacing of 1.0 Å was used to define the possible binding region. The box was large enough to encompass the whole structure of β-LG so that no binding modes were pre-excluded. The exhaustiveness parameter was set to 17 and all other parameters were set as the default. To construct the ternary binding complexes of (β-LG + PLM) + PIC/EGCG and (β-LG + PIC/EGCG) + EGCG/PIC, the second compound was docked to β-LG with the binding region defined by a cubic box that was just enough to encompass the corresponding pocket.

3.3. Molecular Dynamics

Based on the docking-derived binding structures, the binary complexes of β-LG + PIC/EGCG/PLM and the ternary complexes of (β-LG + PLM) + PIC/EGCG and (β-LG + PIC/EGCG) + EGCG/PIC were subjected to MD simulations by using the Amber 20 software [29,30]. Each binding complex was placed in a truncated octahedron box of TIP3P water molecules at a margin distance of 10.0 Å. Environmental Na+ ions were added to maintain electrical neutrality. The Amber ff14SB force field was applied for β-LG. For the compounds, RESP charge together with the second generation of general Amber force field (GAFF2) was applied [31]. Each model was firstly energy minimized by 10,000 steps of steepest descent minimization with a harmonic constraint of 500 kcal·mol^{-1}·Å^{-2} imposed on binding complex (solute), followed by 10,000 steps of conjugated gradient minimization with no constraint. Then, the system was gradually heated from 0 to 300 K under the NVT ensemble for 500 ps, with a weak constraint of 10 kcal·mol^{-1}·Å^{-2} imposed on the solute. The model was subsequently subjected to an equilibrium simulation for 1 ns by removing all constraints. Finally, the production simulation for each model was conducted under the NPT ensemble with a simulation time of 600 ns (replica 1). To investigate the convergence of MD simulations, two additional replicas based on different starting structures for each model were simulated for 200 ns. In all MD simulations, parameters were set according to our previous report [32]. MD trajectories were recorded at an interval of 10 ps for the structural and energetic analyses.

3.4. Principal Component Analysis

Since principal component analysis (PCA) can filter the essential degrees of freedom from a number of local fluctuations [33], PCA was carried out for all MD trajectories using the interactive essential dynamics (IED) method [34]. Based on 5000 frames evenly extracted from the last 100 ns of MD trajectories, PCA of β-LG backbones was carried out for each model by using the CPPTRAJ module of AmberTools. The graphical summaries of motions along the first two eigenvectors are shown in porcupine plots using the VMD software version 1.9.3 [35].

3.5. Noncovalent Interactions

NCIplot calculations were carried out with a step size of 0.10 to visualize the interacting regions between β-LG and the binding compounds [36,37]. The reduced gradients were rendered as an isosurface in VMD, using an isovalue of 0.3 au.

3.6. Binding Free Energy Analysis

The binding free energy between β-LG and EGCG/PIC was obtained from MM/GBSA calculations [38]. A total of 200 snapshots evenly extracted from the last 100 ns of the MD trajectory were used for the calculation of each binding complex. The binding free energy value is equal to the free energy difference between the binding complex ($G_{complex}$) and the sum of β-LG ($G_{receptor}$) and compound (G_{ligand}) as follows:

$$\Delta G_{bind} = G_{complex} - \left(G_{receptor} + G_{ligand}\right) \quad (1)$$

Each of them can be calculated with the equation:

$$\Delta G_{bind} = \Delta H - T\Delta S \approx \Delta E_{MM} + \Delta G_{solv} - T\Delta S \quad (2)$$

where ΔE_{MM} is the molecular mechanical energy of the gas phase, ΔG_{solv} is the solvation-free energy, and $T\Delta S$ is the entropy contribution. ΔE_{MM} comprises contributions from electrostatic energy (ΔE_{ele}), van der Waals interaction energy (ΔE_{vdw}), and internal strain energy (ΔE_{int}). Because we adopted the single simulation approach (only simulating the binding complex), ΔE_{int}, which comprises bonds, angles, and dihedral energies, will cancel out according to Equation (1) [39]:

$$\Delta E_{MM} = \Delta E_{ele} + \Delta E_{vdw} + \Delta E_{int} \quad (3)$$

ΔG_{solv} contains contributions from a polar part (ΔG_{GB}) and a nonpolar (ΔG_{SA}) part:

$$\Delta G_{solv} = \Delta G_{GB} + \Delta G_{SA} \quad (4)$$

ΔG_{GB} was estimated by the generalized Born (GB) model with the interior and exterior dielectric constants set to 4 and 80, respectively [40]. The nonpolar solvation terms were calculated according to the LCPO algorithm:

$$\Delta G_{SA} = \gamma \Delta SASA + \beta \quad (5)$$

where γ and β were set to 0.0072 kcal·mol^{-1}·Å^{-2} and 0, respectively [39,41]. Therefore, the binding free energy was calculated as follows:

$$\Delta G_{bind} = \Delta E_{ele} + \Delta E_{vdw} + \Delta G_{GB} + \Delta G_{SA} - T\Delta S \quad (6)$$

Based on the extracted snapshots, the entropic contribution ($T\Delta S$) was evaluated through normal mode analysis (NMA) [42,43].

4. Conclusions

Through intensive molecular docking and molecular dynamics simulations combined with in-depth analyses, the current work characterized the binding features of β-LG with functional compounds including PIC, EGCG, and PLM. Observations on the interaction modes and binding affinities revealed that polyphenols PIC and EGCG may concomitantly bind to the peripheral and central sites, respectively. The central bound PLM showed a synergistic effect on PIC. Key residues that contributed to the binding were identified, and vdW interactions were discovered to play a pivotal role in the binding of β-LG and all compounds.

The major limitation of the current study comes from the research methods of molecular docking and MD simulation. Though three replicas with distinct starting structures for each model were simulated, differences exist between the models and experimental conditions. However, we believe the limitation should hardly influence the conclusion since the investigation was based on the crystal structure of β-LG, and the docking-derived binding sites of compounds are in agreement with the reported experimental results.

Based on our identified interaction modes and key residues that form direct contacts with the binding compounds, site-directed mutagenesis would certainly lead to improved β-LG variants with higher affinities to the target compound. In all, the study here provides pivotal insights into the binding characteristics of β-LG with PIC, EGCG, and PLM, shedding new light on the development of β-LG-based protection and transportation of functional compounds.

Supplementary Materials: The following supporting information can be downloaded at: https://www.mdpi.com/article/10.3390/molecules29050956/s1. Table S1. Molecular docking results of β-LG and PIC calculated with AutoDock Vina and scored with Dock6. Table S2. Molecular docking results of β-LG and EGCG calculated with AutoDock Vina and scored with Dock6. Table S3. Molecular docking results of β-LG and PLM calculated with AutoDock Vina and scored with Dock6. Table S4. Molecular docking results of β-LG + PIC and EGCG calculated with AutoDock Vina and scored with Dock6. Table S5. Molecular docking results of β-LG + EGCG and PIC calculated with AutoDock Vina and scored with Dock6. Table S6. Detailed information of the constructed models. Figure S1. Superimposition of the best-performed docking conformation of PLM (green) on the crystal configuration (yellow). Figure S2. Initial structures of the β-LG+PIC/EGCG/PLM binary complexes submitted to MD simulations. For PIC and EGCG, three initial docking structures, i.e. replica 1 (orange), replica 2 (blue), and replica 3 (violet), were used in (a) and (b). For PLM, the crystal configuration was used (c). Figure S3. Initial structures for the simulation of the (β-LG + PLM) + PIC/EGCG and (β-LG + PIC/EGCG) + EGCG/PIC tertiary complexes. (a–d) For PIC and EGCG, three initial docking structures, i.e. replica 1 (orange), replica 2 (blue), and replica 3 (violet), were used. Figure S4. Eigenvalue profiles plotted by the first 30 eigenvectors of the compound bound β-LG in PCA analysis. Figure S5. RMSD profiles of the additional replica MD simulations and the conformational comparison of the equilibrated structures. For clarity, only the β-LG structure in replica 1 was shown, with the compounds in replica 1, replica 2, and replica 3 colored in orange, blue, and violet, respectively. Figure S6. RMSF profiles of β-LG with two compounds bound concomitantly. (a–d) RMSFs derived from trajectory analysis and the b-factor of crystal structure are colored in blue and black, respectively. Figure S7. RMSD profiles of the additional replica MD simulations and the conformational comparison of the equilibrated structures. For clarity, only the β-LG structure in replica 1 was shown, with the compounds in replica 1, replica 2, and replica 3 colored in orange, blue, and violet, respectively.

Author Contributions: T.M.: investigation, data curation, writing—original draft. Z.W.: conceptualization, writing—review and editing, methodology. H.Z.: investigation, data curation. Z.Z.: investigation, software, and data curation. W.H.: investigation, data curation. L.X.: investigation, data curation. M.L.: conceptualization, methodology. J.L.: writing—final approval of the version to be published, review and editing, funding acquisition. H.Y.: writing—review and editing, funding acquisition, and project administration. All authors have read and agreed to the published version of the manuscript.

Funding: This research was funded by the Natural Science Foundation of Shandong Province, China (ZR2021MB097) and the Open Project of Shandong Collaborative Innovation Center for Antibody Drugs (CIC-AD1819).

Institutional Review Board Statement: Not applicable.

Informed Consent Statement: Not applicable.

Data Availability Statement: Data are contained within the article and Supplementary Materials.

Acknowledgments: The authors thank the Scientific Research Cloud Platform in the School of Chemistry and Chemical Engineering at Liaocheng University for the support of numerical calculations.

Conflicts of Interest: The authors declare no conflict of interest.

References

1. Ahn, W.S.; Huh, S.W.; Bae, S.M.; Lee, I.P.; Sin, J.I. A major constituent of green tea, EGCG, inhibits the growth of a human cervical cancer cell line, CaSki cells, through apoptosis, G_1 arrest, and regulation of gene expression. *DNA Cell Biol.* **2003**, *22*, 217–224. [CrossRef]
2. Roy, A.M.; Baliga, M.S.; Katiyar, S.K. Epigallocatechin-3-gallate induces apoptosis in estrogen receptor–negative human breast carcinoma cells via modulation in protein expression of p53 and Bax and caspase-3 activation. *Mol. Cancer Ther.* **2005**, *4*, 81–90. [CrossRef]
3. Sen, T.; Dutta, A.; Chatterjee, A. Epigallocatechin-3-gallate (EGCG) downregulates gelatinase-B (MMP-9) by involvement of FAK/ERK/NFκB and AP-1 in the human breast cancer cell line MDA-MB-231. *Anticancer Drugs* **2010**, *21*, 632–644. [CrossRef]
4. Meeran, S.M.; Patel, S.N.; Chan, T.H. A novel prodrug of epigallocatechin-3-gallate: Differential epigenetic hTERT repression in human breast cancer cells. *Cancer Prev. Res.* **2011**, *4*, 1243–1254. [CrossRef]
5. Li, Y.; Shen, X.; Wang, X.; Li, A.; Wang, P.; Jiang, P.; Zhou, J.; Feng, Q. EGCG regulates the cross-talk between JWA and topoisomerase II$_\alpha$ in non-small-cell lung cancer (NSCLC) cells. *Sci. Rep.* **2015**, *5*, 11009. [CrossRef]
6. Farooqi, A.A.; Mansoor, Q.; Ismail, M.; Bhatti, S. Therapeutic Effect of Epigallocatechin-3-gallate (EGCG) and Silibinin on ATM Dynamics in Prostate Cancer Cell Line LNCaP. *World J. Oncol.* **2010**, *1*, 242–246. [CrossRef]
7. Wieder, T.; Prokop, A.; Bagci, B.; Essmann, F.; Bernicke, D.; Schulze-Osthoff, K.; Drken, B.; Schmalz, H.G.; Daniel, P.T.; Henze, G.J. Piceatannol, a hydroxylated analog of the chemopreventive agent resveratrol, is a potent inducer of apoptosis in the lymphoma cell line BJAB and in primary, leukemic lymphoblasts. *Leukemia* **2001**, *15*, 1735–1742. [CrossRef]
8. Siedlecka-Kroplewska, K.; Ślebioda, T.; Kmieć, Z. Induction of autophagy, apoptosis and acquisition of resistance in response to piceatannol toxicity in MOLT-4 human leukemia cells. *Toxicol. Vitr.* **2019**, *59*, 12–25. [CrossRef] [PubMed]
9. Zhang, L.; McClements, D.J.; Wei, Z.; Wang, G.; Liu, X.; Liu, F. Delivery of synergistic polyphenol combinations using biopolymer-based systems: Advances in physicochemical properties, stability and bioavailability. *Crit. Rev. Food Sci. Nutr.* **2019**, *60*, 2083–2097. [CrossRef] [PubMed]
10. Liu, M.; Liu, T.; Shi, Y.; Zhao, Y.; Yan, H.; Sun, B.; Wang, Q.; Wang, Z.; Han, J. Comparative study on the interaction of oxyresveratrol and piceatannol with trypsin and lysozyme: Binding ability, activity and stability. *Food Funct.* **2019**, *10*, 8182–8194. [CrossRef] [PubMed]
11. Wang, L.; Cheng, H.; Ni, Y.; Liang, L. Technology, Encapsulation and protection of bioactive components based on ligand-binding property of β-lactoglobulin. *Zhongguo Shipin Xuebao* **2015**, *15*, 124–129.
12. Cheng, H.; Fang, Z.; Liu, T.; Gao, Y.; Liang, L. A study on β-lactoglobulin-triligand-pectin complex particle: Formation, characterization and protection. *Food Hydrocoll.* **2018**, *84*, 93–103. [CrossRef]
13. Ajuwon, K.M.; Spurlock, M.E. Palmitate Activates the NF-κB Transcription Factor and Induces IL-6 and TNFα Expression in 3T3-L1 Adipocytes. *J. Nutr.* **2005**, *135*, 1841–1846. [CrossRef] [PubMed]
14. Eom, D.-W.; Lee, J.H.; Kim, Y.-J.; Hwang, G.S.; Kim, S.-N.; Kwak, J.H.; Cheon, G.J.; Kim, K.H.; Jang, H.-J.; Ham, J.; et al. Synergistic effect of curcumin on epigallocatechin gallate-induced anticancer action in PC3 prostate cancer cells. *BMB Reports* **2015**, *48*, 461–466. [CrossRef]
15. Liu, C.Z.; Lv, N.; Xu, Y.Q. pH-dependent interaction mechanisms between β-lactoglobulin and EGCG: Insights from multi-spectroscopy and molecular dynamics simulation methods. *J. Food Hydrocoll.* **2022**, *133*, 108022. [CrossRef]
16. Liu, T.; Liu, H.; Ren, Y.; Zhao, Y.; Yan, H.; Wang, Q.; Zhang, N.; Ding, Z.; Wang, Z. Co-encapsulation of (−)-epigallocatechin-3-gallate and piceatannol/oxyresveratrol in β-lactoglobulin: Effect of ligand–protein binding on the antioxidant activity, stability, solubility and cytotoxicity. *Food Funct.* **2021**, *12*, 7126–7144. [CrossRef]
17. Zhang, J.; Liu, X.; Subirade, M.; Zhou, P.; Liang, L. A study of multi-ligand beta-lactoglobulin complex formation. *Food Chem.* **2014**, *165*, 256–261. [CrossRef] [PubMed]
18. Kanakis, C.D.; Hasni, I.; Bourassa, P.; Tarantilis, P.A.; Polissiou, M.G.; Tajmir-Riahi, H.-A. Milk β-lactoglobulin complexes with tea polyphenols. *Food Chem.* **2011**, *127*, 1046–1055. [CrossRef]
19. Graves, A.P.; Shivakumar, D.M.; Boyce, S.E.; Jacobson, M.P.; Case, D.A.; Shoichet, B.K. Rescoring Docking Hit Lists for Model Cavity Sites: Predictions and Experimental Testing. *J. Mol. Biol.* **2008**, *377*, 914–934. [CrossRef]
20. Brozell, S.R.; Mukherjee, S.; Balius, T.E.; Roe, D.R.; Case, D.A.; Rizzo, R.C. Evaluation of dock 6 as a pose generation and database enrichment tool. *J. Comput. Aid. Mol. Des.* **2012**, *26*, 749–773. [CrossRef]
21. Wu, S.Y.; Perez, M.D.; Puyol, P.; Sawyer, L. β-lactoglobulin binds palmitate within its central cavity. *J. Biol. Chem.* **1999**, *274*, 170–174. [CrossRef] [PubMed]
22. Wang, Y.; Li, G.; Meng, T.; Qi, L.; Yan, H.; Wang, Z.G. Molecular insights into the selective binding mechanism targeting parallel human telomeric G-quadruplex. *J. Mol. Graph. Model.* **2022**, *110*, 108058. [CrossRef] [PubMed]
23. Cieplak, P.; Cornell, W.D.; Bayly, C.; Kollman, P.A. Application of the multimolecule and multiconformational RESP methodology to biopolymers: Charge derivation for DNA, RNA, and proteins. *J. Comput. Chem.* **2010**, *16*, 1357–1377. [CrossRef]
24. Díaz-Gómez, D.G.; Galindo-Murillo, R.; Cortés-Guzmán, F. The Role of the DNA Backbone in Minor-Groove Ligand Binding. *Chem. Phys. Chem.* **2017**, *18*, 1909–1915. [CrossRef] [PubMed]

25. Salomon-Ferrer, R.; Case, D.A.; Walker, R.C. An overview of the Amber biomolecular simulation package. *Wiley Interdiscip. Rev. Comput. Mol. Sci.* **2012**, *3*, 198–210. [CrossRef]
26. Eberhardt, J.; Santos-Martins, D.; Tillack, A.F.; Forli, S. AutoDock Vina 1.2.0: New docking methods, expanded force field, and python bindings. *J. Chem. Inf. Model.* **2021**, *61*, 3891–3898. [CrossRef] [PubMed]
27. Olsson, M.H.M.; Søndergaard, C.R.; Rostkowski, M.; Jensen, J.H. PROPKA3: Consistent treatment of internal and surface residues in empirical pKa predictions. *J. Chem. Theory Comput.* **2011**, *7*, 525–537. [CrossRef]
28. Morris, G.M.; Huey, R.; Lindstrom, W.; Sanner, M.F.; Belew, R.K.; Goodsell, D.S.; Olson, A.J. Autodock4 and AutoDockTools4: Automated docking with selective receptor flexiblity. *J. Comput. Chem.* **2009**, *16*, 2785–2791. [CrossRef]
29. Braun, E.; Moosavi, S.M.; Smit, B. Anomalous effects of velocity rescaling algorithms: The flying ice cube effect revisited. *J. Chem. Theory Comput.* **2018**, *14*, 5262–5272. [CrossRef]
30. Case, D.A.; Cheatham, T.E.; Darden, T.; Gohlke, H.; Luo, R.; Merz, K.M.; Onufriev, A.; Simmerling, C.; Wang, B.; Woods, R.J. The Amber biomolecular simulation programs. *J. Comput. Chem.* **2005**, *26*, 1668–1688. [CrossRef]
31. He, X.; Man, V.H.; Yang, W.; Li, T.-S.; Wang, J. A fast and high-quality charge model for the next generation general AMBER force field. *J. Chem. Phys.* **2020**, *153*, 114502. [CrossRef]
32. Wang, Z.; Li, G.; Tian, Z.; Lou, X.; Huang, Y.; Wang, L.; Li, J.; Hou, T.; Liu, J.-P. Insight derived from molecular dynamics simulation into the selectivity mechanism targeting *c-MYC* G-quadruplex. *J. Phys. Chem. B* **2020**, *124*, 9773–9784. [CrossRef]
33. Amadei, A.; Linssen, A.B.M.; Berendsen, H.J.C. Essential dynamics of proteins. *Proteins* **1993**, *17*, 412–425. [CrossRef]
34. Mongan, J. Interactive essential dynamics. *J. Comput. Aid. Mol. Des.* **2004**, *118*, 433–436. [CrossRef]
35. Humphrey, W.; Dalke, A.; Schulten, K.K. VMD—Visual molecular dynamics. *J. Mol. Graph.* **1995**, *14*, 33–38. [CrossRef]
36. Contreras-García, J.; Johnson, E.R.; Keinan, S.; Chaudret, R.; Piquemal, J.-P.; Beratan, D.N.; Yang, W. NCIPLOT: A program for plotting noncovalent interaction regions. *J. Chem. Theory Comput.* **2011**, *7*, 625–632. [CrossRef]
37. Islam, B.; Stadlbauer, P.; Neidle, S.; Haider, S.; Sponer, J. Can we execute reliable MM-PBSA free energy computations of relative stabilities of different guanine quadruplex folds? *J. Phys. Chem. B* **2016**, *120*, 2899–2912. [CrossRef]
38. Hou, T.J.; Wang, J.M.; Li, Y.Y.; Wang, W. Assessing the performance of the MM/PBSA and MM/GBSA methods. 1. The accuracy of binding free energy calculations based on molecular dynamics simulations. *J. Chem. Inf. Model.* **2011**, *51*, 69–82. [CrossRef] [PubMed]
39. Genheden, S.; Ryde, U. How to obtain statistically converged MM/GBSA results. *J. Comput. Chem.* **2010**, *31*, 837–846. [CrossRef] [PubMed]
40. Onufriev, A.; Bashford, D.; Case, D.A. Exploring protein native states and large-scale conformational changes with a modified generalized born model. *Proteins* **2004**, *55*, 383–394. [CrossRef] [PubMed]
41. Chen, F.; Sun, H.; Wang, J.; Zhu, F.; Liu, H.; Wang, Z.; Lei, T.; Li, Y.; Hou, T. Assessing the performance of MM/PBSA and MM/GBSA methods. 8. Predicting binding free energies and poses of protein–RNA complexes. *RNA* **2018**, *24*, 1183–1194. [CrossRef] [PubMed]
42. Hou, J.-Q.; Chen, S.-B.; Tan, J.-H.; Luo, H.-B.; Li, D.; Gu, L.-Q.; Huang, Z.-S. New insights from molecular dynamic simulation studies of the multiple binding modes of a ligand with G-quadruplex DNA. *J. Comput. Aid. Mol. Des.* **2012**, *26*, 1355–1368. [CrossRef] [PubMed]
43. Weiser, J.; Shenkin, P.S.; Still, W.C. Approximate solvent-accessible surface areas from tetrahedrally directed neighbor densities. *J. Biopolym.* **1999**, *50*, 373–380. [CrossRef]

Disclaimer/Publisher's Note: The statements, opinions and data contained in all publications are solely those of the individual author(s) and contributor(s) and not of MDPI and/or the editor(s). MDPI and/or the editor(s) disclaim responsibility for any injury to people or property resulting from any ideas, methods, instructions or products referred to in the content.

Article

Size- and Voltage-Dependent Electron Transport of C₂N-Rings-Based Molecular Chains

Dian Song [1], Jie Li [1,*], Kun Liu [1], Junnan Guo [2], Hui Li [2,*] and Artem Okulov [3]

1. School of Materials Science and Engineering, Jiangsu University of Science and Technology, Zhenjiang 212100, China; sssdian2001@163.com (D.S.); liu_kun@163.com (K.L.)
2. Key Laboratory for Liquid-Solid Structural Evolution and Processing of Materials, Ministry of Education, Shandong University, Jinan 250061, China; guojunnan113005@hotmail.com
3. M.N. Mikheev Institute of Metal Physics, Ural Branch of Russian Academy of Sciences, Ekaterinburg 620077, Russia; okulovartem@imp.uran.ru
* Correspondence: jie_li@just.edu.cn (J.L.); lihuilmy@sdu.edu.cn (H.L.)

Abstract: C_2N-ring-based molecular chains were designed at the molecular level and theoretically demonstrated to show distinctive and valuable electron transport properties that were superior to the parent carbonaceous system and other similar nanoribbon-based molecular chains. This new -type molecular chain presented an exponential attenuation of the conductance and electron transmission with the length. Essentially, the molecular chain retained the electron-resonant tunneling within 7 nm and the dominant transport orbital was the LUMO. Shorter molecular chains with stronger conductance anomalously possessed a larger tunnel barrier energy, attributing to the compensation of a much smaller HOMO–LUMO gap, and these two internal factors codetermined the transport capacity. Some influencing factors were also studied. In contrast to the common O impurity with a tiny effect on electron transmission of the C_2N rings chain, the common H impurity clearly improved it. When the temperature was less than 400 K, the electron transmission varied with temperature within a narrow range, and the structural disorder deriving from proper heating did not greatly modify the transmission possibility and the exponentially decreasing tendency with the length. In a non-equilibrium condition, the current increased overall with the bias but the growth rate varied with size. A valuable negative differential resistance (NDR) effect appeared in longer molecular chains with an even number of big carbon–nitrogen rings and strengthened with size. The emergence of such an effect originated from the reduction in transmission peaks. The conductance of longer molecular chains was enhanced with the voltage but the two shortest ones presented completely different trends. Applying the bias was demonstrated to be an effective way for C_2N-ring-based molecular chains to slow down the conductance decay constant and affect the transport regime. C_2N-ring-based molecular chains show a perfect application in tunneling diodes and controllable molecular devices.

Keywords: new-type molecular chain; electron transport; size dependence; applying the bias; first principles

Citation: Song, D.; Li, J.; Liu, K.; Guo, J.; Li, H.; Okulov, A. Size- and Voltage-Dependent Electron Transport of C₂N-Rings-Based Molecular Chains. *Molecules* 2023, 28, 7994. https://doi.org/10.3390/molecules28247994

Academic Editor: Minghao Yu

Received: 13 November 2023
Revised: 3 December 2023
Accepted: 5 December 2023
Published: 7 December 2023

Copyright: © 2023 by the authors. Licensee MDPI, Basel, Switzerland. This article is an open access article distributed under the terms and conditions of the Creative Commons Attribution (CC BY) license (https://creativecommons.org/licenses/by/4.0/).

1. Introduction

Mostly, the electron transport of molecular wires decays exponentially with length but the structural stability also decreases rapidly with the chain elongation, limiting their applications [1–3]. Therefore, exploring new-type molecular wires with controllable conductance decay and good structural stability to understand electron transport [4,5] and establish well-defined structure–property relationships [6,7] has widely attracted the academic and industrial communities in fields such as field-effect transistors, logic circuits, memory and storage devices [8–10].

Recently, C_2N, which is a new two-dimensional (2D) material with a periodic porous structure, was successfully synthesized [11]. In its structure, sp^2-hybridized carbon (C)

rings are connected through two face-to-face nitrogen (N) bridges that are distributed in a honeycomb lattice, and six electronegative N atoms constitute one periodic hole (see Figure 1a) [12]. Such a unique structure exhibits extremely high thermal stability due to the strong covalent bonding framework formed by the C-C and C-N atom pairs [13]. C_2N displays tunable semiconductor properties and presents a high on/off ratio of ~107 as a field effect transistor [14]. Thus, extensive research about properties such as thermal conductance [15], photoelectronic properties [16], mechanical properties [17], adsorption of molecules [18,19], electrochemical properties [20,21], and gas and impurity separation [22] has been conducted, showing promising applications in the semiconductor [23], nano/optoelectronics [24], sensing technology [25], photovoltaics [26], energy [27], biological [28] and medical industries [29]. However, few studies have been focused on the electrical performance of C_2N at the molecular scale.

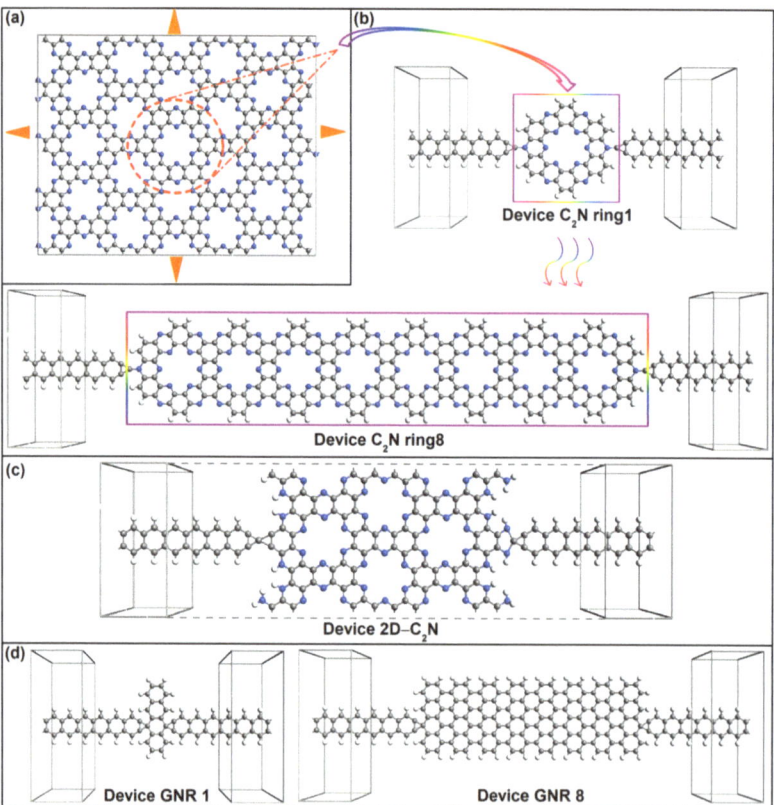

Figure 1. (a) Geometry of C_2N monolayer. The unit cell is indicated by the red dashed circle and the corresponding. Molecular electrostatic potential is shown in Figure S1 of Supporting Information (SI) [17]. (b–d) Structure schematic of C_2N-ring-chain-based devices, 2D-C_2N-based device and GNR-based device, respectively.

Based on the excellent structural stability and electrical properties of 2D C_2N, we theoretically designed a new-type molecular wire, which was one side-by-side single-ring molecular chain from the C_2N monolayer, and studied the equilibrium and non-equilibrium electron transport properties, including the conductance decay and current–voltage characteristics. The corresponding structure–property relationship and inner electron transmission mechanism are well explained.

2. Results and Discussion

The new-style molecular wire was composed of parallel-connected single hexagonal C_2N rings taken from a C_2N monolayer (see Figure 1a). The length of one single C-N garland (denoted as a C_2N ring here) was defined as one unit length. Eight lengths (i.e., 1–8 C_2N rings) of molecular wires were studied. Figure 1b exhibits the schematic view of C_2N-ring-chain-based devices, where the molecular wire was sandwiched between four-atom-wide graphene nanoribbon (GNR) electrodes with a tip C atom as the linked atom. On each side, the tip electrode atom was symmetrically bonded to one N atom on the outer edge of the C_2N ring. Molecular devices were denoted as device C_2N ring i (i = 1–8) depending on the number of C_2N rings.

In addition, we also provide the parent material 2D-C_2N and graphene nanoribbon (GNR)-based chains with eight lengths as a comparison. Corresponding device structures are shown in Figure 1c,d, which all have the same connection style and electrodes to C_2N-ring-molecular-chain-based devices. One column of C rings is defined as one unit length for GNR. Correspondingly, the device with the parent material is written as device 2D C_2N and GNR-based devices are written as device GNR i (i = 1–8) depending on the number of the unit length.

2.1. Equilibrium Conductance Decay and Electronic States

To explore the electron transport of new-type molecular wires, first, Figure 2a presents the equilibrium quantum conductance of the device 2D C_2N and the conductance of the device C_2N rings changing with the number of C_2N rings with an exponential fit. Obviously, the conductance of the shorter C_2N ring chains was stronger than the original material 2D C_2N. The conductance attenuation with the length of a C_2N ring chain (L) satisfies the typical equation [30]:

$$G = A \exp(-\beta L) \quad (1)$$

where β is the decay constant, which is calculated from the slope of the plot of logarithmic conductance (lnG) vs L. More clearly, the lnG–length curve with a linear fit is also given in Figure 2b. A is a constant related to the interaction between the molecular wire and electrodes, reflecting the contact resistance.

Figure 2c gives the β values for different molecular lengths. Noticeably, β varied with length and started to drop when the length was over 4.5 nm (five rings). However, β changed little (from 2.20 to 2.24 nm^{-1}) in the length range [3 nm, 7 nm] and the C_2N ring chain always maintained an exponentially attenuated conductance within 7 nm, indicating that the C_2N ring chain possessed resonant tunneling as the dominant transport mechanism within 7 nm. This was quite different from traditional molecular junctions, which took the direct tunneling regime as the dominant transport mechanism with lengths less than 4 nm and presented an obvious decreased β with lengths over 4 nm, evidencing a transition between the tunneling and hopping mechanism [31,32].

Further, the focus of the following research was the electronic structures of molecular rings, mainly highlighting the electronic states in the transport process, which is closely linked to transport decay [33,34]. The eigenstates of the molecular projected self-consistent Hamiltonian (MPSH) can be considered as molecular orbitals renormalized by the molecule–electrode interaction [35], giving a visual description of the molecular electronic structure [36]. Thus, frontier molecular orbits (LUMOs and HOMOs) are given. Representative states of LUMOs and HOMOs are presented in Figure 2d1,d2, respectively. Others can be found in Figure S2 of the SI. The electronic states of LUMOs were delocalized for C_2N ring chains for all lengths, while the wavefunctions started to be localized and were mainly distributed in the rings on both sides from device C_2N ring 3. This indicates stronger electronic states of the LUMO, in stark contrast to the HOMO. Thus, the dominant transport orbital was the LUMO for C_2N-ring-chain-based molecular devices in the tunneling mechanism.

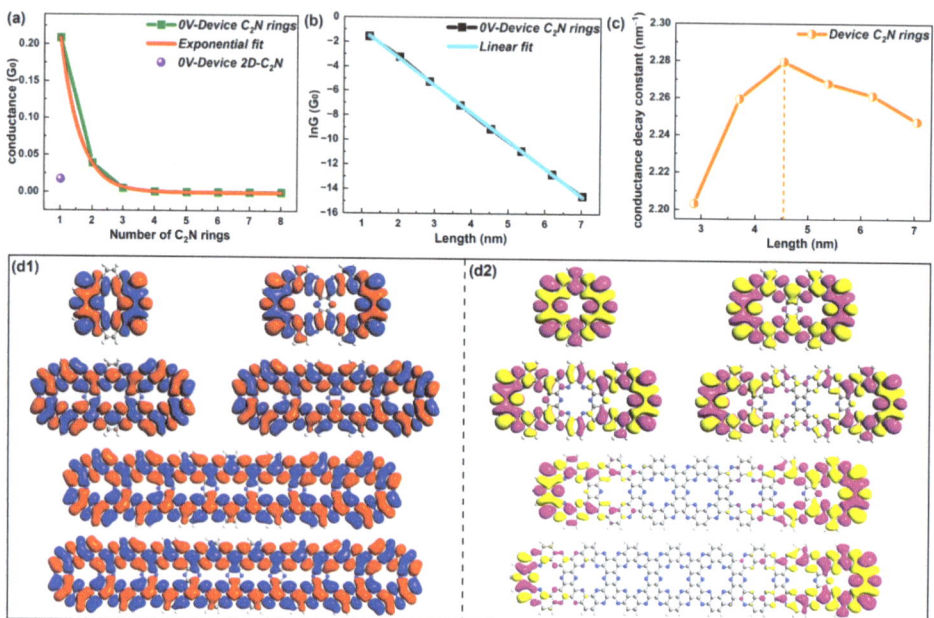

Figure 2. (**a**) Equilibrium quantum conductance—number of C_2N rings curve with an exponential fit and conductance of 2D-C_2N-based device. (**b**) lnG as a function of length with a linear fit. (**c**) Changes in the conductance decay constant within the length range of [2.85, 7] nm. (**d1,d2**) LUMOs and HOMOs for representative device C_2N rings 1, 2, 3, 4, 7 and 8, respectively.

2.2. Mechanisms: Electron Transmission, HOMO–LUMO Gap and Tunnel Barrier Energy

The transmission at the Fermi level (E_F) expresses the transport capacity of molecular devices [37]. Thus, Figure 3a shows the trend of the transmission coefficient at E_F changing with the number of C_2N rings and the transmission of the device 2D-C_2N. Similar to the conductance, the electron transmission coefficient at E_F was much larger for shorter device C_2N rings in contrast to the parent-material-based device, signifying stronger electron transport abilities of shorter C_2N-ring-based chains. The electron transmission probability took on an exponential decline, directly explaining the reduction in the conductance with length. On a deeper level, the tunnel barrier energy, i.e., the energy difference between the dominant transport orbital and the electrode Fermi level, is the internal determinant for the electron transport of molecular devices [38]. Usually, the tunnel barrier energy of the molecular wire would increase as the length increases, essentially causing the conductance decay [39]. Unexpectedly, the tunnel barrier energy of the C_2N ring chains did not show the overall uptrend with length (see Figure 3b). Device C_2N rings 1 and 2 with stronger transport capacity actually possessed higher tunnel barrier energy compared with the longer devices. When the number of rings grew from 1 to 2 or 3 to 8, the tunnel barrier energy followed the general rule.

Furthermore, such an abnormal phenomenon of the intrinsic factor is directly reflected in the transmission pathways, which are in essence transition paths of electrons between atoms [40]. Transmission pathways of representative devices are exhibited in Figure 3c1–c4. Others are shown in Figure S3 of the SI. Specifically, the thickness of the arrow indicates the magnitude of the local transmission between each pair of atoms, and the arrowhead and the color designate the direction of the electron flow [40]. Obviously, electrons transferred along little C/C_2N rings up and down the outer edges, not passing the joint rings. In particular, stronger transmission appeared in little C_2N rings, judging by the color and width of the transmission arrows. No matter whether the number increased from 1 to 2 or 3 to 8, the transferring arrows gradually reduced and became thinner, indicating that the electron transport weakened, but the transmission power of device C_2N ring 3 was stronger than device C_2N ring 2. All these were consistent with the tendency of the tunnel barrier energy.

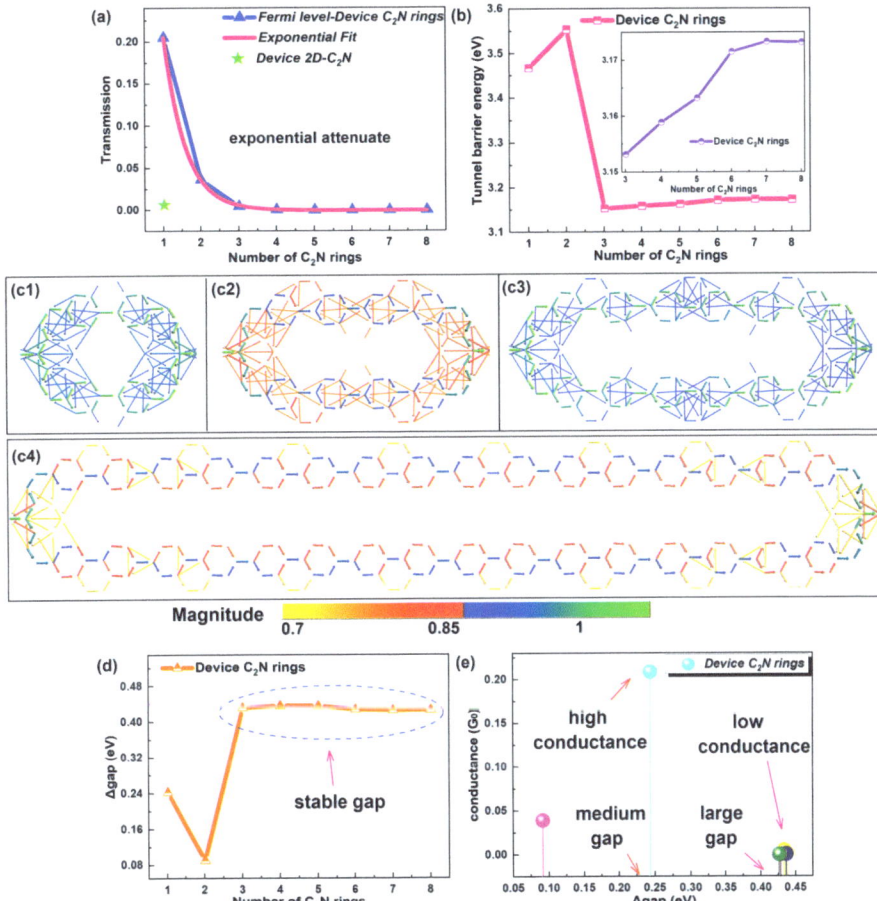

Figure 3. (a) Equilibrium electron transmission for E_F number of C_2N rings curve with an exponential fit and the transmission of a 2D-C_2N-based device. (b) Tunnel barrier energy as a function of number of C_2N rings in [1, 8], and the inset shows the corresponding changes in a narrower range of [3, 8]. (c1–c4) Transmission pathways of representative device C_2N rings 1, 2, 3 and 8, respectively. (d) HOMO–LUMO gap as a function of number of C_2N rings. (e) Conductance as a function of HOMO–LUMO gap of C_2N rings.

To investigate the anomalous behavior, the tendency of the HOMO–LUMO gap with the number of C_2N rings was studied (see Figure 3d). Shorter chains presented smaller gaps. Larger and stable gaps were found for longer chains. The conductance as a function of the gap is also provided in Figure 3e. For the new-style chain, a large gap did not bring about strong conductance, and high conductance occurred in the chain with a medium gap. Generally, a wider HOMO–LUMO gap produced higher conductance [1]. Clearly, a narrower gap of device C_2N rings 1 and 2 compensated for their larger tunnel barrier energy. These two internal factors codetermined the conductance and transport capacity of the C$_2$N-ring-chain-based devices.

2.3. Comparisons with Other Similar Systems and Influencing Factors

For comparison, Figure 4 gives the equilibrium electron transport performance of the GNR-based chains with different lengths. Unlike the device C$_2$N rings, the device GNR presented a fluctuating tendency of conductance with the number of columns of C rings (called the number of rings), as shown in Figure 4a and the corresponding inset (shown in a narrower range). For more clarity, the lnG–length graph provided in Figures 4b and 4c exhibits the comparison of the conductance as a function of the length for C$_2$N-ring- and GNR-based chains on a log scale. Obviously, the conductance of the device GNR showed non-exponential attenuation with the length of the chain. Furthermore, the device C$_2$N had a distinct advantage over the device GNR on the conductance strength.

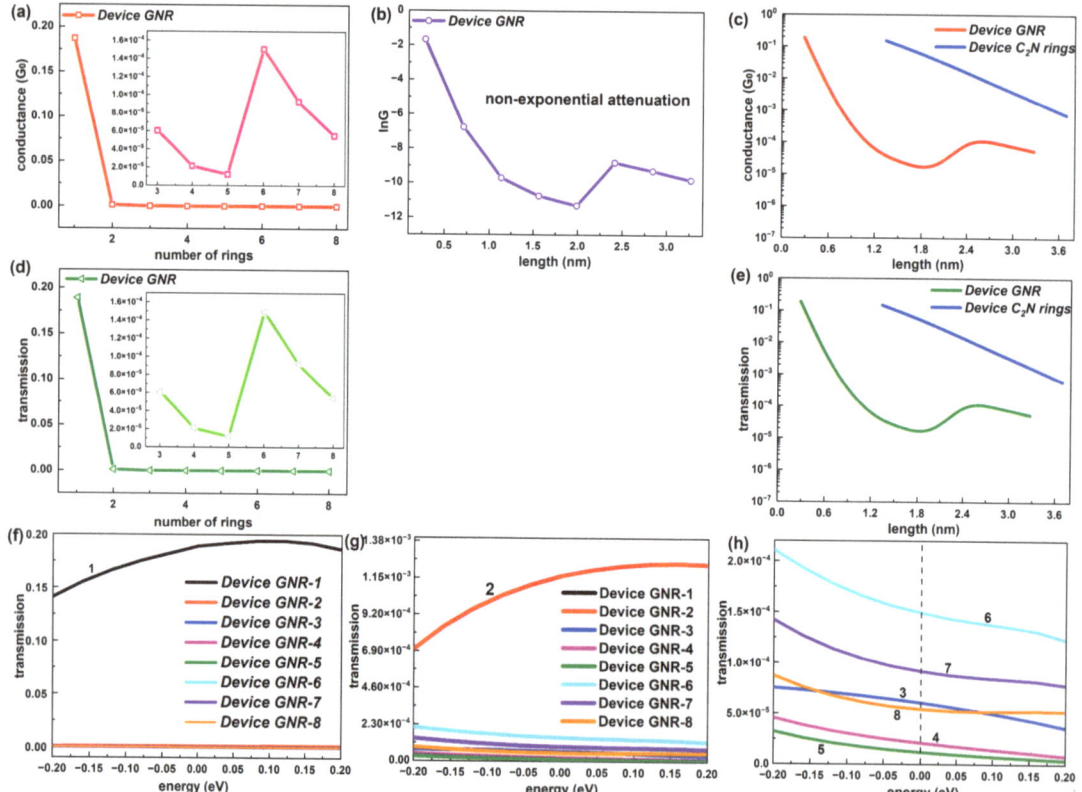

Figure 4. (a) Conductance at E_F as a function of number of rings (1–8) for GNR-based devices. The inset shows the number range [3, 8]. (b) lnG as a function of length for GNR-based devices. (c) Comparison of conductance between C$_2$N-ring-chain-based devices and GNR-based devices changing with the length. (d) Electron transmission at E_F as a function of number of rings (1–8) for GNR-based devices. The inset shows the number range [3, 8]. (e) Comparison of electron transmission at E_F between C$_2$N-ring-chain-based devices and GNR-based devices changing with the length. (f–h) Transmission spectrum around E_F of GNR-based devices in transmission ranges [0, 0.2], [0, 1.38 × 10^{-3}] and [0, 2 × 10^{-4}], respectively.

Consistent with the conductance, the electron transmission coefficient at E_F of the GNR-based chain took on an undulating state that varied with the number of rings, which can be found in Figure 4d. A contrast of the transmission–length relationship between the device C_2N rings and the device GNR is also presented on a log scale in Figure 4e. It follows that such a trend was also not an exponential decline and the transmission coefficient of the device C_2N rings was far higher than the device GNR at the same length. Thus, the new-type C_2N-ring-based chains possessed better electronic transport capacity compared with the GNR-based chains.

Further, Figure 4f,g exhibit the comparison of the transmission spectrum within the energy interval [−0.2 eV, 0.2 eV], i.e., near E_F among the GNR-based chains with different lengths. From Figure 4f, the electron transmission of the shortest GNR-based chain was significantly stronger than other lengths of GNR chains in the whole energy range. Device GNR-2 also displayed a much larger transmission coefficient than longer GNR chains near E_F, according to the inset of Figure 4f. For these two shortest GNR-based chains, the transmission probability increased gradually from −0.2 eV to 0.2 eV. The other six lengths of GNR-based chains all presented quite low electron transmission probabilities, which were distributed in the same order of magnitude (10^{-5}). Unlike C_2N-ring-based chains with the law of decline, there were no specific rules in the relative strength of the electron transmission among these six GNR-based chains (GNR-3 to GNR-8). Different from the two shortest GNR-based chains, these six longer chains all showed slowly decreasing transmission values from −0.2 eV to 0.2 eV. Overall, the C_2N-ring-based chains demonstrated a more regular transmission spectrum with the length.

In addition, some factors were also considered to investigate the corresponding influence on the electron transport of the new-type molecular chain, including the defects, disorder, temperature and electrode materials. The results of the synthesis and characterization show that there were varieties of point defects in the C_2N-based materials, mainly including the C and N vacancies, as well as the substitutional and interstitial impurities of C, N, O and H atoms [11]. Here, we considered two common defects, that is, H and O impurities in the big ring, which were shown to possess relatively lower formation energies in defective C_2N-based materials [16]. For each length of the C_2N ring chain, one defect atom (H or O) was considered here. The insets of Figure 5a exhibit the positions of H or O impurities in the C_2N ring chain.

To study the influence of the defects on electron transport, first, Figure 5a presents the electron transmission at E_F changing with the number of C_2N rings for these two kinds of defective C_2N ring chains. One obvious and unexpected increase could be found in device C_2N ring 2 with a defect H atom, which broke the decreasing trend of electron transmission with the length of the perfect C_2N ring chain. However, as the length continued to increase, the electron transmission still went down with the length for the C_2N ring chain with an H impurity. As a result, the value of each C_2N ring was higher than the perfect C_2N rings but still showed an exponential decline with the length, which could be seen in the comparison of electron transmission at E_F changing with the length among the devices with no defects and the two defects (Figure 5b). The C_2N-ring-based chain with an O impurity presented a similar electron transmission value and exponential downward trend to the perfect C_2N-ring-based chain, indicating the slight influence of the O impurity on the electron transmission at E_F of the C_2N-ring-based chain. Further, the electron transmission spectrum around E_F is given. Figure 5c shows the comparison of the spectrum for device C_2N ring 1 with no defect and H and O impurities, while others are exhibited in Figure S4 of the SI. The O impurity also slightly affected the distribution and value of electron transmission near E_F but changed the height and position of the transmission peaks for larger energies. In contrast, the H impurity greatly modified the transmission curve around E_F.

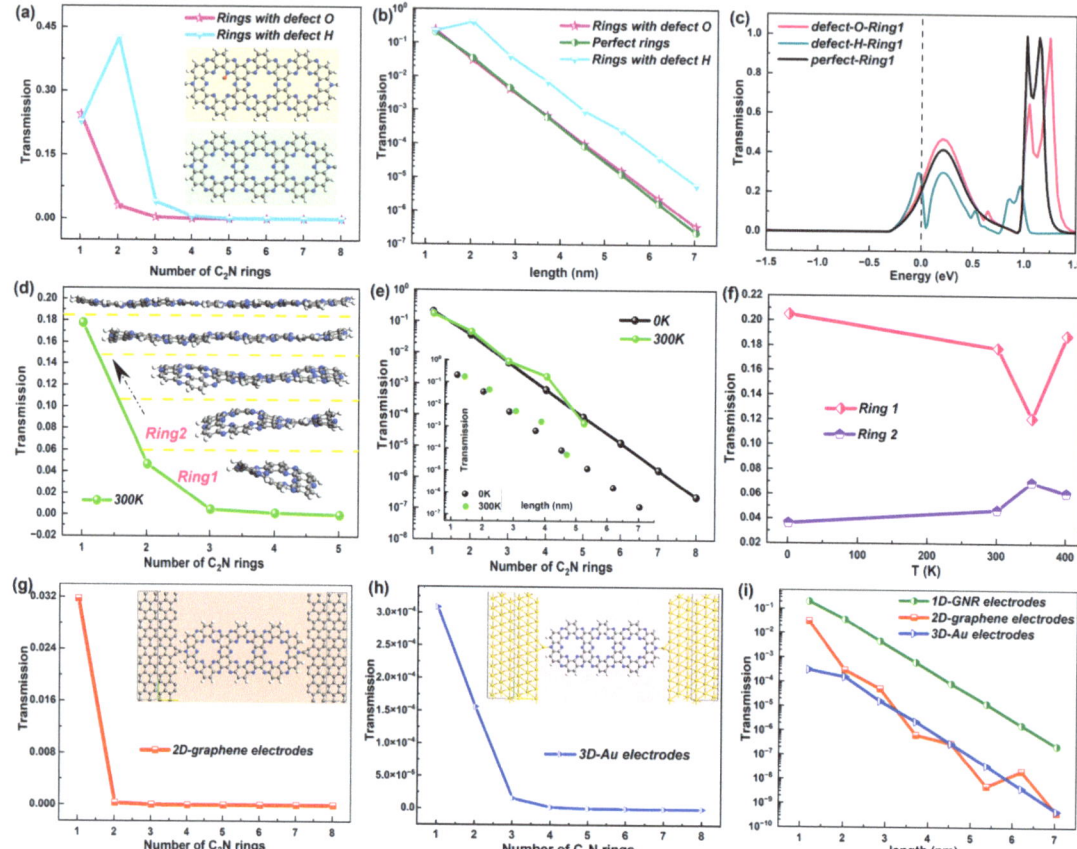

Figure 5. (**a**) Electron transmission at E_F of the C_2N ring chains with O and H impurities, changing with the number of C_2N rings. The insets show the positions of two such impurities. (**b**) Comparison of electron transmission at E_F of the C_2N ring chains with no defect and O and H impurities, changing with the length of the chains. (**c**) Electron transmission spectrum around E_F of device C_2N ring 1 with no defect and O and H impurities. (**d**) Electron transmission at E_F as a function of the number of C_2N rings at 300 K. The insets show the corresponding 5 structures of the C_2N ring chains at 300 K. (**e**) Comparison of electron transmission at E_F of C_2N rings at 0 K and 300 K changing with the number of C_2N rings and the length (shown in the inset), respectively. (**f**) Electron transmission at E_F of device C_2N rings 1 and 2 as a function of the temperature T. (**g**,**h**) Electron transmission at E_F as a function of the number of C_2N rings for the C_2N-ring-chain-based devices with 2D graphene electrodes and 3D Au electrodes, respectively. The insets show the device structures. (**i**) Comparison of electron transmission at E_F of C_2N ring chains based on 3 different electrodes changing with the length of the chains. For clarity, all comparisons are shown on a logarithmic scale.

When the temperature rises, structural distortions, deformation and disorder occur in C_2N rings, especially shorter chains, and may affect the electron transport properties. When the temperature rose to 300 K, significant structural distortions could be observed in rings 1–3 with ups and downs, which can be seen in the insets of Figure 5d. As the length increased, the C_2N ring chain showed less structural fluctuation, which is also consistent with the stable structure of 2D-C_2N [25,41], but possessed a little higher electron transmission coefficient at E_F than the rings at 0 K, and one prominent enhancement in the transmission value appeared in ring 4, which is reflected in Figure 5e. Furthermore, when the C$_2$N ring chain was at 300 K, the length increased slightly compared with the chain at 0 K. However, such structural disorder originating from a higher temperature did not change the exponentially decreasing tendency of electron transmission at E$_F$ with the length.

To further investigate the effect of the temperature on the electron transport, Figure 5f displays the electron transmission at E$_F$ as a function of the temperature T for C$_2$N ring 1 and C$_2$N ring 2 as representatives based on the larger structural changes of shorter chains. They do not show the monotonical changing trend of the electron transmission coefficient at E$_F$ with T. For the shortest C$_2$N ring 1, the electron transmission coefficient at E$_F$ decreased with T until 350 K and then increased again. Differently, C$_2$N ring 2 showed enhanced electron transmission at E$_F$ until 350 K and then decreased again, but the changing range of electron transmission was not significant and was smaller than that of C$_2$N ring 1. No matter how the temperature changed, the electron transmission possibility at E$_F$ of ring 1 was bigger than that of C$_2$N ring 2 and the value did not vary greatly with T for the two rings.

Furthermore, as an important part of the molecular device, the electrodes may greatly affect the electron transport properties. Therefore, we chose different types of materials as the electrodes to study such an influence, namely, 2D graphene and 3D Au electrodes in contrast to the original 1D GNR electrodes. The corresponding structures of the two devices are presented in the insets of Figure 5g,h, which also show the electron transmission at E$_F$ changing with the number of C2N rings. It is not hard to find the sharp drop in the electron transmission value with the number of C$_2$N rings in these two devices. For clarity, Figure 5i presents the comparison of the electron transmission at E$_F$ of the C$_2$N ring chains based on these three different electrodes changing with the length of the chains on a logarithmic scale. The transmission possibility of 3D-Au-electrode-based device maintained the exponential attenuation with length with a similar speed to the 1D-GNR-based device but was much weaker than the same length of C$_2$N ring chain. The outcome was different for the 2D-graphene-based device. The electron transmission still became weaker with increasing length but was no longer strictly exponential. When the length of the C$_2$N ring chain was the same, the transmission coefficient was always noticeably lower than the 1D-GNR-based device, but sometimes a little higher than the 3D-Au-based device.

2.4. Non-Equilibrium Electron Transport Properties: IV Characteristics

Non-equilibrium electron transport properties were also studied for the C$_2$N ring chains. The current–voltage characteristics are related to practical and important non-equilibrium performance [42]. Figure 6 presents typical current–voltage curves with derivatives at biases, representing changing rates. Others are given in Figure S5 of the SI. Overall, the current of the C$_2$N-ring-based chains increased with the voltage but showed different rates based on different lengths. When the number of big carbon–nitrogen rings did not exceed three, the growth rate of the current went down monotonically with increased voltage. The current of device C$_2$N rings 4 and 5 tended to grow linearly at a rate with a gentle variation. When the number exceeded five, the speed of the current rose with voltage. Moreover, at the same bias, the growth rate of current decreased as the length increased.

Remarkably, as the length of the C$_2$N ring chain increased, the I–V curve no longer strictly maintained the increasing trend and started to show a negative differential resistance (NDR) effect, which is indispensable for several electronic components, such as the Esaki and resonant tunneling diodes [4]. A slight NDR feature only appeared in the longer devices with an even number (6 and 8) of C$_2$N rings. As presented in Figure 6d,e, device C$_2$N ring 6 showed a peak-to-valley current ratio (PVCR) of 1.01 in the bias range [0.3 V, 0.33 V], while device C$_2$N ring 8 had a PVCR of 1.27 between 0.19 V and 0.24 V. Here, the PVCR is equal to the ratio of the maxima and the minima of the current in the voltage range where the current drops [43]. It follows that NDR would enhance, that is, the voltage range where NDR occurs would enlarge and the PVCR would rise, as the number of C$_2$N rings increases. In contrast, the I–V performance of the device 2D-C$_2$N showed a non-linear increasing trend and had a long plateau of slow growth, which was more like the overall trends of longer C$_2$N-ring-chain-based devices but did not show a valuable NDR effect.

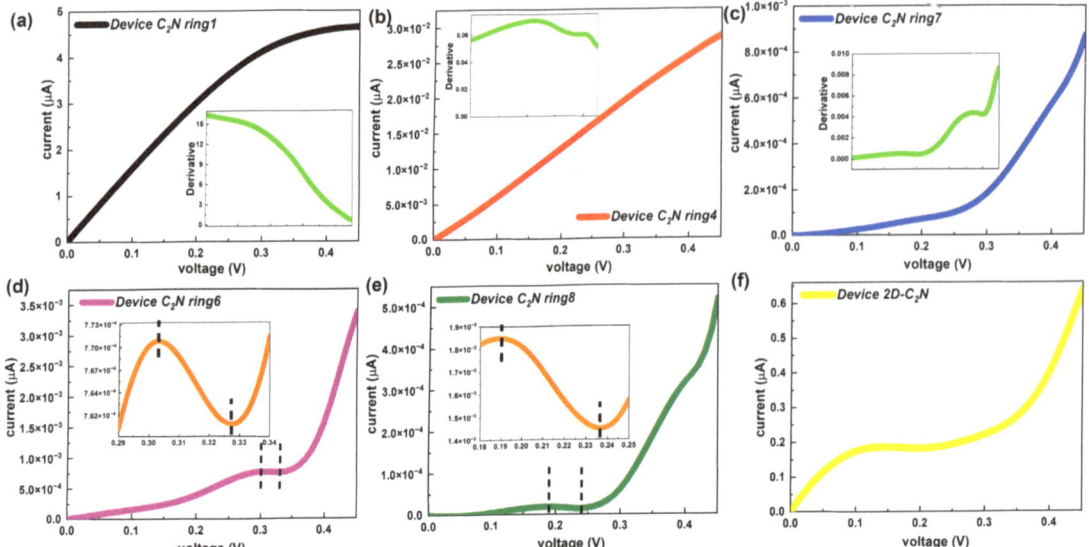

Figure 6. (a–c) I–V curves of representative device C$_2$N rings 1, 4 and 7, respectively. Each inset shows the first derivative of corresponding I–V curve. (d,e) I–V curves of device C$_2$N rings 6 and 7, respectively. I–V curves within the specific bias interval are exhibited in the insets. (f) I–V curve of the device 2D-C$_2$N.

Here, the current was calculated using the Landauer–Büttiker formula, indicating that the current through devices was inseparable from the transmission coefficients of the devices. To explain the current–voltage characteristics of the C$_2$N-ring-based chain, Figure 7 presents the transmission functions of the device C$_2$N rings under representative biases. In our calculations, the average Fermi level, which was the average chemical potential of left and right electrodes, was set as zero. The current was determined using the integral area of the transmission curve within the bias window, i.e., $[-V/2, +V/2]$ [44], shown as dashed lines. For all C$_2$N-ring-based devices, 0.1 V, 0.2 V, 0.3 V and 0.4 V were selected as representative biases. Actually, there was just the left side of the transmission peak inside each bias window for all C$_2$N-ring-based devices and the summit of the transmission peak was far away from E$_F$ for most C$_2$N-based devices, except the shortest device C$_2$N ring 1, which is demonstrated by the transmission functions under four representative biases within larger energy intervals in Figure S6 of the SI. Therefore, to display the changes of the integral area within the bias window more clearly, transmission curves under non-equilibrium conditions are presented in narrower energy ranges. Several typical devices (device C$_2$N rings 1, 3, 6 and 8) are shown in Figure 7a–d and others are displayed in Figure S7 of the SI. There were differences in the growth rates of transmission curves between C$_2$N-based devices from -0.2 eV to 0.2 eV. Apparent enlargement can be found in the integral area of the transmission curve inside the bias window for these devices as the bias increased based on the variations in the height and width of the specific area (marked with a dashed line in the figure), which explains the overall increasing trend of the I–V curves well.

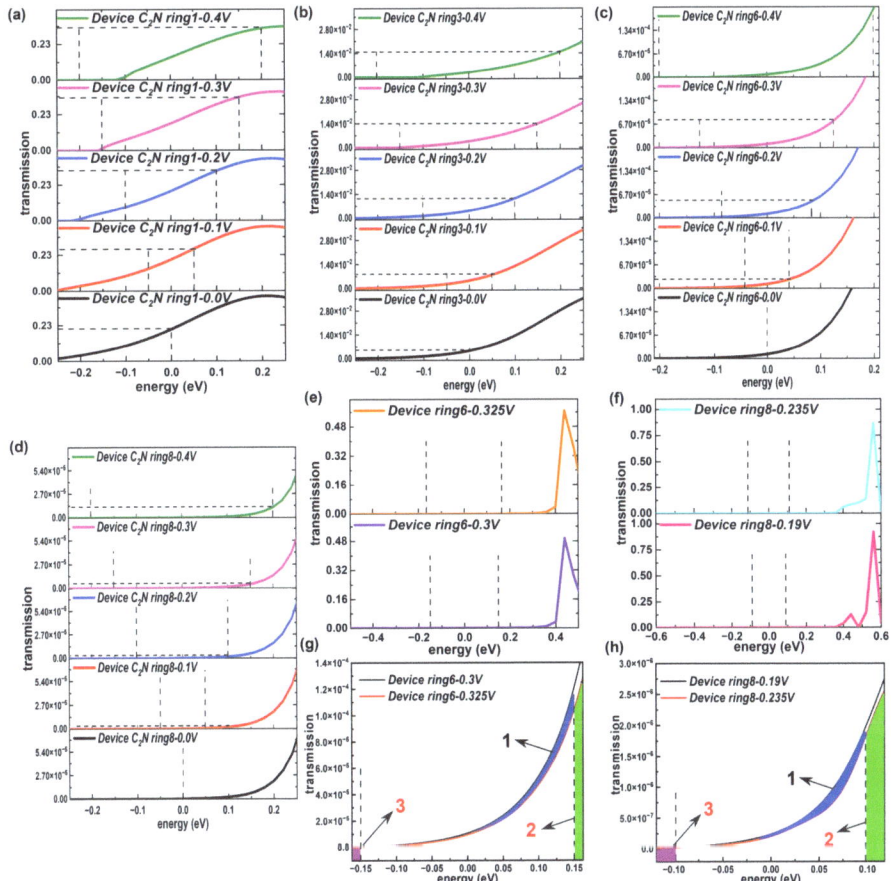

Figure 7. (**a–d**) Transmission spectrum within narrower energy ranges at 0 V, 0.1 V, 0.2 V, 0.3 V and 0.4 V of representative device C_2N rings 1, 3, 6 and 8, respectively. (**e**) Transmission spectrum at 0.3 V and 0.325 V of device C_2N ring 6. (**f**) Transmission spectrum at 0.19 V and 0.235 V of device C_2N ring 8. (**g,h**) Transmission spectrum in (**e,f**) within narrower energy ranges.

However, there was a noted drop in the current emerged over a small voltage interval for device C_2N ring 6 and device C_2N ring 8. To clearly explain the origins of the NDR effect appearing in the longer devices with an even number of C_2N ring chains, we first provide the transmission functions in larger energy ranges under the voltages at both ends of the decreasing voltage intervals (0.3 V and 0.325 V for device C_2N ring 6, 0.19 V and 0.235 V for device C_2N ring 8), as displayed in Figure 7e,f. Similar to the typical biases above, under the new biases, the summits of the transmission peaks still remained away from E_F. Because the peak value was far larger than the coefficient at E_F, the changes in the area inside the bias window could not be directly observed. Thus, each energy range on display was narrowed to a minimum (see Figure 7g,h), corresponding to the selected voltages, i.e., [−0.325/2, 0.325/2] for device C_2N ring 6 and [−0.235/2, 0.235/2] for device C_2N ring 8. The integral area of the transmission peak under 0.325 V was smaller than that under 0.3 V by comparing the area 1 sandwiched between curves and the sum of extra areas 2 and 3, as judged by a huge order of magnitude difference between the horizontal and vertical coordinates. Areas 1, 2 and 3 are marked with different colors in Figure 7g,h.

The same was true for device C_2N ring 8. Therefore, the NDR effect in essence emerged from the reduction in the transmission peaks.

2.5. Non-Equilibrium Electron Transport Properties: Conductance–Voltage Relations

Non-equilibrium changes in the conductance were associated with the length and voltage. Relative relationships for the device 2D-C_2N and device C_2N rings are exhibited in Figure 8. Taking significant differences in conductance values into consideration, the conductance–voltage curve is given separately for the device 2D-C_2N, device C_2N ring 1 and device C_2N ring 2, corresponding to Figure 8a–c, and the other six C_2N rings-based chains are compared on a linear and log scale respectively, corresponding to Figure 8d,e. As observed, the conductance fluctuated with the bias and the values were relatively small under low biases for the parent-material-based device, which was much weaker than the two shortest C_2N-ring-chain-based devices under any voltage.

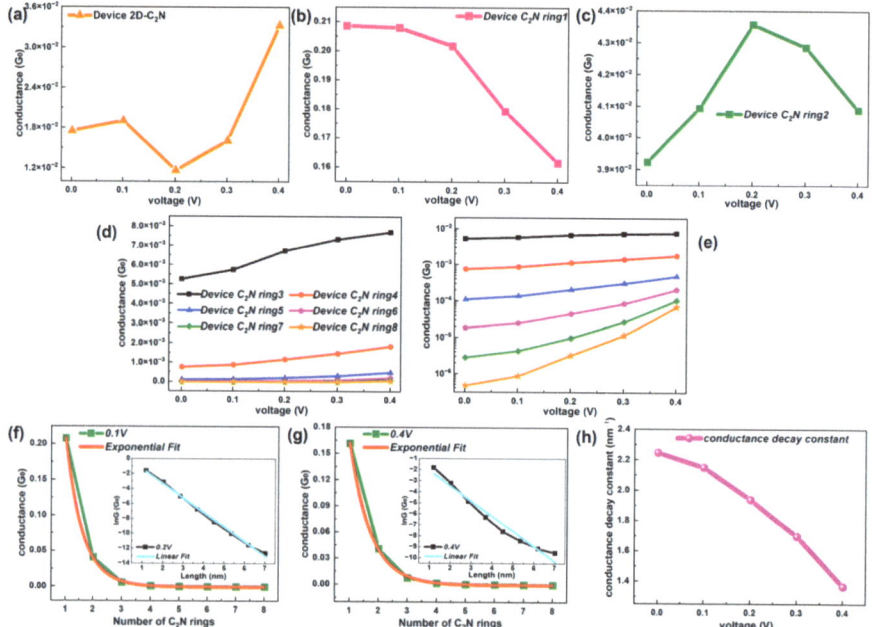

Figure 8. (**a–c**) Conductance–voltage functions of device 2D-C_2N, device C_2N ring 1 and device C_2N ring 2, respectively. (**d,e**) Comparison of the conductance changing with bias for device C_2N rings 3, 4, 5, 6, 7 and 8 on linear and log scales, respectively. (**f,g**) Conductance as a function of number of C_2N rings with an exponential fit at 0.1 V and 0.4 V, respectively. Each inset exhibits the corresponding lnG as a function of length with a linear fit. (**h**) Changing trend of the conductance decay constant with voltage.

As the bias enhances, the two shortest C_2N-ring-based chains show peculiar conductance–voltage characteristics, which did not show any regularity in the length. The conductance of device C_2N ring 1 dropped with the increased bias, while one single peak appeared in the conductance–voltage curve of device C_2N ring 2. Different from the above two, the longer six C_2N-ring-chain-based devices presented regularly enhanced conductance with voltage and the growth accelerated in larger biases, which was more prominent in longer chains. However, at each bias, the conductance still decayed with the increasing length.

To investigate the influence of the voltage on the trend of the conductance changing with the length, Figure 8f,g and Figure S8 in the SI show the conductance–length curve at several biases. When the voltage was applied, the conductance of the C_2N-ring-chain-based devices still decayed rapidly with the length, but the trend gradually deviated from the exponential fit and the extent of this deviation deepened as the voltage increased. Further, Figure 8h gives the conductance decay constant (β) as a function of the voltage. One significant drop can be observed in the conductance decay constant with the increased voltage, suggesting a transition between resonant tunneling and the intrachain hopping mechanism under the bias. This provides an effective way for the C_2N ring chain to control the attenuation speed of the conductance with the length or to affect the inner electron transport regime.

3. Computational Methods

The simulated electronic devices were constructed in Virtual NanoLab within a supercell with over 15 Å of vacuum space to allow for electrostatic interactions to decay for a system. All structures were structurally optimized before calculating the electron transport properties. All calculations used the first principles theory based on density functional theory (DFT) and non-equilibrium Green's function (NEGF) and were performed in the Atomistix Toolkit (ATK) [45,46]. Numerical LCAO basis sets and norm-conserving pseudopotentials were adopted. For high accuracy, GGA-PBE formulation was selected and the K-point sampling was set to $1 \times 1 \times 100$. The double-zeta plus polarization (DZP) basis for all atoms was adopted. The density mesh cut-off for the electrostatics potential was 75 Ha and the electron temperature was set as 300 K. Structural optimizations used the quasi-Newton method until all residual forces on each atom were smaller than 0.05 eV/Å. The convergence criterion for the total energy was set to 10^{-5} via the mixture of the Hamiltonian.

There is no doubt that DFT-GGA underestimates the band gap when calculating electron transmission. Here, we also chose the HSE06 function as a comparison. For device ring 1 as a representative, the calculated HOMO–LUMO gap was 0.2432472 eV with the GGA function, while the gap was 0.2700527 eV using the HSE06 function. The difference in the gap was relatively small. Furthermore, we also calculated the transmission spectrum around the E_F with the two functionals, which is shown in Figure 9. The electron transmission curves calculated with the two functionals are very close, especially for smaller energies, which is reflected in the similar transmission peaks and transmission coefficients. For larger energies, there were some little differences in the number and height of transmission peaks. Thus, there was a very small gap between the results calculated with GGA and HSE06 functionals. More importantly, such small differences did not affect the comparison between the molecular chains with different lengths, which was our research focus.

The conductance G can be expressed in terms of the transmission function within the Landauer–Büttiker formalism [47,48]:

$$G = G_0 \int_{-\infty}^{+\infty} dE\, T(E,V) \left(-\frac{\partial f(E)}{\partial E} \right) \quad (2)$$

The current through a molecular junction is calculated from the Landauer–Büttiker equation [49]:

$$I = \frac{2e}{h} \int_{-\infty}^{+\infty} dE\, T(E,V)(f_1(E) - f_2(E)) \quad (3)$$

where $G_0 = 2e^2/h$ is the quantum unit of conductance, h is the Planck's constant, e is the electron charge, $f(E)$ is the Fermi distribution function, $f_{1,2}(E)$ are the Fermi functions of source and drain electrodes, and $T(E,V)$ is the quantum mechanical transmission probability of electrons, which can be given as [49]

$$T(E,V) = tr\left[\Gamma_L(E,V) G^R(E,V) \Gamma_R(E,V) G^A(E,V) \right] \quad (4)$$

where G^R and G^A are the retarded and advanced Green functions of the conductor part respectively and Γ_L and Γ_R are the coupling functions to the left and right electrodes, respectively.

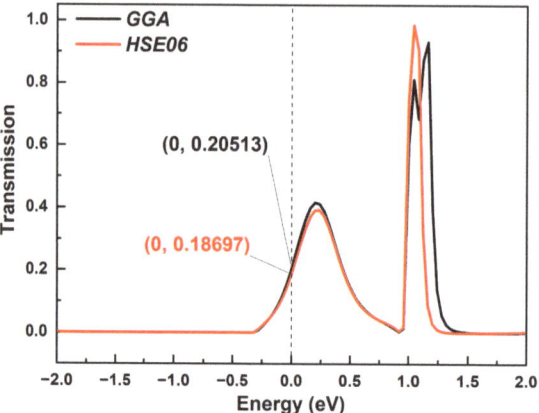

Figure 9. The comparison of electron transmission spectrum calculated with GGA and HSE06 functionals.

4. Conclusions

We have designed new C_2N ring chains in different sizes and theoretically investigated comprehensive electron transport properties. The results show that the conductance of the new-type chain decayed exponentially with the length and so did the electron transmission. The length affected the conductance decay constant, which decreased when the length was greater than 4.5 nm. However, the electron tunneling mechanism still dominated the electron transport by the LUMO, which showed delocalized electronic states for each length. This new-type molecular wire showed one abnormal tunnel barrier energy-length curve, where shorter C_2N ring chains with strong conductance exhibited higher barrier energy thanks to the compensation of their much narrower HOMO–LUMO gap. When a bias was applied, the currents of different wires grew at different rates. Furthermore, the NDR effect occurred in longer chains with an even number of C_2N rings and increased with size. The reduction in the electron transmission peak brought about such a valuable effect. The two shortest C_2N-ring-based chains showed peculiar conductance–voltage characteristics, which did not show any regularity in the length. Meanwhile, the six longer C_2N-ring-chain-based devices presented regularly enhanced conductance with the voltage at larger rates for longer chains. The conductance attenuation speed and transport regime could be regulated effectively by applying the voltage. Furthermore, the new-type chain was demonstrated to be better than the parent material and other similar nanoribbon-based chains in electron transport. Some factors were also considered to investigate the corresponding influence on the new-type molecular chain. The common H impurity obviously enhanced the electron transmission of the C_2N ring chain, while the common O impurity had a minimal effect on it. The structural disorder originating from the higher temperature did not change the exponential decreasing tendency of electron transmission with the length of the C_2N ring chain and the electron transmission fluctuated with the temperature T within a small range when T did not exceed 400 K. The C_2N ring chain with 1D GNR electrodes possessed noticeably stronger electron transmission than those with 2D graphene and 3D Au electrodes. These findings offer a solid backing for the application of C_2N ring chains in tunneling diodes and controllable molecular devices.

Supplementary Materials: The following supporting information can be downloaded from https://www.mdpi.com/article/10.3390/molecules28247994/s1—Figure S1: Molecular electrostatic potential for the pore (unit cell) of C_2N monolayer; Figure S2: (a,b) LUMOs and HOMOs of device C_2N rings 5 and 6, respectively; Figure S3: (a,b) Transmission pathways of device C_2N rings 4, 5, 6 and 7, respectively; Figure S4: (a–g) Comparison of electron transmission spectrum around E_F of the C_2N ring chains with no defect and O and H impurities for device C_2N rings 2, 3, 4, 5, 6, 7 and 8, respectively; Figure S5: (a–e) I–V curves of device C_2N rings 2, 3, 5, 6 and 8, respectively. Each inset shows the first derivative of the corresponding I-V curve at biases, representing the growth rate of the current with the bias; Figure S6: (a–h) Transmission spectrum within wider energy ranges at 0 V, 0.1 V, 0.2 V, 0.3 V and 0.4 V of device C_2N rings 1–8, respectively; Figure S7: (a–d) Transmission spectrum within narrower energy ranges at 0 V, 0.1 V, 0.2 V, 0.3 V and 0.4 V of device C_2N rings 2, 4, 5 and 7, respectively; Figure S8: (a,b) Conductance as a function of number of C_2N rings with an exponential fit at 0.2 V and 0.3 V, respectively. Each inset exhibits the corresponding lnG as a function of length with a linear fit.

Author Contributions: Conceptualization, J.L.; methodology, J.G.; software, H.L.; formal analysis, J.L. and D.S.; investigation, A.O.; resources, K.L.; writing—original draft preparation, D.S.; writing—review and editing, K.L. and J.L.; funding acquisition, K.L. and J.L. All authors have read and agreed to the published version of the manuscript.

Funding: This research was funded by the National Natural Science Foundation of China (grant no. 52105351), Natural Science Foundation of Jiangsu Province (grant nos. BK20210890 and BK20210873), High-end Foreign Experts Recruitment Plan of China (grant nos. G2023014009L and G2023014004), Postdoctoral Science Foundation of China (grant no. 2022M722928), Jiangsu Provincial Double-Innovation Doctor Program (grant nos. JSSCBS20210991 and JSSCBS20210985), Natural Science Research Projects in Universities of Jiangsu Province (grant no. 21KJB460015) and state assignment of the Ministry of Science and Higher Education of the Russian Federation (theme "Additivity" no. 121102900049-1).

Institutional Review Board Statement: Not applicable.

Informed Consent Statement: Not applicable.

Data Availability Statement: The data presented in this study are available on request from the corresponding author. The data are not publicly available due to the requirements of funding projects.

Conflicts of Interest: The authors declare no conflict of interest.

References

1. Bryce, M.R. A review of functional linear carbon chains (oligoynes, polyynes, cumulenes) and their applications as molecular wires in molecular electronics and optoelectronics. *J. Mater. Chem. C* **2021**, *9*, 10524–10546. [CrossRef]
2. Zhang, B.; Ryan, E.; Wang, X.; Song, W.; Lindsay, S. Electronic Transport in Molecular Wires of Precisely Controlled Length Built from Modular Proteins. *ACS Nano* **2022**, *16*, 1671–1680. [CrossRef] [PubMed]
3. Patrick, C.W.; Woods, J.F.; Gawel, P.; Otteson, C.E.; Thompson, A.L.; Claridge, T.D.W.; Jasti, R.; Anderson, H.L. Polyyne [3]Rotaxanes: Synthesis via Dicobalt Carbonyl Complexes and Enhanced Stability. *Angew. Chem. Int. Edit.* **2022**, *61*, e202116897. [CrossRef] [PubMed]
4. Li, P.; Jia, C.; Guo, X. Molecule-Based Transistors: From Macroscale to Single Molecule. *Chem. Rec.* **2021**, *21*, 1284–1299. [CrossRef]
5. Yang, S.-Y.; Qu, Y.-K.; Liao, L.-S.; Jiang, Z.-Q.; Lee, S.-T. Research Progress of Intramolecular π-Stacked Small Molecules for Device Applications. *Adv. Mater.* **2022**, *34*, 2104125. [CrossRef] [PubMed]
6. Gu, M.-W.; Peng, H.H.; Chen, I.W.P.; Chen, C.-H. Tuning surface d bands with bimetallic electrodes to facilitate electron transport across molecular junctions. *Nat. Mater.* **2021**, *20*, 658–664. [CrossRef] [PubMed]
7. Naher, M.; Milan, D.C.; Al-Owaedi, O.A.; Planje, I.J.; Bock, S.; Hurtado-Gallego, J.; Bastante, P.; Abd Dawood, Z.M.; Rincón-García, L.; Rubio-Bollinger, G.; et al. Molecular Structure–(Thermo)electric Property Relationships in Single-Molecule Junctions and Comparisons with Single- and Multiple-Parameter Models. *J. Am. Chem. Soc.* **2021**, *143*, 3817–3829. [CrossRef]
8. Li, L.; Low, J.Z.; Wilhelm, J.; Liao, G.; Gunasekaran, S.; Prindle, C.R.; Starr, R.L.; Golze, D.; Nuckolls, C.; Steigerwald, M.L.; et al. Highly conducting single-molecule topological insulators based on mono- and di-radical cations. *Nat. Chem.* **2022**, *14*, 1061–1067. [CrossRef]
9. Guldi, D.M.; Nishihara, H.; Venkataraman, L. Molecular wires. *Chem. Soc. Rev.* **2015**, *44*, 842–844. [CrossRef]
10. Robertson, N.; McGowan, C.A. A comparison of potential molecular wires as components for molecular electronics. *Chem. Soc. Rev.* **2003**, *32*, 96–103. [CrossRef]
11. Mahmood, J.; Lee, E.K.; Jung, M.; Shin, D.; Jeon, I.-Y.; Jung, S.-M.; Choi, H.-J.; Seo, J.-M.; Bae, S.-Y.; Sohn, S.-D.; et al. Nitrogenated holey two-dimensional structures. *Nat. Commun.* **2015**, *6*, 6486. [CrossRef]

12. Liang, Z.; Yang, D.; Tang, P.; Zhang, C.; Jacas Biendicho, J.; Zhang, Y.; Llorca, J.; Wang, X.; Li, J.; Heggen, M.; et al. Atomically dispersed Fe in a C2N Based Catalyst as a Sulfur Host for Efficient Lithium–Sulfur Batteries. *Adv. Energy Mater.* **2021**, *11*, 2003507. [CrossRef]
13. Gu, Z.; Zhao, L.; Liu, S.; Duan, G.; Perez-Aguilar, J.M.; Luo, J.; Li, W.; Zhou, R. Orientational Binding of DNA Guided by the C2N Template. *ACS Nano* **2017**, *11*, 3198–3206. [CrossRef] [PubMed]
14. Zhang, J.; Fang, C.; Li, Y.; An, W. Tetrahedral W_4 cluster confined in graphene-like C_2N enables electrocatalytic nitrogen reduction from theoretical perspective. *Nanotechnology* **2022**, *33*, 245706. [CrossRef] [PubMed]
15. Rajabpour, A.; Bazrafshan, S.; Volz, S. Carbon-nitride 2D nanostructures: Thermal conductivity and interfacial thermal conductance with the silica substrate. *Phys. Chem. Chem. Phys.* **2019**, *21*, 2507–2512. [CrossRef] [PubMed]
16. Zhang, H.; Zhang, X.; Yang, G.; Zhou, X. Point Defect Effects on Photoelectronic Properties of the Potential Metal-Free C2N Photocatalysts: Insight from First-Principles Computations. *J. Phys. Chem. C* **2018**, *122*, 5291–5302. [CrossRef]
17. Deng, S.; Hu, H.; Zhuang, G.; Zhong, X.; Wang, J. A strain-controlled C2N monolayer membrane for gas separation in PEMFC application. *Appl. Surf. Sci.* **2018**, *441*, 408–414. [CrossRef]
18. Yong, Y.; Cui, H.; Zhou, Q.; Su, X.; Kuang, Y.; Li, X. C2N monolayer as NH_3 and NO sensors: A DFT study. *Appl. Surf. Sci.* **2019**, *487*, 488–495. [CrossRef]
19. Panigrahi, P.; Sajjad, M.; Singh, D.; Hussain, T.; Andreas Larsson, J.; Ahuja, R.; Singh, N. Two-dimensional Nitrogenated Holey Graphene (C2N) monolayer based glucose sensor for diabetes mellitus. *Appl. Surf. Sci.* **2022**, *573*, 151579. [CrossRef]
20. Zhong, W.; Qiu, Y.; Shen, H.; Wang, X.; Yuan, J.; Jia, C.; Bi, S.; Jiang, J. Electronic Spin Moment As a Catalytic Descriptor for Fe Single-Atom Catalysts Supported on C_2N. *J. Am. Chem. Soc.* **2021**, *143*, 4405–4413. [CrossRef]
21. He, M.; An, W.; Wang, Y.; Men, Y.; Liu, S. Hybrid Metal–Boron Diatomic Site Embedded in C_2N Monolayer Promotes C–C Coupling in CO_2 Electroreduction. *Small* **2021**, *17*, 2104445. [CrossRef]
22. Liu, B.; Law, A.W.-K.; Zhou, K. Strained single-layer C_2N membrane for efficient seawater desalination via forward osmosis: A molecular dynamics study. *J. Membr. Sci.* **2018**, *550*, 554–562. [CrossRef]
23. Gao, E.; Yang, H.; Guo, Y.; Nielsen, S.O.; Baughman, R.H. The Stiffest and Strongest Predicted Material: C_2N Atomic Chains Approach the Theoretical Limits. *Adv. Sci.* **2023**, *10*, 2204884. [CrossRef]
24. Ma, Z.; Luo, Y.; Wu, P.; Zhong, J.; Ling, C.; Yu, Y.; Xia, X.; Song, B.; Ning, L.; Huang, Y. Unique Geometrical and Electronic Properties of TM_2—B_2 Quadruple Active Sites Supported on C_2N Monolayer Toward Effective Electrochemical Urea Production. *Adv. Funct. Mater.* **2023**, *33*, 2302475. [CrossRef]
25. Tian, Z.; López-Salas, N.; Liu, C.; Liu, T.; Antonietti, M. C2N: A Class of Covalent Frameworks with Unique Properties. *Adv. Sci.* **2020**, *7*, 2001767. [CrossRef]
26. Guan, Z.; Lian, C.-S.; Hu, S.; Ni, S.; Li, J.; Duan, W. Tunable Structural, Electronic, and Optical Properties of Layered Two-Dimensional C_2N and MoS_2 van der Waals Heterostructure as Photovoltaic Material. *J. Phys. Chem. C* **2017**, *121*, 3654–3660. [CrossRef]
27. Xu, J.; Mahmood, J.; Dou, Y.; Dou, S.; Li, F.; Dai, L.; Baek, J.-B. 2D Frameworks of C_2N and C_3N as New Anode Materials for Lithium-Ion Batteries. *Adv. Mater.* **2017**, *29*, 1702007. [CrossRef] [PubMed]
28. Huang, M.; Meng, H.; Luo, J.; Li, H.; Feng, Y.; Xue, X.-X. Understanding the Activity and Design Principle of Dual-Atom Catalysts Supported on C2N for Oxygen Reduction and Evolution Reactions: From Homonuclear to Heteronuclear. *J. Phys. Chem. Lett.* **2023**, *14*, 1674–1683. [CrossRef] [PubMed]
29. Ahsan, F.; Yar, M.; Gulzar, A.; Ayub, K. Therapeutic potential of C_2N as targeted drug delivery system for fluorouracil and nitrosourea to treat cancer: A theoretical study. *J. Nanostruct. Chem.* **2023**, *13*, 89–102. [CrossRef]
30. Li, H.; Garner, M.H.; Su, T.A.; Jensen, A.; Inkpen, M.S.; Steigerwald, M.L.; Venkataraman, L.; Solomon, G.C.; Nuckolls, C. Extreme conductance suppression in molecular siloxanes. *J. Am. Chem. Soc.* **2017**, *139*, 10212–10215. [CrossRef]
31. Kim; Beebe, J.M.; Jun, Y.; Zhu, X.Y.; Frisbie, C.D. Correlation between HOMO Alignment and Contact Resistance in Molecular Junctions: Aromatic Thiols versus Aromatic Isocyanides. *J. Am. Chem. Soc.* **2006**, *128*, 4970–4971. [CrossRef] [PubMed]
32. Ho Choi, S.; Kim, B.; Frisbie, C.D. Electrical Resistance of Long Conjugated Molecular Wires. *Science* **2008**, *320*, 1482–1486. [CrossRef] [PubMed]
33. Bande, A.; Michl, J. Conformational Dependence of σ-Electron Delocalization in Linear Chains: Permethylated Oligosilanes. *Chem.— Eur. J.* **2009**, *15*, 8504–8517. [CrossRef] [PubMed]
34. Schepers, T.; Michl, J. Optimized ladder C and ladder H models for sigma conjugation: Chain segmentation in polysilanes. *J. Phys. Org. Chem.* **2002**, *15*, 490–498. [CrossRef]
35. Li, Z.; Kosov, D.S. Orbital interaction mechanisms of conductance enhancement and rectification by dithiocarboxylate anchoring group. *J. Phys. Chem. B* **2006**, *110*, 19116–19120. [CrossRef] [PubMed]
36. Liu, K.; Li, J.; Liu, R.; Li, H.; Okulov, A. Non-equilibrium electronic properties of ultra-thin SiC NWs influenced by the tensile strain. *J. Mater. Res. Technol.* **2023**, *26*, 6955–6965. [CrossRef]
37. Liu, K.; Li, Y.; Liu, Q.; Song, D.; Xie, X.; Duan, Y.; Wang, Y.; Li, J. Superior electron transport of ultra-thin SiC nanowires with one impending tensile monatomic chain. *Vacuum* **2022**, *199*, 110950. [CrossRef]
38. Song, D.; Liu, K.; Li, J.; Zhu, H.; Sun, L.; Okulov, A. Mechanical tensile behavior-induced multi-level electronic transport of ultra-thin SiC NWs. *Mater. Today Commun.* **2023**, *36*, 106528. [CrossRef]

39. Zhang, L.; Li, T.; Li, J.; Jiang, Y.; Yuan, J.; Li, H. Perfect Spin Filtering Effect on Fe_3GeTe_2-Based Van der Waals Magnetic Tunnel Junctions. *J. Phys. Chem. C* **2020**, *124*, 27429–27435. [CrossRef]
40. Solomon, G.C.; Herrmann, C.; Hansen, T.; Mujica, V.; Ratner, M.A. Exploring local currents in molecular junctions. *Nat. Chem.* **2010**, *2*, 223–228. [CrossRef]
41. Longuinhos, R.; Ribeiro-Soares, J. Stable holey two-dimensional C_2N structures with tunable electronic structure. *Phys. Rev. B* **2018**, *97*, 195119. [CrossRef]
42. Cao, J.; Dong, J.; Saglik, K.; Zhang, D.; Solco, S.F.D.; You, I.J.W.J.; Liu, H.; Zhu, Q.; Xu, J.; Wu, J.; et al. Non-equilibrium strategy for enhancing thermoelectric properties and improving stability of $AgSbTe_2$. *Nano Energy* **2023**, *107*, 108118. [CrossRef]
43. Wang, D.; Chen, Z.; Su, J.; Wang, T.; Zhang, B.; Rong, X.; Wang, P.; Tan, W.; Guo, S.; Zhang, J.; et al. Controlling Phase-Coherent Electron Transport in III-Nitrides: Toward Room Temperature Negative Differential Resistance in AlGaN/GaN Double Barrier Structures. *Adv. Funct. Mater.* **2021**, *31*, 2007216. [CrossRef]
44. Lu, W.; Meunier, V.; Bernholc, J. Nonequilibrium Quantum Transport Properties of Organic Molecules on Silicon. *Phys. Rev. Lett.* **2005**, *95*, 206805. [CrossRef] [PubMed]
45. Brandbyge, M.; Mozos, J.-L.; Ordejón, P.; Taylor, J.; Stokbro, K. Density-functional method for nonequilibrium electron transport. *Phys. Rev. B* **2002**, *65*, 165401. [CrossRef]
46. Taylor, J.; Guo, H.; Wang, J. Ab initio modeling of quantum transport properties of molecular electronic devices. *Phys. Rev. B* **2001**, *63*, 245407. [CrossRef]
47. Landauer, R. Electrical resistance of disordered one-dimensional lattices. *Philos. Mag.* **1970**, *21*, 863–867. [CrossRef]
48. Reed, M.A.; Zhou, C.; Muller, C.; Burgin, T.; Tour, J. Conductance of a molecular junction. *Science* **1997**, *278*, 252–254. [CrossRef]
49. Datta, S. *Electronic Transport in Mesoscopic Systems*; Cambridge University Press: Cambridge, UK, 1997.

Disclaimer/Publisher's Note: The statements, opinions and data contained in all publications are solely those of the individual author(s) and contributor(s) and not of MDPI and/or the editor(s). MDPI and/or the editor(s) disclaim responsibility for any injury to people or property resulting from any ideas, methods, instructions or products referred to in the content.

Article

Mechanistic Studies on Aluminum-Catalyzed Ring-Opening Alternating Copolymerization of Maleic Anhydride with Epoxides: Ligand Effects and Quantitative Structure-Activity Relationship Model

Xiaowei Xu [1,*], Hao Li [1], Andleeb Mehmood [2], Kebin Chi [1], Dejun Shi [1], Zhuozheng Wang [1], Bin Wang [3,*], Yuesheng Li [3] and Yi Luo [1,*]

[1] PetroChina Petrochemical Research Institute, Beijing 102206, China
[2] College of Physics and Optoelectronic Engineering, Shenzhen University, Shenzhen 518000, China
[3] Tianjin Key Laboratory of Composite & Functional Materials, School of Materials Science and Engineering, Tianjin University, Tianjin 300350, China
* Correspondence: xuxiaowei010@petrochina.com.cn (X.X.); binwang@tju.edu.cn (B.W.); luoyi010@petrochina.com.cn (Y.L.)

Abstract: Previous work has indicated that aluminum (Al) complexes supported by a bipyridine bisphenolate (BpyBph) ligand exhibit higher activity in the ring-opening copolymerization (ROCOP) of maleic anhydride (MAH) and propylene oxide (PO) than their salen counterparts. Such a ligand effect in Al-catalyzed MAH-PO copolymerization reactions has yet to be clarified. Herein, the origin and applicability of the ligand effect have been explored by density functional theory, based on the mechanistic analysis for chain initiation and propagation. We found that the lower LUMO energy of the (BpyBph)AlCl complex accounts for its higher activity than the (salen)AlCl counterpart in MAH/epoxide copolymerizations. Inspired by the ligand effect, a structure-energy model was further established for catalytic activity (TOF value) predictions. It is found that the LUMO energies of aluminum chloride complexes and their average NBO charges of coordinating oxygen atoms correlate with the catalytic activity (TOF value) of Al complexes (R^2 value of 0.98 and '3-fold' cross-validation Q^2 value of 0.88). This verified that such a ligand effect is generally applicable in anhydride/epoxide ROCOP catalyzed by aluminum complex and provides hints for future catalyst design.

Keywords: aluminum complex; density functional theory; ring-opening alternating copolymerization; ligand effect; multivariate linear regression

1. Introduction

Aliphatic polyesters have drawn a lot of attention as a potentially viable choice for petroleum-based polymers because of their numerous renewable sources, hydrolytic degradability, and excellent biocompatibility [1–3]. In this context, the ring-opening copolymerization (ROCOP) of epoxides with anhydrides is a burgeoning technology to produce aliphatic polyesters [4–6]. The ROCOP of epoxides with anhydrides most commonly uses binary catalyst systems comprising metal complexes and nucleophilic cocatalyst components. The earliest epoxide/anhydride copolymerizations were initiated with amines and metal alkoxides, yielding low-molecular-weight polymers with broad dispersity and, in many cases, significant polyether contamination from epoxide homopolymerization [7,8].

The first well-controlled ROCOP of propylene oxide with phthalic anhydride catalyzed by aluminum porphyrin complex and a tetralkyl ammonium halide was reported in 1985 [9]. Subsequently, Coates and co-workers found that a β-diiminate zinc acetate [(BDI)ZnOAc] complex is an effective catalyst for various epoxide/cyclic anhydrides copolymerization and terpolymerizations of epoxide/cyclic anhydrides/CO_2 [10,11]. This was followed in 2011 by another two catalysts for epoxide/anhydride copolymerization, N,N'-

bis(salicylidene)-cyclohexanediamine chromium-(III) chloride [(salcy)CrCl], which have been successfully used in epoxide/CO_2 copolymerizations [12–14]. Since then, (salcy)MX complexes with similar backbones have become some of the most widely used complexes for epoxide/anhydride copolymerization, including Cr [13–17], Co [13–15,17–19], Al [13,15], Mn [13,20], and Fe [21] complexes. Among reported catalysts, Co and Cr-based complexes with salen-type ligands are still the most effective organometallic complexes for MAH/PO copolymerization. However, toxic Co and Cr elements are difficult to remove completely and their residuals will hinder the further application of aliphatic polyesters. Thus, it is necessary to develop metal-based complex with a low toxicity and high efficiency.

For this purpose, aluminum complexes are mainly explored owing to their low toxicity, easy preparation, and the high degree of control over the molecular weight and microstructure. In 2012, Duchateau's group applied metal salen (salen = N,N-bis(3,5-di-tert-butylsalicylidene)diimine) chloride complexes with different diimine linkages to catalyze the MAH/PO copolymerization and found that Al-salophen complexes clearly outperformed the other three Al-complexes [15]. Recently, Wang et al. have synthesized and characterized a series of bipyridine bisphenolate Al complexes, (BpyBph)AlX (X = axial group, such as Cl, OOCCH3, OOCCF3). They assessed the catalytic activity of these complexes for MAH/PO copolymerization and investigated the impact of the ligand's steric and electronic nature on their catalytic activity in the absence of cocatalyst. There is a negligible effect on activity (TOF = 5.9 h^{-1} vs. 5.6 h^{-1}, respectively) when replacing the *para* t-Bu groups with the electron-donating –OCH3 group. However, with the bulky cumyl substituents both on the *para* and *ortho* position of the bisphenolate Al complexes, the corresponding complex was proved to be the least active (TOF = 4.0 h^{-1}), possibly because of the steric bulk around the metal center and leading to hindered monomer coordination and insertion. The catalytic activity of bipyridine bisphenolate Al complexes with *para* fluorine atoms was the highest (TOF = 8.3 h^{-1}), which is attributed to the increased Lewis acidity of the complex from the electron-withdrawing group facilitating the activation of PO. It is noteworthy that the ligand backbone has a significant effect on the polymerization performance for MAH/PO copolymerization. Additionally, they found the anion group in the axial group affected the catalytic activity. The catalytic activity decreased with the decrease in the leaving ability of the axial group (leaving ability: $Cl^- > TFA^- > OAc^-$). As Table 1 shows, **A1**/PPNCl binary catalytic system demonstrated a remarkably higher catalytic activity (TOF = 36 h^{-1}) and could produce higher M_n polyesters with a narrow and unimodal distribution (PDI = 1.28). By contrast, **B1**/PPNCl binary catalytic system demonstrated a lower activity (TOF = 6.0 h^{-1}) and lower M_n unsaturated polyesters [22]. From this, it is clear that the bipyridine bisphenolate Al-complex (**A1**) provides superior results to the salen Al complex (**B1**). However, the origin of such activity discrepancies induced by the ligand backbone has remained unclear, which is fundamentally crucial for the further development of such catalyst systems.

The regression analysis technique has been successfully applied to construct a quantitative structure-property relationship (QSPR) of polymerization catalysts, which is a meaningful method to analyze the main factor governing catalytic performance on the basis of the molecular descriptor and QSPR model [23,24]. For this purpose, some data sets for QSPR analysis consisted of a known catalyst structure and the catalytic performance measured by experimental and computational approaches [25]. Sun's group investigated the relationship between the structure of late transition metal complexes and their experimental activity in ethylene oligo/polymerization by molecular modeling and the QSAR method [24]. Our group explored the origin of the stereoselectivity in the yttrium-catalyzed polymerization of 2-vinylpyridine and the poisoning effect of a polar monomer toward Brookhart-type catalysts, by combining DFT calculations and multivariate regression analysis [26]. As for the copolymerization of anhydride and cyclohexene oxide, previous studies have focused mostly on the ROCOP of phthalic anhydride and cyclohexene oxide catalyzed by the Cr-complexes. It was revealed that the maximal rate was achieved with one equivalent of [PPN]Cl as the cocatalyst and the polymerization rate is first-order dependence on [epoxide]. The ring-opening of epoxide was the rate-determining step [27]. Tolman and co-workers reported a mechanistic study on the ROCOP of epoxides and cyclic anhydrides using a (salph)AlCl/PPNCl (PPN = bis(triphenylphosphine)iminium) catalytic pair [28]. However, there is a lack of systematic studies into the structure effect of the ligand on the copolymerization activity.

Table 1. ROAC of MAH/PO by Al complexes bearing different ligand backbones in the presence of PPNCl.

Catalyst	t (h)	Conv. (%)	TOF (h^{-1})	Ester Linkage (%)	$M_{n,thero}$ (kDa)	$M_{n,sec}$ (kDa)	PDI
A1/PPNCl	2	71	36	>99	5.6	4.6	1.28
B1/PPNCl	2	12	6.0	>99	0.9	n.d.	n.d.

In the present work, to elucidate the relationship between the ligand backbone of Al complexes and their catalytic activity in MAH/PO copolymerization (Table 1), DFT calculations have been conducted to disclose the polymerization mechanism and the origin of activity difference. Then, some key descriptors were calculated to correlate the catalyst activity (TOF value), and a QSPR model is thus constructed. On the basis of these theoretical studies, it has been found that the higher activity of **A1** with bipyridine bisphenolate than the **B1** with a traditional salen-type ligand can be explained by the lower LUMO energy and more negatively charged coordinating oxygen atoms.

2. Results and Discussion

To discriminate the distinct effects of the BpyBph ligand from typical salen analogues, we comparatively explored the possible mechanism of ROCOP of MAH/PO catalyzed by **A1** or **B1** with the aid of DFT calculations in the present paper. The complexes **A1** and **B1** were simplified to **A** and **B**, respectively, in which the t-Bu substituents were simplified by H atoms. To prove the rationality of the simplified models, the key steps involving PO insertion (**INT3** → **TS2** and **INT10** → **TS4**) mediated by the original catalyst (**A1** and **B1**) were calculated for comparison. As shown in Table S1, the energy differences for the abovementioned steps are almost unchanged, in spite of the full or simplified model, suggesting a neglectable effect of the t-Bu substituents on the relative energies.

2.1. Chain Initiation Stage

The computed energy profiles for the chain initiation step mediated by complexes **A** and **B** are shown in Figure 1. Upon an attack of PO to **A** (or **B**), the PO could be activated by the initial five-coordination Al complex and undergo ring-opening upon the nucleophilic attack of free Cl$^-$ via the concerted **TS1**, leading to the **INT2**. Alternatively, it is possible to convert the initial aluminum complex (**A** or **B**) to 6-coordination bis(chlorinated) aluminate complex **INT1'** by its reaction with exogenous Cl$^-$ derived from [PPN]Cl. Owing to the fact that the weakly nucleophilic Cl$^-$ is an easier leaving group than alkoxide with a strong nucleophilicity, the resulting six-coordination aluminate species **INT2** is feasible to release the chloride anion, leaving a vacant site available for coordination of the coming PO. A coordination of PO to **INT3** generated the neutrally 6-coordinated **INT4** which could be converted to the bis(alkoxide) Al complex **INT5** via **TS2** by the attack of another nucleophilic anion. It is noteworthy that the ring-opening of PO in the chain initiation stage proceeded at the two sides of the [ONNO-Al] plane, and the bis(alkoxide) Al complex **INT5** was the real active species that could initiate the subsequent chain propagation. As shown in Figure 1, the activation energies ($\Delta G_1^{\ddagger}$) of the first PO insertion were very similar for complexes **A** and **B**; however, the $\Delta G_2^{\ddagger}$ for complex **A** (24.4 kcal/mol) is higher than that for **B** (19.3 kcal/mol), suggesting the slower chain initiation by complex **A**.

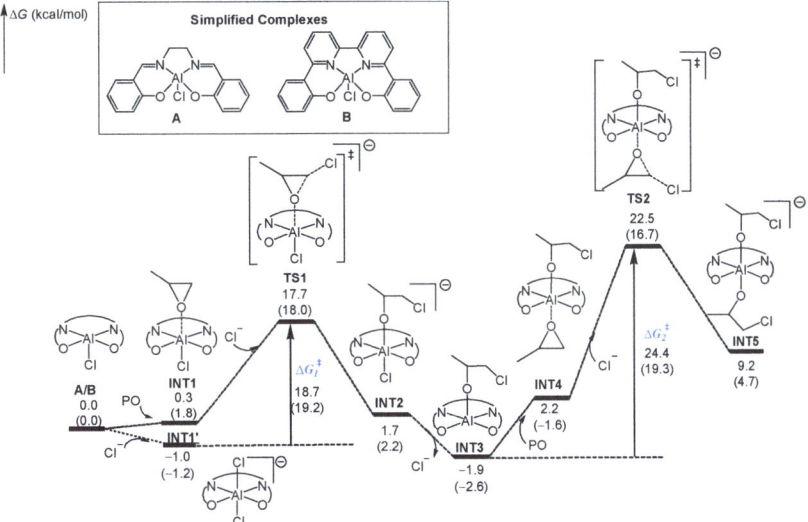

Figure 1. The calculated reaction routes for chain initiation mediated by **A** or **B**. The relative free energies in the **A**-catalyzed system and **B**-catalyzed system (in parentheses) are given in kcal/mol. The symbol '$\ddagger$' means transition state.

To elucidate the origin of the slower chain initiation in **A**-mediated polymerization, the distortion/interaction analysis was comparatively performed for the key TSs, **A_TS2** and **B_TS2**, respectively [29–32]. Both TSs were divided into three fragments, catalyst alkoxide (Fragment A), monomer moiety (Fragment B), and chloride ion (Fragment C) (Table 2), and these energies were evaluated through single-point calculations. The interaction energy ΔE_{int} was estimated by the single-point energies of TS and the three fragments. These single-point energies, together with the energies of the respective fragments in their optimal geometry, allow the estimation of the distortion energies of the three fragments, $\Delta E_{dist}(A)$, $\Delta E_{dist}(B)$, and $\Delta E_{dist}(C)$. The distortion energy of the fragment is defined as the energy difference between its distorted geometry in the transition state and its optimized structure. Therefore, the energy of the transition state, $\Delta E^{\ddagger}$, is $\Delta E^{\ddagger} = \Delta E_{int} + \Delta E_{dist}(A) + \Delta E_{dist}(B) + \Delta E_{dist}(C)$.

As shown in Table 2, although the interaction energies (ΔE_{int}) of the three fragments are almost equal (−33.9 vs. −34.1 kcal/mol) in the two TSs, the disfavored items of $\Delta E_{dist}(A)$ and $\Delta E_{dist}(B)$ in **A_TS2** (17.9 and 20.0 kcal/mol) are much larger than in **B_TS2** (15.0 and 18.6 kcal/mol), which indicates that the larger distortion could account for the lower stability of **A_TS2**. To gain more insight, the geometries of fragment **A** are carefully compared. The change in the dihedral angle of the [ONNO] equatorial ligand plane in fragment **A** of **A_TS2** (13.3° for the fragment A of **A_TS2** relative to **B_INT3**) is much bigger than in the case of **B_TS2** (6.3° for the fragment A of **B_TS2** relative to **B_INT3**). These geometrical changes mainly account for the larger distortion of **A_TS2** and thus its lower stability.

Table 2. Distortion/interaction analysis of (**a**) **A_TS2** and (**b**) **B_TS2**. Energies are given in kcal/mol. Green circle denotes catalyst fragment. Blue circle denotes monomer fragment. Pink circle denotes anion. The symbol '‡' means transition state.

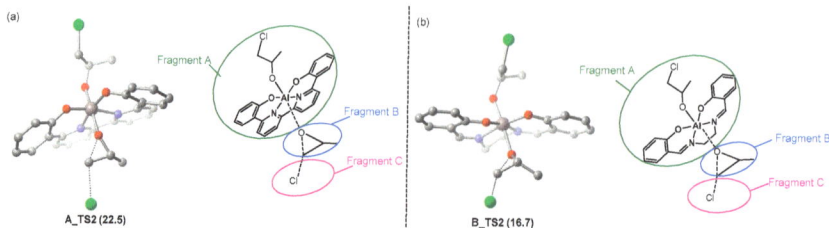

TS2	$\Delta G^{\ddagger}$	ΔE_{int}(A-B-C)	ΔE_{dist}(A)	ΔE_{dist}(B)	ΔE_{dist}(C)	ΔE_{dist}	$\Delta E^{\ddagger}$
A_TS2	24.4	−33.9	17.9	20.0	0.0	37.9	3.9
B_TS2	19.3	−34.1	15.0	18.6	0.0	33.6	−0.5

2.2. Chain Propagation Stage

To further explore the effect of the ligand backbone on the polymerization activity, the chain propagation was also calculated, as shown by the following process (Figure 2). Based on the bis(alkoxide) Al complex **INT5**, the enchainment of MAH could take place through **TS3** and yield anionic six-coordination **INT6** with alkoxide at one side and carboxylate at the other side of the ONNO-Al plane. The release of the carboxylate anion (an easier leaving group than alkoxide) resulted in five-coordination neutral **INT7** with an alkoxide and a vacant coordination site in the axial position, which could coordinate with either PO or RCOO$^-$ (dissociated from **INT6**) and thus has two possible reaction pathways. One possible route is the formation of **INT8**, in which PO was activated upon coordination to **INT7**. Then, the bis(alkoxide) Al complex **INT9** was generated by the nucleophilic attack of RCOO$^-$ via transition state **TS4**. **INT9** is structurally similar to the bis(alkoxide) aluminate complex **INT5**, and the repeat of the process of **INT5** to **INT9** could grow the polymer chain. The alternative route is an association of RCOO$^-$ with **INT7** and the formation of the anionic six-coordination **INT10**. Upon the approach of MAH, the alkoxide in **INT10** could nucleophilically attach the carboxyl carbon of the MAH and generate the bi(carboxylate) Al complex **INT11**, which could convert to **INT12** via the release of a carboxylate anion. The RCOO$^-$ anion (dissociated from **INT11**) is easily coordinated with **INT12** in another fashion and forms more stable intermediate (**INT12′**). Similar to **INT7**, PO could coordinate to **INT12** and was ring-opening under the action of a dissociated carboxylate anion from **INT11** to afford **INT14** (analogues to **INT10**). The chain propagation could occur via a similar process from **INT10** to **INT14**.

As shown in Figure 2, the free energy barrier for epoxide opening (**INT12′** → **TS4**) is the highest ($\Delta G_4^{\ddagger}$: 33.3 and 42.5 kcal/mol for the **A**- and **B**-catalyzed systems, respectively) among the reaction steps. Therefore, the ring-opening of PO serves as a rate-determining step of the copolymerization reaction. This is in agreement with the experimental finding that the polymerization rate is first-order in [PO] and zero-order in [MAH] [22]. Meanwhile, it was found that the overall energy barrier of the **B** system is $\Delta G_4^{\ddagger}$ = 42.5 kcal/mol, which is 9.2 kcal/mol higher than that in the **A** system. This is in line with the considerable difference in reactivity between complexes **A1** and **B1** [22].

Furthermore, distortion/interaction analyses for turn-over limiting TSs (**A_TS4** and **B_TS4**) were conducted as well, in order to elucidate the ligand framework effect on catalytic activity. As shown in Table 3, in the case of **A_TS4**, the total deformation energy ΔE_{dist} is 42.4 kcal/mol, which could be balanced out by its ΔE_{int} (−46.1 kcal/mol) leading to an ΔE of −3.7 kcal/mol. By contrast, the less negative interaction energy (−39.5 kcal/mol) in **B_TS4** could not compensate for its total deformation energy (ΔE_{dist} = 41.2 kcal/mol), thus producing a higher $\Delta E^{\ddagger}$ (1.7 kcal/mol). Therefore, the stronger interaction between three fragments could account for the greater stability of **A_TS4**. Meanwhile, it has been found that the interaction (−29.6 kcal/mol) between the fragment A and B in **A_TS4** is stronger than that in **B_TS4** (−24.8 kcal/mol) and mainly contributes to the stability of **A_TS4**. Such a discrepancy probably originates from the higher electrophilic ability of **A_INT7** compared with **B_INT7**. As expected, the calculated LUMO energies for **A_INT7** and **B_INT7** were −3.21 and −2.73 eV (Figure 3a), respectively. With the LUMO population on the ligand backbone, the LUMO energy of **A_INT7** was closer to the HOMO energy of PO, which resulted in an increase in the reactivity of the corresponding metal complex toward PO (Figure 3b). Essentially, these differences in LUMO energy were also observed for **A** and **B** (−3.22 vs. −2.74 eV, Figure 3a). Thus, the lower LUMO energy of **A1**, which bears the BpyBph ligand, may account for the higher activity of **A1** than the typical (salen)Al analogues (**B1**).

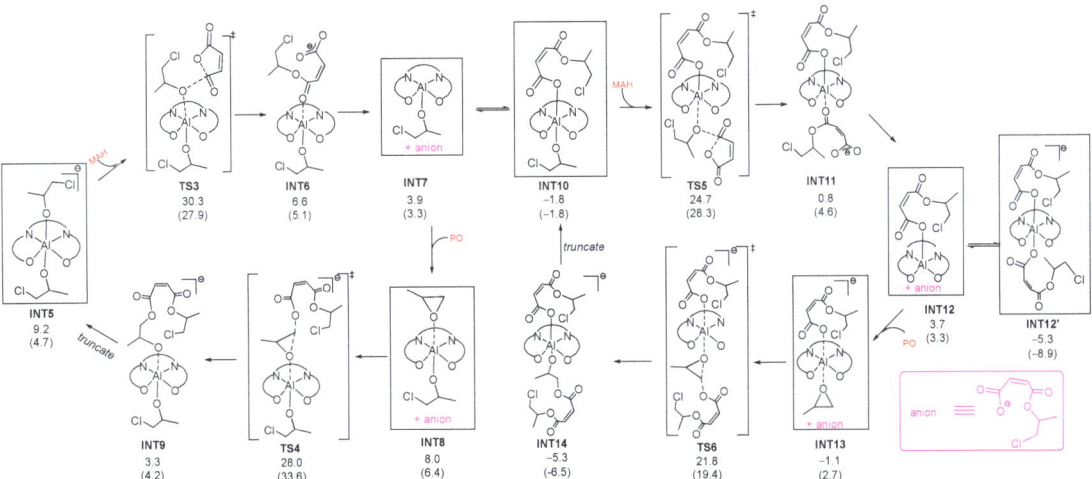

Figure 2. The calculated reaction routes for chain propagation mediated by **A** or **B**. The symbol "‡" denotes transition state. The pink and red ones represent anion and monomers, respectively. The relative free energies in the **A**-catalyzed system and **B**-catalyzed system (in parentheses) are given in kcal/mol. The energy reference point is the same as that in Figure 1.

Table 3. Distortion/interaction analysis of (a) **A_TS4** and (b) **B_TS4**. Energies are given in kcal/mol. Green circle denotes catalyst fragment. Blue circle denotes monomer fragment. Pink circle denotes anion. The symbol '‡' means transition state.

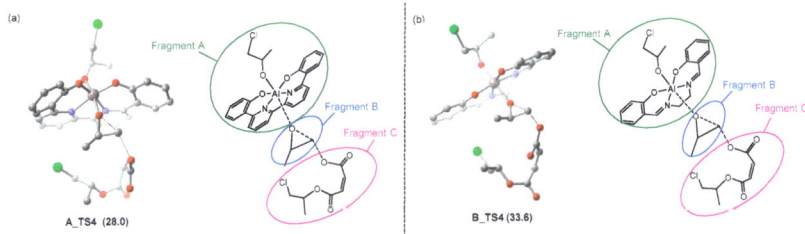

TS4	$\Delta G^{\ddagger}$	ΔE_{int}(A-B-C)	ΔE_{dist}(A)	ΔE_{dist}(B)	ΔE_{dist}(C)	ΔE_{dist}	$\Delta E^{\ddagger}$
A_TS4	33.3	−46.1	18.0	23.3	1.0	42.4	−3.7
B_TS4	42.5	−39.5	14.8	26.1	0.2	41.2	1.7

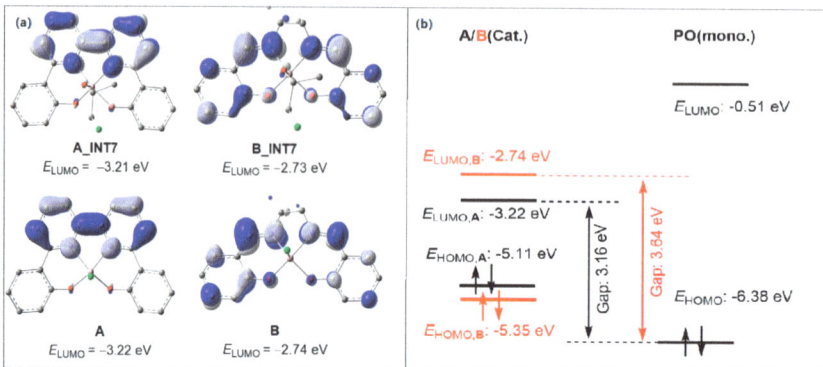

Figure 3. (a) LUMO plots (isosurface value = 0.02 a.u.) and corresponding orbital energies. The blue and grey clouds in the figure represent the positive and negative orbits. (b) Energy electronic levels diagrams.

2.3. Multivariate Linear Regression Analysis

To further validate the above ligand effect, we further explored the six complexes (**A1~F1**, Figure 4) that have been reported in the original paper [22] on the basis of the elucidated reaction mechanisms in the **A** and **B** systems, and constructed their quantitative structure-activity relationship through multivariate linear regression (MLR) analysis. The catalytic activity toward PO and MAH copolymerization was estimated by the turnover frequency (TOF, h^{-1}) value. Seven parameters of each aluminum chloride complex were chosen to characterize the copolymerization activity of these complexes (**A1~F1**, Figure 4), including the LUMO energy (E_{LUMO}, in eV), Wiberg bond index of Al-Cl (WBI), dihedral angle of [ONNO] (D, in degree), the NBO charges of central metal (q_{Al}) and coordinating heteroatoms, viz., chloride atom (q_{Cl}), nitrogen atoms (q_N, average charge on the N atoms), and oxygen atoms (q_O, the average charge on the O atoms).

Taking the catalysts **A1~F1** as the training set, we constructed the catalyst activity regression model for CHO and MAH copolymerization, with the help of the stepwiselm function in Matlab to perform multivariate linear regression and remove the insignificant descriptors (p-value > 0.05). The TOF and parameters (seven descriptors aforementioned, viz., E_{LUMO}, WBI, D, q_{Al}, q_{Cl}, q_N, and q_O) are shown in Table S2. The accuracy of the model was good, with the training determination coefficient (R^2) of 0.98 and the '3-fold' cross-validation determination coefficient (Q^2) of 0.88. As expected, the predicted TOF values derived from the MLR model are consistent with the experimental ones.

In the regression model, only the LUMO energy (E_{LUMO}) and the average NBO charges of oxygen atoms (q_O) were retained (Figure 5). According to the negative values of E_{LUMO} and q_O, a more negative E_{LUMO} and a more negative q_O could lead to a higher TOF value. In addition, it has been found that electron-withdrawing groups or conjugated backbone endowed a higher activity (higher TOF value) of the corresponding metal complexes. This demonstrates that the electron-withdrawing group or conjugated backbone is beneficial to decreasing such a LUMO energy or the average NBO charges of oxygen atoms. Such an electronic effect therefore increases the activity of the corresponding metal complex toward PO/MAH copolymerization. This strategy was promising to utilize in the future design of metal-based catalysts and expand their application by advancing catalytic properties.

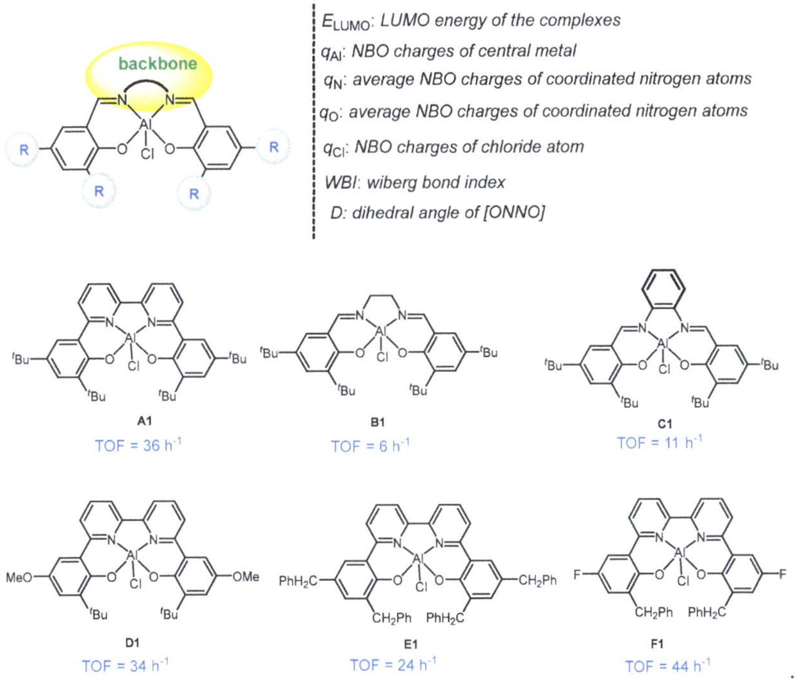

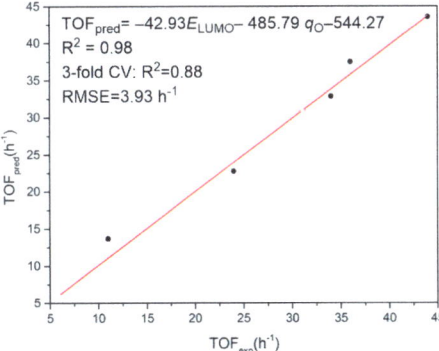

Figure 4. Some descriptors and TOF values (in h^{-1}) for the six complexes (**A1**~**F1**).

Figure 5. Plot of the experimental activity TOF$_{exp}$ (h^{-1}) vs. corresponding prediction values of TOF$_{pred}$ (h^{-1}) derived from the multivariate linear regression model. Black circles are six complexes (**A1**~**F1**) and two black circles (**A1** and **D1**) are overlapped in the figure. The red line to the predicted TOF values has the Pearson correlation coefficient R = 0.98.

3. Methods

The M06L functional was used for geometrical optimization and subsequent frequency calculations without any symmetry or geometrical constraints [33]. Frequency calculations were performed to ensure that the structures found were stationary points (no imaginary frequencies for minima and one imaginary frequency for transition-state structures). Intrinsic reaction coordinate (IRC) calculations were also carried out to investigate whether each of the transition structures actually connected the reactant and product [34,35]. The optimization of the transition state was carried out using the Berny algorithm [36]. In these calculations, the double-ζ 6-31+G(d,p) basis set was used for all atoms [37,38]. Such basis sets are referred to BSI. To obtain more accurate energies, single-point energy calculations were performed at the level of M06-2X/BSII together with the SMD model for considering the solvation effect of THF [39], which was used for modeling epoxide as the solvent [28,40]. In the BSII, the 6-311+G(d,p) basis set was used for all of the atoms. To justify the use of the above XC-functional and basis sets, the key intermediates and transition states were calculated at the level of SMD(THF)/M06-2X/6-311++G(2d,p)//B3LYP-D3/6-311+G** [41–44]. The comparative data with different computational methods are collected in Table S3. The three-dimensional images of the optimized structures were prepared using CYLview [45]. All the calculations were performed by the Gaussian 16 program [46]. As indicated by previous calculations for the separated and anion-paired [PPN]$^+$ cations, the [PPN]$^+$ cation could be viewed as a noncoordinating and separated counterion owing to its large bulk [28].

4. Conclusions

The ligand effect in the copolymerization of maleic anhydride (MAH) and epoxides (PO) catalyzed by Al-based complexes has been investigated by DFT calculations. Having achieved an agreement in activity discrepancy between BpyBph and salen-ligated Al complexes, it is found that the BpyBph ligand endows its Al complex with a lower LUMO energy than its salen counterpart, thus leading to the higher catalytic activity. Inspired by this ligand effect, a quantitative structure-activity relationship has been constructed via multivariate linear regression (MLR) analysis. The trained MLR model features an R^2 value of 0.98 and a '3-fold' cross-validation determination coefficient (Q^2) of 0.88. Strong correlations were found between the catalyst activity of aluminum chloride complexes and their LUMO energies (E_{LUMO}), as well as the average NBO charge of coordinating oxygen atoms (q_O). When the E_{LUMO} is lower and the ligation oxygen atoms are more negatively charged (more negative q_O), the Al complexes have a higher catalytic activity and lead to more rapid PO/MAH copolymerization. This is in agreement with the experimental results in which the complexes (**F1**) with an electron-withdrawing group or conjugated backbone demonstrate the highest TOF value, compared with other aluminum analogues (**A1~E1**). The mechanistic insight gained in this study could help to better interpret the catalytic activity and contribute to the rational design of novel metal catalysts.

Catalysts are the core competence of determining the industrial production efficiency and advanced product development of polyester. The development of traditional catalysts relies on chance encounters, chemical intuition, and large-scale experimental behavior. Meanwhile, the measurement of catalytic performance is high cost and needs a lot of resources as well. In order to meet the requirements of current social and economic development, it is urgent to change the outmoded development pattern to a predictable and designable model. With the rapid evolution of high-performance computing technology, the era of artificial intelligence is inexorably surging. Relying on the rapid development of different algorithms and computer hardware, it is the right time to harvest the potential of machine learning in the field of catalysis across academy and industry. Despite a substantial number of successful applications of machine learning, this exciting topic is still largely in its nascent stage and it is believed that machine learning will play an increasingly important role in accelerating the development of various kinds of functional materials in the foreseeable future. Herein, we will focus on the development of novel polyester and polyolefin catalysts by using machine-learning methods combined with essential DFT calculations, and provide insight into the underlying mechanism of the relationship between the microstructure of the catalyst and its polymerization performance by constructing the prediction models of catalytical activity and selectivity at the molecular and electronic level in the near future. This may provide a catalyst design guideline for developing high-performance polymerization catalysts and accelerate new high-performance catalyst development and materials discovery.

Supplementary Materials: The following supporting information can be downloaded at: https://www.mdpi.com/article/10.3390/molecules28217279/s1, Table S1. The computed free energy barrier for the key steps with full and simplified modes (Energy in kcal/mol); Table S2. Computed descriptors and TOF values (in h^{-1}) for the six complexes (**A1~F1**). Table S3. The comparative data with different computational methods. The Cartesians of all explored structures as well as a list of the energies and imaginary frequencies were present in the Supplementary Materials.

Author Contributions: X.X.: data curation, formal analysis, software, writing-original draft. H.L.: validation, data curation. A.M.: validation, visualization. K.C.: validation. D.S.: visualization. Z.W.: visualization. B.W.: conceptualization, project administration, writing—review and editing, supervision. Y.L. (Yuesheng Li): writing—review and editing. Y.L. (Yi Luo): conceptualization, funding acquisition, project administration, writing—review and editing. All authors have read and agreed to the published version of the manuscript.

Funding: The authors are grateful for financial support by the National Natural Science Foundation of China; this work is funded under the grant numbers 22071015, 52222302, and 51973156.

Institutional Review Board Statement: Not applicable.

Informed Consent Statement: Not applicable.

Data Availability Statement: The data presented in this study are available on request from the corresponding author.

Conflicts of Interest: The authors declare no conflict of interest.

Sample Availability: Samples of the compounds are not available from the authors.

References

1. Müller, R.J.; Kleeberg, I.; Deckwer, W.D. Biodegradation of polyesters containing aromatic constituents. *J. Biotechnol.* **2001**, *86*, 87. [CrossRef] [PubMed]
2. Olson, D.A.; Gratton, S.E.A.; Desimone, J.M.; Sheares, V.V. Amorphous Linear Aliphatic Polyesters for the Facile Preparation of Tunable Rapidly Degrading Elastomeric Devices and Delivery Vectors. *J. Am. Chem. Soc.* **2006**, *128*, 13625. [CrossRef] [PubMed]
3. Brown, A.H.; Sheares, V.V. Amorphous Unsaturated Aliphatic Polyesters Derived from Dicarboxylic Monomers Synthesized by Diels-Alder Chemistry. *Macromolecules* **2007**, *40*, 4848. [CrossRef]
4. Paul, S.; Zhu, Y.; Romain, C.; Brooks, R.; Saini, P.K.; Williams, C.K. Ring-opening copolymerization (ROCOP): Synthesis and properties of polyesters and polycarbonates. *Chem. Commun.* **2015**, *51*, 6459. [CrossRef] [PubMed]
5. Longo, J.M.; Sanford, M.J.; Coates, G.W. Ring-Opening Copolymerization of Epoxides and Cyclic Anhydrides with Discrete Metal Complexes: Structure–Property Relationships. *Chem. Rev.* **2016**, *116*, 15167. [CrossRef]
6. Stößer, T.; Chen, T.T.D.; Zhu, Y.; Williams, C.K. 'Switch' catalysis: From monomer mixtures to sequence-controlled block copolymers. *Philos. Trans. Royal. Soc. A Math. Phys. Eng. Sci.* **2017**, *376*, 20170066. [CrossRef]
7. Lustoň, J.; Maňasek, Z. Copolymerization of Epoxides with Cyclic Anhydrides Catalyzed by Tertiary Amines in the Presence of Proton-Donating Compounds. *J. Macromol. Sci. Part A Chem.* **1979**, *13*, 853. [CrossRef]
8. Lustoň, J.; Vašš, F. Anionic Copolymerization of Cyclic Ethers with Cyclic Anhydrides. *Adv. Polym. Sci.* **1984**, *54*, 91.
9. Aida, T.; Inoue, S. Catalytic Reaction on Both Sides of a Metalloporphyrin Plane. Alternating Copolymerization of Phthalic Anhydride and Epoxypropane with an Aluminum Porphyrin-Quaternary Salt System. *J. Am. Chem. Soc.* **1985**, *107*, 1358. [CrossRef]
10. Jeske, R.C.; DiCiccio, A.M.; Coates, G.W. Alternating Copolymerization of Epoxides and Cyclic Anhydrides: An Improved Route to Aliphatic Polyesters. *J. Am. Chem. Soc.* **2007**, *129*, 11330. [CrossRef]
11. Jeske, R.C.; Rowley, J.M.; Coates, G.W. Pre-rate-determining selectivity in the terpolymerization of epoxides, cyclic Anhydrides, and CO_2: A one-step route to diblock copolymers. *Angew. Chem. Int. Ed.* **2008**, *47*, 6041. [CrossRef] [PubMed]
12. Huijser, S.; HosseiniNejad, E.; Sablong, R.; de Jong, C.; Koning, C.E.; Duchateau, R. Ring-Opening Co- and Terpolymerization of an Alicyclic Oxirane with Carboxylic Acid Anhydrides and CO_2 in the Presence of Chromium Porphyrinato and Salen Catalysts. *Macromolecules* **2011**, *44*, 1132. [CrossRef]
13. Robert, C.; de Montigny, F.; Thomas, C.M. Tandem synthesis of alternating polyesters from renewable resources. *Nat. Commun.* **2011**, *2*, 586. [CrossRef]
14. DiCiccio, A.M.; Coates, G.W. Ring-Opening Copolymerization of Maleic Anhydride with Epoxides: A Chain-Growth Approach to Unsaturated Polyesters. *J. Am. Chem. Soc.* **2011**, *133*, 10724. [CrossRef] [PubMed]
15. Hosseini Nejad, E.; van Melis, C.G.W.; Vermeer, T.J.; Koning, C.E.; Duchateau, R. Alternating Ring-Opening Polymerization of Cyclohexene Oxide and Anhydrides: Effect of Catalyst, Cocatalyst, and Anhydride Structure. *Macromolecules* **2012**, *45*, 1770. [CrossRef]

16. Liu, F.P.; Li, J.; Liu, Y.; Ren, W.M.; Lu, X.B. Alternating Copolymerization of trans-Internal Epoxides and Cyclic Anhydrides Mediated by Dinuclear Chromium Catalyst Systems. *Macromolecules* **2019**, *52*, 5652. [CrossRef]
17. Winkler, M.; Romain, C.; Meier, M.A.R.; Williams, C.K. Renewable polycarbonates and polyesters from 1,4-cyclohexadiene. *Green Chem.* **2015**, *17*, 300. [CrossRef]
18. Jeon, J.Y.; Eo, S.C.; Varghese, J.K.; Lee, B.Y. Copolymerization and terpolymerization of carbon dioxide/propylene oxide/phthalic anhydride using a (salen)Co(III) complex tethering four quaternary ammonium salts. *Beilstein J. Org. Chem.* **2014**, *10*, 1787. [CrossRef]
19. Duan, Z.; Wang, X.; Gao, Q.; Zhang, L.; Liu, B.; Kim, I. Highly active bifunctional cobalt-salen complexes for the synthesis of poly(ester-block-carbonate) copolymerviaterpolymerization of carbon dioxide, propylene oxide, and norbornene anhydride isomer: Roles of anhydride conformation consideration. *J. Polym. Sci. A Polym. Chem.* **2014**, *52*, 789. [CrossRef]
20. Liu, D.; Zhang, Z.; Zhang, X.; Lü, X. Alternating Ring-Opening Copolymerization of Cyclohexene Oxide and Maleic Anhydride with Diallyl-Modified Manganese(III)–Salen Catalysts. *Aust. J. Chem.* **2016**, *69*, 47. [CrossRef]
21. Mundil, R.; Hošťálek, Z.; Šeděnková, I.; Merna, J. Alternating ring-opening copolymerization of cyclohexene oxide with phthalic anhydride catalyzed by iron(III) salen complexes. *Macromol. Res.* **2015**, *23*, 161. [CrossRef]
22. Chen, X.L.; Wang, B.; Pan, L.; Li, Y.S. Synthesis of Unsaturated (Co)polyesters from Ring-Opening Copolymerization by Aluminum Bipyridine Bisphenolate Complexes with Improved Protonic Impurities Tolerance. *Macromolecules* **2022**, *55*, 3502. [CrossRef]
23. Parveen, R.; Cundari, T.R.; Younker, J.M.; Rodriguez, G.; McCullough, L. DFT and QSAR Studies of Ethylene Polymerization by Zirconocene Catalysts. *ACS Catal.* **2019**, *9*, 9339. [CrossRef]
24. Yang, W.; Ma, Z.; Yi, J.; Ahmed, S.; Sun, W.H. Catalytic performance of bis(imino)pyridine Fe/Co complexes toward ethylene polymerization by 2D-/3D-QSPR modeling. *J. Comput. Chem.* **2019**, *40*, 1374. [CrossRef] [PubMed]
25. Xiang, S.; Wang, J. Quantum chemical descriptors based QSAR modeling of neodymium carboxylate catalysts for coordination polymerization of isoprene. *Chin. J. Chem. Eng.* **2020**, *28*, 1805. [CrossRef]
26. Zhao, Y.; Lu, H.; Luo, G.; Kang, X.; Hou, Z.; Luo, Y. Origin of stereoselectivity and multidimensional quantitative analysis of ligand effects on yttrium-catalysed polymerization of 2-vinylpyridine. *Catal. Sci. Technol.* **2019**, *9*, 6227. [CrossRef]
27. Darensbourg, D.J.; Poland, R.R.; Escobedo, C. Kinetic Studies of the Alternating Copolymerization of Cyclic Acid Anhydrides and Epoxides, and the Terpolymerization of Cyclic Acid Anhydrides, Epoxides, and CO_2 Catalyzed by (salen)CrIIICl. *Macromolecules* **2012**, *45*, 2242. [CrossRef]
28. Fieser, M.E.; Sanford, M.J.; Mitchell, L.A.; Dunbar, C.R.; Mandal, M.; Van Zee, N.J.; Urness, D.M.; Cramer, C.J.; Coates, G.W.; Tolman, W.B. Mechanistic Insights into the Alternating Copolymerization of Epoxides and Cyclic Anhydrides Using a (Salph)AlCl and Iminium Salt Catalytic System. *J. Am. Chem. Soc.* **2017**, *139*, 15222. [CrossRef] [PubMed]
29. Bickelhaupt, F.M.; Houk, K.N. Analyzing Reaction Rates with the Distortion/Interaction-Activation Strain Model. *Angew. Chem. Int. Ed.* **2017**, *56*, 10070. [CrossRef]
30. Kitaura, K.; Morokuma, K. A new energy decomposition scheme for molecular interactions within the Hartree-Fock approximation. *Int. J. Quantum Chem.* **1976**, *10*, 325. [CrossRef]
31. Liu, F.; Liang, Y.; Houk, K.N. Bioorthogonal cycloadditions: Computational analysis with the distortion/interaction model and predictions of reactivities. *Acc. Chem. Res.* **2017**, *50*, 2297. [CrossRef] [PubMed]
32. Fernández, I.; Bickelhaupt, F.M. The activation strain model and molecular orbital theory: Understanding and designing chemical reactions. *Chem. Soc. Rev.* **2014**, *43*, 4953. [CrossRef] [PubMed]
33. Zhao, Y.; Truhlar, D.G. The M06 Suite of Density Functionals for Main Group Thermochemistry, Thermochemical Kinetics, Noncovalent Interactions, Excited States, and Transition Elements: Two New Functionals and Systematic Testing of Four M06-Class Functionals and 12 Other Functionals. *Theor. Chem. Acc.* **2007**, *120*, 215.
34. Fukui, K. The Path of chemical reactions-The IRC approach. *Acc. Chem. Res.* **1981**, *14*, 363. [CrossRef]
35. Fukui, K. Formulation of the reaction coordinate. *J. Phys. Chem.* **1970**, *74*, 4161. [CrossRef]
36. Schlegel, H.B. Optimization of equilibrium geometries and transition structures. *J. Comput. Chem.* **1982**, *3*, 214. [CrossRef]
37. Hehre, W.J.; Ditchfield, R.; Pople, J.A. Self—Consistent Molecular Orbital Methods. XII. Further Extensions of Gaussian—Type Basis Sets for Use in Molecular Orbital Studies of Organic Molecules. *J. Chem. Phys.* **1972**, *56*, 2257. [CrossRef]
38. Hariharan, P.C.; Pople, J.A. The Influence of Polarization Functions on Molecular Orbital Hydrogenation Energies. *Theoret. Chim. Acta* **1973**, *28*, 213. [CrossRef]
39. Marenich, A.V.; Cramer, C.J.; Truhlar, D.G. Universal Solvation Model Based on Solute Electron Density and on a Continuum Model of the Solvent Defined by the Bulk Dielectric Constant and Atomic Surface Tensions. *J. Phys. Chem. B* **2009**, *113*, 6378. [CrossRef]
40. Zhang, D.D.; Feng, X.; Gnanou, Y.; Huang, K.W. Theoretical Mechanistic Investigation into Metal-Free Alternating Copolymerization of CO_2 and Epoxides: The Key Role of Triethylborane. *Macromolecules* **2018**, *51*, 5600. [CrossRef]
41. Perdew, J.P.; Burke, K.; Wang, Y. Generalized gradient approximation for the exchange-correlation hole of a many-electron system. *Phys. Rev. B* **1996**, *54*, 16533. [CrossRef] [PubMed]
42. Becke, A.D. Density-functional thermochemistry. III. The role of exact exchange. *J. Chem. Phys.* **1993**, *98*, 5648. [CrossRef]
43. Perdew, J.P.; Wang, Y. Accurate and simple analytic representation of the electron-gas correlation energy. *Phys. Rev. B Condens. Matter Mater. Phys.* **1992**, *45*, 13244. [CrossRef] [PubMed]

44. Grimme, S.; Antony, J.; Ehrlich, S.; Krieg, H.A. A consistent and accurate ab initio parametrization of density functional dispersion correction (DFT-D) for the 94 elements H-Pu. *J. Chem. Phys.* **2010**, *132*, 154104. [CrossRef] [PubMed]
45. Legault, C.Y. *Cylview*; Version 1.0 b; Universitéde Sherbrooke: Sherbrooke, QC, Canada, 2009; Available online: https://www.cylview.org/ (accessed on 23 October 2023).
46. Frisch, M.J.; Trucks, G.W.; Schlegel, H.B.; Scuseria, G.E.; Robb, M.A.; Cheeseman, J.R.; Scalmani, G.; Barone, V.; Mennucci, B.; Petersson, G.A.; et al. *Gaussian 16*; Revision B.01; Gaussian, Inc.: Wallingford, CT, USA, 2016.

Disclaimer/Publisher's Note: The statements, opinions and data contained in all publications are solely those of the individual author(s) and contributor(s) and not of MDPI and/or the editor(s). MDPI and/or the editor(s) disclaim responsibility for any injury to people or property resulting from any ideas, methods, instructions or products referred to in the content.

Article

Tunable Electronic Transport of New-Type 2D Iodine Materials Affected by the Doping of Metal Elements

Jie Li [1,*], Yuchen Zhou [1], Kun Liu [1], Yifan Wang [1], Hui Li [2,*] and Artem Okulov [3]

[1] School of Materials Science and Engineering, Jiangsu University of Science and Technology, Zhenjiang 212100, China; zhouyuchen23@163.com (Y.Z.); liu_kun@163.com (K.L.); 18001759763@163.com (Y.W.)
[2] Key Laboratory for Liquid-Solid Structural Evolution and Processing of Materials, Ministry of Education, Shandong University, Jinan 250061, China
[3] M.N. Mikheev Institute of Metal Physics, Ural Branch of Russian Academy of Sciences, Ekaterinburg 620077, Russia; okulovartem@imp.uran.ru
* Correspondence: jie_li@just.edu.cn (J.L.); lihuilmy@sdu.edu.cn (H.L.)

Citation: Li, J.; Zhou, Y.; Liu, K.; Wang, Y.; Li, H.; Okulov, A. Tunable Electronic Transport of New-Type 2D Iodine Materials Affected by the Doping of Metal Elements. *Molecules* 2023, 28, 7159. https://doi.org/10.3390/molecules28207159

Academic Editor: Minghao Yu

Received: 26 September 2023
Revised: 13 October 2023
Accepted: 16 October 2023
Published: 19 October 2023

Copyright: © 2023 by the authors. Licensee MDPI, Basel, Switzerland. This article is an open access article distributed under the terms and conditions of the Creative Commons Attribution (CC BY) license (https://creativecommons.org/licenses/by/4.0/).

Abstract: 2D iodine structures under high pressures are more attractive and valuable due to their special structures and excellent properties. Here, electronic transport properties of such 2D iodine structures are theoretically studied by considering the influence of the metal-element doping. In equilibrium, metal elements in Group 1 can enhance the conductance dramatically and show a better enhancement effect. Around the Fermi level, the transmission probability exceeds 1 and can be improved by the metal-element doping for all devices. In particular, the device density of states explains well the distinctions between transmission coefficients originating from different doping methods. Contrary to the "big" site doping, the "small" site doping changes transmission eigenstates greatly, with pronounced electronic states around doped atoms. In non-equilibrium, the conductance of all devices is almost weaker than the equilibrium conductance, decreasing at low voltages and fluctuating at high voltages with various amplitudes. Under biases, K-big doping shows the optimal enhancement effect, and Mg-small doping exhibits the most effective attenuation effect on conductance. Contrastingly, the currents of all devices increase with bias linearly. The metal-element doping can boost current at low biases and weaken current at high voltages. These findings contribute much to understanding the effects of defects on electronic properties and provide solid support for the application of new-type 2D iodine materials in controllable electronics and sensors.

Keywords: new-type 2D iodine materials; electron transport; doping of metal elements; applying the bias; first principles

1. Introduction

Currently, one-atom-thick 2D nanomaterials are catching researchers' attention because of their fascinating structures, outstanding properties and promising applications [1,2]. The elemental 2D materials are surely striking among the 2D family due to their simples chemical composition and amazing characteristics [3,4]. For example, the most typical one is graphene (group-IVA), with high electron mobility, Young's modulus, and thermal conductivity [5,6]. Unlike the honeycomb structure based on six-membered carbon rings of graphene, silicene, germanene, and stanene (group-IVA) possess a buckled hexagonal honeycomb lattice with variable electronic structures [7,8]. Borophene (group-IIIA) shows metallic features and highly anisotropic electronic properties [9]. Unlike the above, black phosphorene (group-VA) is a semiconductor with a tunable band gap and shows anisotropic properties on the surface [10]. In recent years, arsenene [11], anitmonene [12], and bismuthene [13] (group-VA), as well as selinene [14] and tellurene [15] (group-VIA), have been reported theoretically and experimentally. Elemental 2D nanomaterials have shown great potential in electronics [16], sensing [17], energy storage [18], photothermal therapy [19], and other applications [20].

Noticeably, elemental 2D nanomaterials in the group-VIIA have begun to receive attention. Two-dimensional binary iodides, including BiI_3 [21], PbI_2 [22], and RhI_3 [23], have been proved to behave excellent performance in photonics. Like these binary iodides, iodine is a layered semiconductor. Atom-thin 2D iodine materials have now been successfully prepared via liquid phase stripping [24]. Such 2D iodine overcomes the deficiencies of slow battery kinetics, poor rate capacity, low conductivity, etc., from the crystal, showing great potential for use in high-performance rechargeable batteries [25]. It should be noted that iodine is a molecular system in which iodine molecules are connected to the layer plane via halogen bonds rather than covalent bonds, making it a fundamental point of interest [26]. Therefore, the special structure of 2D iodine material combined with the nanometer size effect shows great application value in terms of high-pressure physics, the energy field, the optoelectronics field, etc. [27–31]. Regardless, 2D iodine-based research on emerging materials is in its infancy and still requires a great deal of research.

High pressure can effectively regulate the physical and chemical properties of materials [32–34]. The iodine molecule, as one of the seven homonuclear diatomic molecules, shows stability under ambient conditions and is high-profile as a result of successive transitions (a superconductive state [35] or metallization [36]) induced by the pressure. At ambient pressure, solid iodine is an orthorhombic molecular crystal (space group, Cmca; phase I). It has been demonstrated that there are no signs of structural phase transition when the pressure is below 23.2 GPa [37]. The conductance strengthens by eight orders of magnitude with the pressure to 12 GPa [36]. When the pressure is high enough, iodine molecules are close to each other and cause the conduction and valence bands to overlap, and as a result, iodine is converted into molecular metals. Pressure-induced photon-generated carriers increase rapidly, leading to the increase in the photoelectric properties of iodine [38]. The special structure and excellent properties caused by high pressure make the 2D iodine structure more attractive and valuable for use under high pressures.

However, there has not been nearly enough research performed on 2D iodine structures under high pressure, as well as their electronic transport properties and applications. Here, a 2D iodine monolayer under a certain high pressure is the research subject. We theoretically study the influence of metal-element doping on the equilibrium and non-equilibrium electron transport properties, including the conductance, electron transmission, and current–voltage characteristics. Electronic structures and charge transferring states are also well explained.

2. Results and Discussion

Here, we studied a new-type 2D iodine structure, which is under a high pressure of 15 GPa (see Figure 1). Just as Li et al. reported, the pressure would make iodine atoms more clustered together, so that 2D iodine shows much shorter atomic distances of less than 3 Å and no band gap under the pressure of 15 GPa, different to the general condition [39]. However, iodine is a 2D layered molecular crystal with covalent intramolecular bonds, so atoms are regularly bonded in pairs. Here, four iodine atoms form a rectangle, where each iodine atom is bonded to one iodine atom in another rectangle. As a result, iodine atoms from four rectangles form an eight-membered ring. Significant electronic interactions occur with the next-nearest neighbors of each atom. Taking into account the difficulty of the bonding, the doped metallic elements are selected from groups 1, 2, and 3, which are Na, K, Mg, Ca, Al, and Ga, respectively. There are two types of doping sites for 2D iodine depending to its structural characteristics. As Figure 1 shows, one is in the big 8-I ring, and the other is in the small 4-I ring.

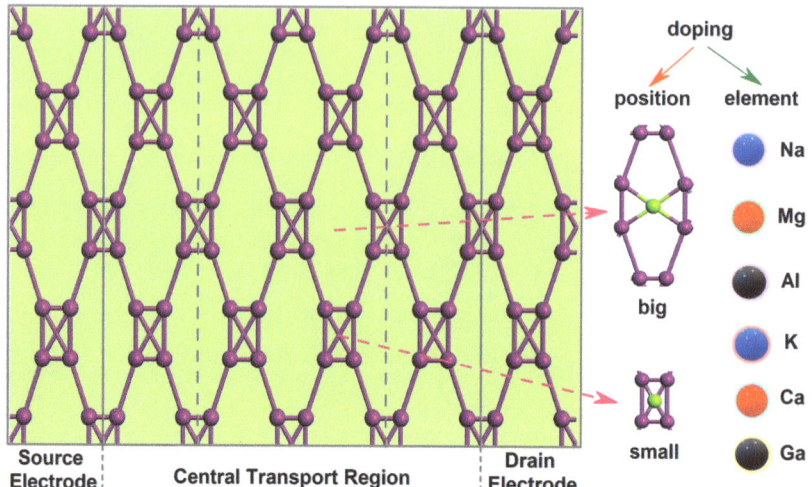

Figure 1. Structural illustration of the devices based on no-doped and doped iodine monolayers.

To study the electrical sensitivity to metal atoms, electronic devices are constructed in Virtual NanoLab, as shown in Figure 1. The device actually has a two-probe configuration. For the left and right electrode regions, the system should be periodic along the transport direction (the C direction). Besides the transferring material, the central region includes the left and right electrode extensions in order to screen out the perturbations from the scatterer and solve a bulk problem for the fully periodic electrode cell. Due to their being no band gap [39], the material itself is used as the electrode. Simulated electronic devices are within a supercell with over 15 Å of vacuum space to allow electrostatic interactions to decay for the system [40]. The device with the original new-type 2D iodine is denoted as Device no-doped. The doped 2D iodine-based devices are written as Device Na-big, Device Na-small, Device Mg-big, Device Mg-small, Device Al-big, Device Al-small, Device K-big, Device K-small, Device Ca-big, Device Ca-small, Device Ga-big and Device Ga-small, respectively, according to the type and sites (big 8-I ring and small 4-I ring) of the doped element.

2.1. Equilibrium Electron Transport

Firstly, we study the sensitivity of equilibrium conductance to metal elements. Figure 2a shows the comparison between equilibrium conductance among all new-type 2D iodine-based devices before and after doping. Obviously, the conductance of doped devices is much stronger than the original device. The doping position has a different influence on conductance for different doped metal elements. The devices doped in "big" site (called "big" doped devices) exhibit stronger conductance than the devices doped in "small" site (called "small" doped devices) with the elements Al and K. Moreover, the doping with these two elements in medium atomic numbers contributes to the most pronounced enhancement in conductance. The "small" doped devices show higher conductance values than the "big" doped devices for the devices with the doped metal elements Na and Ca in larger atomic numbers, and the influence of these two elements is also strong on the conductance value, though not as obviously as Al and K. Markedly different from the above, the doping site exhibits little effect on conductance for Mg-doped devices and Ga-doped devices, and the two types of elemental doping also show the least noticeable increases in conductance. Overall, both of the elements in Group 1 can enhance the conductance dramatically and show better doping effects than the four other elements in groups 2 and 3, which present significant influencing differences within the same main group.

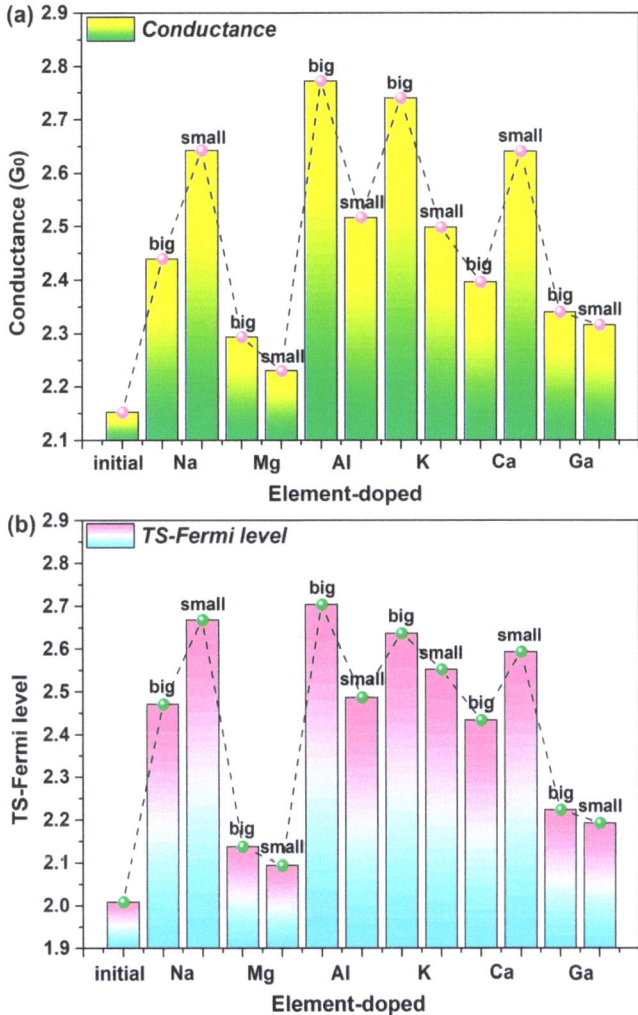

Figure 2. (**a**,**b**) Comparisons between equilibrium quantum conductance and electron transmission at the Fermi level among the undoped and doped iodine-based devices.

To investigate the differences in conductance, Figure 2b shows the transmission coefficient of all the devices at the Fermi level (E_F), which expresses the transport capacity of electronic devices [41]. In contrast to the original device, doped devices have a significantly strong electron transmission. On the whole, most doped metal elements can dramatically boost the electron transmission, and there is a small difference between the transmission coefficients of doped devices. The doping site also exhibits varying influences on the electron transmission for various doped metal elements, which is consistent with the conductance, but this difference in impact is not as obvious as that on the conductance.

In detail, for the devices with the doped elements Al and K, the "big" site shows a greater advantage than the "small" site regarding the improvement in the electron transmission. When doped with metal elements Na and Ca, "small" doped devices tend to have higher electron transmission values than "large" doped ones. Unlike the above, the Mg-doped device exhibits the close value in the "big" and "small" sites, and the same is true for the Ca-doped device. Furthermore, these two elements demonstrate the minimum increase in the electron transmission coefficient. In general, both elements of Group 1 have a significant increase in electron transmission and show a superior doping effect compared to the other four elements of groups 2 and 3, which exhibit extreme electron transmission values. Thus, the electron transmission at E_F can well explain the conductance in the equilibrium for these new-type 2D iodine-based devices.

Further, the equilibrium transmission spectrum in [−1 eV, 1 eV] is given and plotted in solid lines in Figure 3. Remarkably, the K-doped devices present much stronger electron transmission than the Device no-doped in the whole energy range. The electron transmission coefficients of Device Al-big and Device Ca-small are also obviously larger than Device no-doped, and Device Ga-small shows slightly stronger electron transmission than Device no-doped near E_F, which fully demonstrates the advantage of the doping position. However, compared to Device no-doped, other types of doping devices present weaker electron transmissions in a quite small part of [−1 eV, 0 eV] while showing notably greater coefficients in the rest of the energy interval. It signifies that metal-element doping can improve the entire transmission spectrum around E_F. For the role of the doping position, the "big" site is superior to the "small" site for Al-doped and K-doped devices, while the "small" site has an advantage over the "big" site for Na-doped and Ca-doped devices, which is consistent with the effect on the conductance and electron transmission at E_F. For the special Mg-doped and Ga-doped devices, the influence of the doping position is uncertain, reflected in the variational contrast relationships of transmission values at different energy intervals around E_F. Moreover, it can be seen that the transmission probability around E_F exceeds 1 for new-type 2D iodine-based devices because more than one obviously strong electron transmission channel contributes to electron transport near E_F for such new devices [40].

The electronic structures of the devices are calculated to analyze the variations in the transmission spectrum resulting from the doping [42]. Figure 3 displays the device density of states (DDOS) of all devices in dashed lines. It can be observed that the shapes of DDOS–energy curves are pretty similar to those of the transmission (TS)–energy curves for all devices, except the Ca-doped and Ga-doped devices with tiny differences. It indicates that the DDOS is the determining inner factor of electron transport for the new-type 2D iodine-based devices [43]. The relative size relationships of DDOS values among Device no-doped, Device Na-big, and Device Na-small are consistent with those of the transmission coefficients in most of the energy range, ignoring very small negative energy intervals. The same is true with the Mg-doped, Al-doped, and K-doped devices. However, some anomalies appear in devices based on the doped elements of Ca and Ga with higher atomic numbers. Smaller DDOS brings about a higher transmission value, which mainly occurs in the comparison between two different doping positions. On the whole, the DDOS can well explain the differences between the electron transmission probability resulting from the doping and different doping sites.

In order to more intuitively describe the changes in the equilibrium electron transport characteristics, Table 1 presents the variations in and variation rates of all doped devices in terms of the equilibrium conductance, electron transmission possibility at E_F, and the device density of states at E_F. Clearly, in the equilibrium condition, the most noticeable change in conductance occurs for Device Al-big, followed by Device K-big and Device Na-small, compared to the undoped device. The devices doped with Mg show the smallest changes in conductance. The changes in electron transmission possibility are basically consistent with those of conductance among the various doped devices. So, Device Al-big, Device K-big, and Device Na-small exhibit a significant advantage in terms of enhancing electron transport capacity, which also possess stronger electron transport. In contrast, the doping with Mg goes against such enhancement, which also shows weaker electron transport.

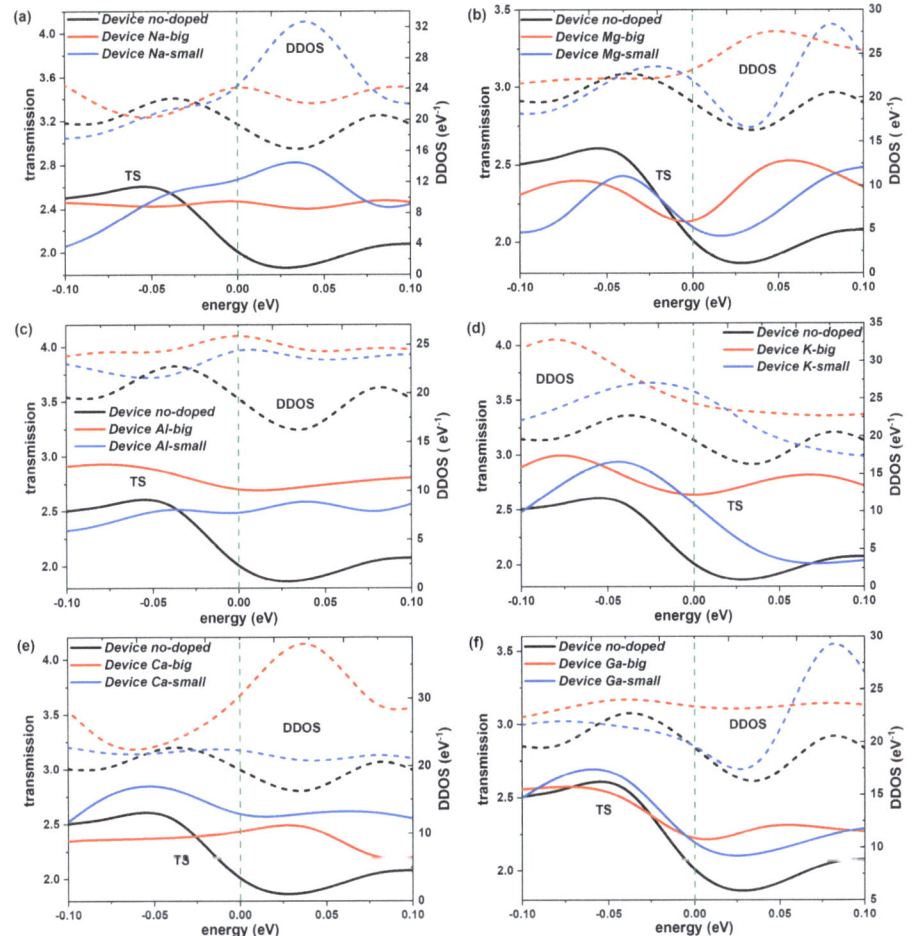

Figure 3. (a–f) Equilibrium electron transmission and device density of states (DDOS) of the comparison between Device no-doped and the iodine devices doped with Na, Mg, Al, K, Ca, and Ga, respectively.

In contrast, Device Ca-big displays the biggest change in DDOS, which does not show strong electron transport capacity. What's more, a remarkable improvement in the DDOS can be found in the devices doped with K and Al. However, the three devices above, with excellent electron transport and corresponding enhancement effects, also exhibit such improvements in DDOS. Moreover, the doping with Mg and Ga brings on slight changes in DDOS, especially for Device Ga-small.

In addition to the analysis of intrinsic electronic structures, the spatial distribution of charge transfer is investigated to study the electron transport of the new-type 2D iodine-based devices. The dominant eigenstate of the transmission matrix at E_F is calculated and displayed in Figure 4. The transmission eigenstate is a complex wave function, corresponding to a scattering state that comes from the left electrode and travels towards the right [44]. There are two continuous distributions of electronic states from the left electrode to the right electrode for Device no-doped, which qualitatively demonstrates that the transmission coefficient exceeds 1.

Table 1. Changes in and variation rates of equilibrium electron transport properties for doped new 2D iodine-based devices compared to the undoped 2D iodine-based device. Here, Con, TS, and DDOS are short for equilibrium conductance, electron transmission possibility at the Fermi level (E_F), and device density of states at E_F, respectively.

Dopants	Doping Site	Changes in Con (G0)	Variation Rate (%)	Changes in TS	Variation Rate (%)	Changes in DDOS (eV^{-1})	Variation Rate (%)
Na	big	0.2874	13.36%	0.4623	23.01%	4.6973	24.15%
	small	0.4903	22.78%	0.6594	32.82%	4.8661	25.01%
K	big	0.5877	27.31%	0.6279	31.25%	4.8935	25.15%
	small	0.3463	16.09%	0.5428	27.02%	6.4726	33.27%
Mg	big	0.1417	6.58%	0.1293	6.44%	3.7493	19.27%
	small	0.0775	3.60%	0.0849	4.22%	2.5120	12.91%
Ca	big	0.2442	11.35%	0.4250	21.15%	10.9708	56.39%
	small	0.4883	22.69%	0.5842	29.08%	2.8705	14.76%
Al	big	0.6207	28.84%	0.6949	34.59%	6.4266	33.04%
	small	0.3648	16.95%	0.4781	23.80%	4.9305	25.35%
Ga	big	0.1880	8.73%	0.2142	10.66%	3.9206	20.15%
	small	0.1637	7.61%	0.1835	9.14%	0.1126	0.58%

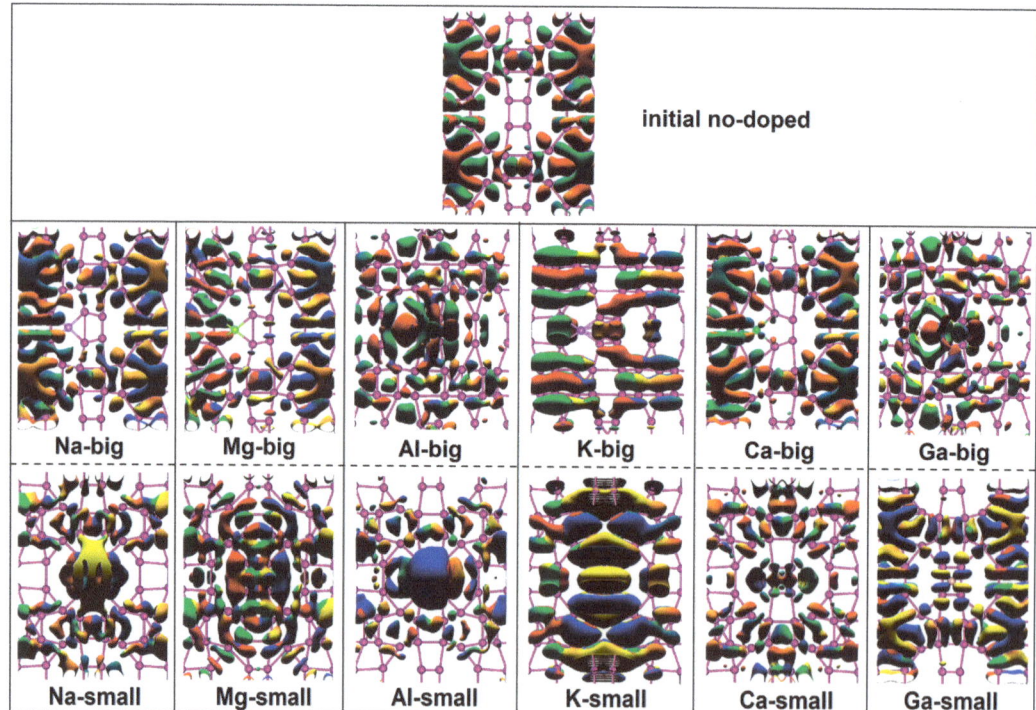

Figure 4. Transmission eigenstates of the undoped iodine-based device and all doped iodine devices with the same isovalue.

Intuitively, the transmission eigenstates are slightly changed by "big" site doping, apart from the elements in Group 3, which show noticeable electronic states around the doping positions. For the other four elements, the transmission eigenstates are quite similar to the undoped device for the three doped elements of Na, Mg, and Ca while influenced by the doped K, with more than two transferring channels found throughout the entire device, albeit without conspicuous electronic states around the doping site, similar to the impacts of the elements in Group 3, which contribute to the stronger electron transport capacity of Device K-big. As a whole, for the "big" doping, the distributions of transmission states are seemingly affected by the elements in the same main group.

Unlike the above, the transmission eigenstate presents pronounced changes influenced by the "small" doping of metal elements. Apparent electronic states can be observed in the surrounding areas of the doped atom for all metal elements and are particularly strong for the four elements with larger atomic numbers. Around the doped atom, the morphology and distribution of electronic states vary with the doped element. Compared to the "big" doped device, the "small" doped one possesses more distributions of electronic states throughout the entire device, despite having weaker electronic states for the doped Na, internally causing stronger electron transmission of Device Na-small. The same applies to the Ca-doped devices. The Mg-doped devices show a similar number of transferring channels and strengths of transmission eigenstates on the two doping sites, and so does the Ga-doped devices, explaining the almost large transmission coefficients at the two doping positions. For Al-doped and K-doped devices, the "big" doping site promotes more transferring channels, stronger electronic states, and further stronger electron transmission compared to the "small" doping site.

2.2. Non-Equilibrium Electron Transport

Non-equilibrium electron transport properties are also studied for the new-type 2D iodine-based devices. Figure 5 exhibits the conductance of all studied devices in the bias region of [0 V, 3 V]. For Device no-doped, a slight drop of conductance can be observed in [0 V, 0.5 V]. Then, the conductance grows gradually to the initial value and later keeps small fluctuations. So, the conductance value is less affected by the voltage for the no-doped device. It is clear that all devices have things in common: the conductance decreases under lower biases (less than about 0.6 V) and fluctuates as the bias increases. However, the amplitude of the variation differs. A small change in conductance (just 0.1 G_0) can only be seen in the "big" doped devices with the elements from Group 3 in the initial descent stage. For the wave phase that follows, Device Na-big and Device K-small present the minimum changing range (about 0.3 G_0), and both Ca-doping devices show relatively slight fluctuations in conductance with the bias, while the conductance alters the most (about 0.7 G_0) with the change in the voltage for Device Al-big and Device K-big.

A significant difference of the conductance compared to the doping position can be found in the Mg-doped, Al-doped, and K-doped devices at some voltages. For the doping of the other types of metal elements, there is little difference in conductance between the "big" doping and "small" doping that is influenced by the bias. In addition, the conductance at all voltages is lower than the equilibrium conductance for the doped device for the elements in Group 1 and the "small" doping devices with the elements in Group 2. In contrast, the conductance at some voltages is a little higher than the equilibrium conductance for the doped device with the elements in Group 3 and the "big" doped devices with the elements in Group 2. Therefore, under the influence of the voltage, the conductance is almost reduced for the undoped and doped devices.

To measure the doping effect quantitatively, Figure 5 also illustrates the variation rate of conductance relative to the undoped condition for the doped devices. Here, the variation rate is defined as the ratio of the conductance difference between one doped device and undoped device relative to the conductance of the undoped device. Obviously, all doped devices present positive variation rates, indicating the promotion of the doping to the conductance at initial small voltages, and show widely negative rates, representing the weakening effect of the doping on the conductance at medium biases. When applied by the bias, the best doping for enhancing conductance is Device Al-big and Device K-big with strength rates of over 30% (up to 37%), mainly at low voltages. The most effective attenuation for the conductance appears in Device Mg-small, with a variation rate of −23%. The effect of the other doping methods is relatively weak.

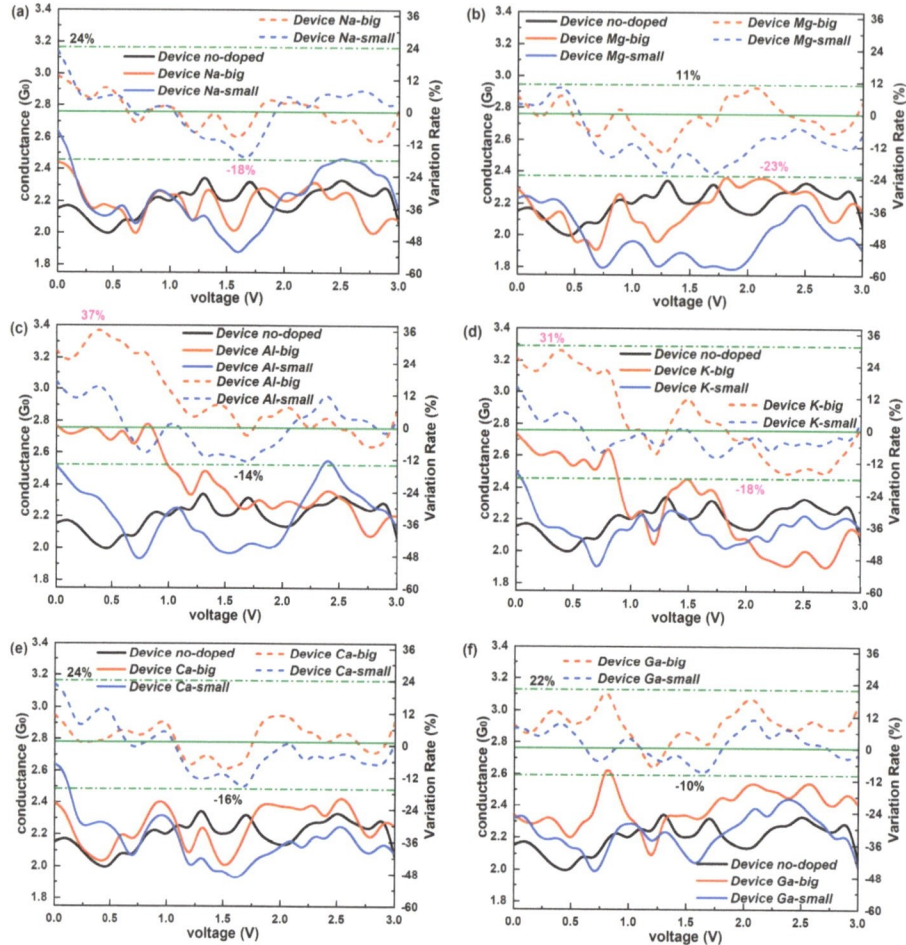

Figure 5. (a–f) Conductance–voltage curves of the undoped iodine-based device and the iodine devices doped with Na, Mg, Al, K, Ca, and Ga, respectively. The dashed lines show the variation rate of the conductance at various voltages for the corresponding doped devices relative to the undoped iodine-based device.

The current–voltage characteristics are one element of practical and important non-equilibrium performance [45]. Figure 6 presents the current–voltage curves of all devices, as well as the corresponding variation rates. Similar to the variation rate of the conductance, the variation rate of the current is the ratio of the difference in current between one doped device and one undoped device divided by the current of Device no-doped. Unlike the fluctuating trends of the conductance, the current increases linearly with the bias for the doped and undoped new-type 2D iodine-based device. It is not easy to find the differences in the current between the two doping positions or the doping effect from the comparisons of values alone, because the changing range of the current is too wide at the voltage range of [0 V, 3 V]. So, the following analysis is mainly based on the variation rate of the current.

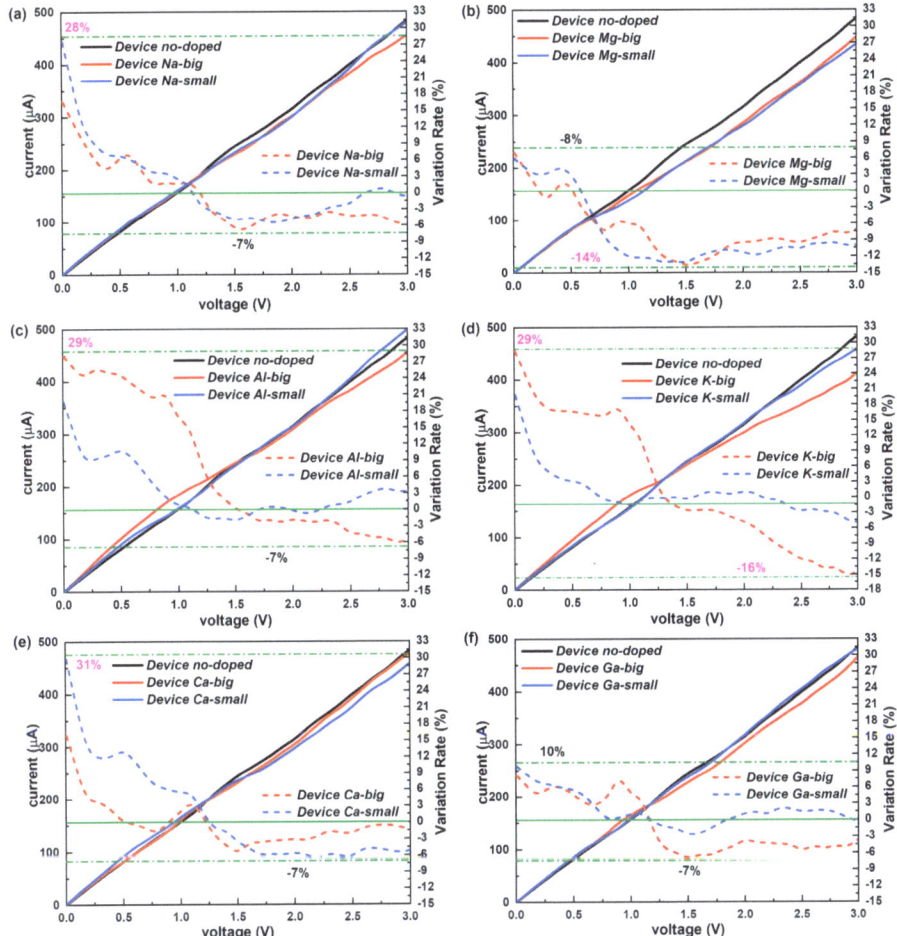

Figure 6. (a–f) Current–voltage curves of the undoped iodine-based device and the iodine devices doped with Na, Mg, Al, K, Ca, and Ga, respectively. The dashed lines show the variation rate of the current at various voltages for the corresponding doped devices relative to the undoped iodine-based device.

It is not hard to find that the current of the new-type device can achieve growth at low voltages and reduction at high voltages via the doping of metal elements. The specific doping effects of different sites vary with the type of the element. For the doped devices with Na and Mg in smaller atomic numbers, the effect (strengthened or weakened) on the current of the "big" doped device synchronizes roughly with the "small" doped device, and there is a small difference between the two in terms of the value of the current. Though the Ga-doped devices also present a slight gap between the two, Device Ga-small possesses a negative variation rate, while Device Ga-big shows a positive one, which means that the effect on the current of the "big" doped device does not synchronize with the "small" doped device. In contrast, huge discrepancies in the variation rates can be observed between the "big" doped device and the "small" doped device for the other three kinds of doped devices, especially for the Al-doped and K-doped devices at lower biases. Like the doped device with Ga in Group 3, the out-of-sync phenomenon of the doping effect also exists in the Al-doped devices due to different doping positions. Such an obvious out-of-sync phenomenon only appears in the doped devices with metal elements in Group 3.

However, in contrast, with the equilibrium condition, the reinforcement effect on the current fades at voltages for the doped devices, but the reduction effect can be achieved by applying higher biases. The doping of most elements can reach excellent results, with a variation rate of about 30% in terms of the raising impact on the current except Mg and Ga, and the choice of the doping position depends on the element type. The variation rate of −7% is easily accessible to the doped devices by applying a voltage, and Device K-big, Device Mg-big, and Device Mg-small all present the best weakening effect on the current, with a rate of about −15%. So, clearly, the Mg-doped doping can weaken the current by applying a certain bias whatever the doping site. The optimum intensifying and dampening impact can both be possible with the K-doped doping by varying the voltage.

3. Computational Methods

All structures were optimized before calculating electron transport properties. All calculations based on the first-principles theory, combining density functional theory (DFT) [46] with non-equilibrium Green's function (NEGF) [47], was implemented in the Atomistix Toolkit (ATK) package [48]. The exchange–correlation function was a generalized gradient approximation (GGA) of Perdew–Burke–Ernzerhof (PBE), while the double-zeta plus polarization (DZP) basis was chosen for all atoms. K-point sampling was set to $1 \times 5 \times 50$, and the transport direction was along the C axis. The density mesh cut-off for the electrostatics potential was 75 Ha. The electron temperature was set as 300 K. Structural optimizations use the quasi-Newton method until all residual forces on each atom are smaller than 0.05 eV/Å. The convergence criterion for the total energy was 10^{-5} via the mixture of the Hamiltonian.

The conductance G could be expressed in terms of the transmission function within the Landauer–Büttiker formalism [49,50]:

$$G = G_0 \int_{-\infty}^{+\infty} dE T(E,V) \left(-\frac{\partial f(E)}{\partial E} \right) \tag{1}$$

The current through a molecular junction was calculated using the Landauer–Büttiker equation [51]

$$I = \frac{2e}{h} \int_{-\infty}^{+\infty} dE T(E,V)(f_1(E) - f_2(E)) \tag{2}$$

where $G_0 = 2e^2/h$ is the quantum unit of conductance, h is Planck's constant, e is the electron charge, $f_{1,2}(E)$ are the Fermi functions of source and drain electrodes, and $T(E,V)$ is the quantum mechanical transmission probability of electrons. Thus, it can be given as depicted in [51]

$$T(E,V) = tr\left[\Gamma_L(E,V) G^R(E,V) \Gamma_R(E,V) G^A(E,V) \right] \tag{3}$$

where G^R and G^A are the stunted and advanced Green functions of the conductor part, respectively, and Γ_L and Γ_R are the coupling functions to the left and right electrodes, respectively.

4. Conclusions

We have systematically studied the doping influence of metal elements on the equilibrium and non-equilibrium electron transport properties of the devices based on the new-type 2D iodine structure. Our results show that the equilibrium conductance of doped devices is much stronger than those of the undoped device, and the doping position has varying influence on the conductance depending on the element type. The elements in Group 1 can both enhance the conductance dramatically and show better doping effects than the other elements in groups 2 and 3, well explained based on the electron transmission at E_F. The transmission probability around E_F exceeds 1 for the new devices due to their being more than one obviously strong electron transferring channel. The metal-element doping can improve the entire transmission spectrum around E_F. Essentially, the device density of states (DDOS) is the determining inner factor in electron transport for these devices and can well illustrate the differences of transmission coefficients from doping and the doping site. For the spatial distribution of charge transferring, the transmission eigenstates are slightly changed by the "big" site doping for most elements, and they are a semblable for the elements in the same group. In contrast, the transmission eigenstates change dramatically when influenced by the "small" doping of metal elements, presenting pronounced electronic states in the surrounding of the doped atom. In non-equilibrium conditions, the conductance decreases under lower biases and fluctuates as the bias increases, and it is reduced for almost all devices under voltages. However, the amplitude of the variations differs. The best doping for enhancing conductance can reach 37%, mainly at low voltages, while the most effective attenuation appears in the Mg-small doping, with a variation rate of −23%. In contrast, the current increases linearly with the bias for all devices. The current of the new-type device can achieve growth (up to 31%) at low voltages and reduction (up to 16%) at high voltages via the doping of metal elements. These results help us to understand the effects of defects on the electronic properties of new 2D iodine materials and provide a solid support for the applications of new 2D iodine materials in controllable electronic devices and sensors.

Author Contributions: Conceptualization, J.L.; software, H.L.; formal analysis, J.L. and Y.Z.; investigation, A.O.; resources, K.L.; writing—original draft preparation, J.L. and Y.W.; writing—review and editing, K.L. and J.L.; funding acquisition, K.L. and J.L. All authors have read and agreed to the published version of the manuscript.

Funding: This research was funded by the National Natural Science Foundation of China (Grant No. 52105351), the Natural Science Foundation of Jiangsu Province (Grant No. BK20210873 and BK20210890), the High-End Foreign Experts Recruitment Plan of China (Grant No. G2023014009L and G2023014004), the China Postdoctoral Science Foundation (Grant No. 2022M722928), the Jiangsu Provincial Double-Innovation Doctor Program (Grant No. JSSCBS20210985 and JSSCBS20210991), the Natural Science Research Projects in the Universities of Jiangsu Province (Grant No. 21KJB460015), and the State Assignment of the Ministry of Science and Higher Education of the Russian Federation (theme "Additivity" No. 121102900049-1).

Institutional Review Board Statement: Not applicable.

Informed Consent Statement: Not applicable.

Data Availability Statement. Data will be made available on request.

Conflicts of Interest: The authors declare no conflict of interest.

Sample Availability: Not applicable.

References

1. Glavin, N.R.; Rao, R.; Varshney, V.; Bianco, E.; Apte, A.; Roy, A.; Ringe, E.; Ajayan, P.M. 2D Materials: Emerging Applications of Elemental 2D Materials (Adv. Mater. 7/2020). *Adv. Mater.* **2020**, *32*, 2070052. [CrossRef]
2. Liu, Y.; Huang, Y.; Duan, X. Van der Waals integration before and beyond two-dimensional materials. *Nature* **2019**, *567*, 323–333. [CrossRef] [PubMed]
3. Li, M.; Kang, J.S.; Nguyen, H.D.; Wu, H.; Aoki, T.; Hu, Y. Anisotropic Thermal Boundary Resistance across 2D Black Phosphorus: Experiment and Atomistic Modeling of Interfacial Energy Transport. *Adv. Mater.* **2019**, *31*, 1901021. [CrossRef]
4. Kim, S.; Cui, J.; Dravid, V.P.; He, K. Orientation-Dependent Intercalation Channels for Lithium and Sodium in Black Phosphorus. *Adv. Mater.* **2019**, *31*, 1904623. [CrossRef]
5. Novoselov, K.S.; Fal'ko, V.I.; Colombo, L.; Gellert, P.R.; Schwab, M.G.; Kim, K. A roadmap for graphene. *Nature* **2012**, *490*, 192–200. [CrossRef]
6. Novoselov, K.S.; Andreeva, D.V.; Ren, W.; Shan, G. Graphene and other two-dimensional materials. *Front. Phys.* **2019**, *14*, 13301. [CrossRef]
7. Molle, A.; Goldberger, J.; Houssa, M.; Xu, Y.; Zhang, S.-C.; Akinwande, D. Buckled two-dimensional Xene sheets. *Nat. Mater.* **2017**, *16*, 163–169. [CrossRef] [PubMed]
8. Tan, C.; Cao, X.; Wu, X.-J.; He, Q.; Yang, J.; Zhang, X.; Chen, J.; Zhao, W.; Han, S.; Nam, G.-H.; et al. Recent Advances in Ultrathin Two-Dimensional Nanomaterials. *Chem. Rev.* **2017**, *117*, 6225–6331. [CrossRef]
9. Mannix, A.J.; Zhou, X.-F.; Kiraly, B.; Wood, J.D.; Alducin, D.; Myers, B.D.; Liu, X.; Fisher, B.L.; Santiago, U.; Guest, J.R.; et al. Synthesis of borophenes: Anisotropic, two-dimensional boron polymorphs. *Science* **2015**, *350*, 1513–1516. [CrossRef]

10. Carvalho, A.; Wang, M.; Zhu, X.; Rodin, A.S.; Su, H.; Castro Neto, A.H. Phosphorene: From theory to applications. *Nat. Rev. Mater.* **2016**, *1*, 16061. [CrossRef]
11. Hu, Y.; Liang, J.; Xia, Y.; Zhao, C.; Jiang, M.; Ma, J.; Tie, Z.; Jin, Z. 2D Arsenene and Arsenic Materials: Fundamental Properties, Preparation, and Applications. *Small* **2022**, *18*, 2104556. [CrossRef]
12. Niu, T.; Meng, Q.; Zhou, D.; Si, N.; Zhai, S.; Hao, X.; Zhou, M.; Fuchs, H. Large-Scale Synthesis of Strain-Tunable Semiconducting Antimonene on Copper Oxide. *Adv. Mater.* **2020**, *32*, 1906873. [CrossRef]
13. Reis, F.; Li, G.; Dudy, L.; Bauernfeind, M.; Glass, S.; Hanke, W.; Thomale, R.; Schäfer, J.; Claessen, R. Bismuthene on a SiC substrate: A candidate for a high-temperature quantum spin Hall material. *Science* **2017**, *357*, 287–290. [CrossRef] [PubMed]
14. Qin, J.; Qiu, G.; Jian, J.; Zhou, H.; Yang, L.; Charnas, A.; Zemlyanov, D.Y.; Xu, C.-Y.; Xu, X.; Wu, W.; et al. Controlled Growth of a Large-Size 2D Selenium Nanosheet and Its Electronic and Optoelectronic Applications. *ACS Nano* **2017**, *11*, 10222–10229. [CrossRef] [PubMed]
15. Wang, Y.; Qiu, G.; Wang, R.; Huang, S.; Wang, Q.; Liu, Y.; Du, Y.; Goddard, W.A.; Kim, M.J.; Xu, X.; et al. Field-effect transistors made from solution-grown two-dimensional tellurene. *Nat. Electron.* **2018**, *1*, 228–236. [CrossRef]
16. Xia, F.; Wang, H.; Jia, Y. Rediscovering black phosphorus as an anisotropic layered material for optoelectronics and electronics. *Nat. Commun.* **2014**, *5*, 4458. [CrossRef]
17. Cui, S.; Pu, H.; Wells, S.A.; Wen, Z.; Mao, S.; Chang, J.; Hersam, M.C.; Chen, J. Ultrahigh sensitivity and layer-dependent sensing performance of phosphorene-based gas sensors. *Nat. Commun.* **2015**, *6*, 8632. [CrossRef]
18. Zhou, J.; Chen, J.; Chen, M.; Wang, J.; Liu, X.; Wei, B.; Wang, Z.; Li, J.; Gu, L.; Zhang, Q.; et al. Few-Layer Bismuthene with Anisotropic Expansion for High-Areal-Capacity Sodium-Ion Batteries. *Adv. Mater.* **2019**, *31*, 1807874. [CrossRef]
19. Chimene, D.; Alge, D.L.; Gaharwar, A.K. Two-Dimensional Nanomaterials for Biomedical Applications: Emerging Trends and Future Prospects. *Adv. Mater.* **2015**, *27*, 7261–7284. [CrossRef]
20. Gusmão, R.; Sofer, Z.; Bouša, D.; Pumera, M. Pnictogen (As, Sb, Bi) Nanosheets for Electrochemical Applications Are Produced by Shear Exfoliation Using Kitchen Blenders. *Angew. Chem. Int. Edit.* **2017**, *56*, 14417–14422. [CrossRef]
21. Wei, Q.; Chen, J.; Ding, P.; Shen, B.; Yin, J.; Xu, F.; Xia, Y.; Liu, Z. Synthesis of Easily Transferred 2D Layered BiI_3 Nanoplates for Flexible Visible-Light Photodetectors. *ACS Appl. Mater. Interfaces* **2018**, *10*, 21527–21533. [CrossRef] [PubMed]
22. Zhong, M.; Huang, L.; Deng, H.-X.; Wang, X.; Li, B.; Wei, Z.; Li, J. Flexible photodetectors based on phase dependent PbI_2 single crystals. *J. Mater. Chem. C* **2016**, *4*, 6492–6499. [CrossRef]
23. Wang, F.; Zhang, Z.; Zhang, Y.; Nie, A.; Zhao, W.; Wang, D.; Huang, F.; Zhai, T. Honeycomb RhI_3 Flakes with High Environmental Stability for Optoelectronics. *Adv. Mater.* **2020**, *32*, 2001979. [CrossRef]
24. Hulkko, E.; Kiljunen, T.; Kiviniemi, T.; Pettersson, M. From Monomer to Bulk: Appearance of the Structural Motif of Solid Iodine in Small Clusters. *J. Am. Chem. Soc.* **2009**, *131*, 1050–1056. [CrossRef] [PubMed]
25. Qian, M.; Xu, Z.; Wang, Z.; Wei, B.; Wang, H.; Hu, S.; Liu, L.M.; Guo, L. Realizing Few-Layer Iodinene for High-Rate Sodium-Ion Batteries. *Adv. Mater.* **2020**, *32*, 2004835. [CrossRef] [PubMed]
26. Huran, A.W.; Wang, H.-C.; San-Miguel, A.; Marques, M.A.L. Atomically Thin Pythagorean Tilings in Two Dimensions. *J. Phys. Chem. Lett.* **2021**, *12*, 4972–4979. [CrossRef]
27. Liu, P.; Zhang, G.; Yan, Y.; Jia, G.; Liu, C.; Wang, B.; Yin, H. Strain-tunable phase transition and doping-induced magnetism in iodinene. *Appl. Phys. Lett.* **2021**, *119*, 102403. [CrossRef]
28. Bafekry, A.; Stampfl, C.; Faraji, M.; Mortazavi, B.; Fadlallah, M.M.; Nguyen, C.V.; Fazeli, S.; Ghergherehchi, M. Monoelemental two-dimensional iodinene nanosheets: A first-principles study of the electronic and optical properties. *J. Phys. D Appl. Phys.* **2022**, *55*, 135104. [CrossRef]
29. Kuang, C.; Zeng, W.; Qian, M.; Liu, X. Liquid-Phase Exfoliated Few-Layer Iodine Nanosheets for High-Rate Lithium-Iodine Batteries. *ChemPlusChem* **2021**, *86*, 865–869. [CrossRef]
30. You, Y.; Zhu, Y.-X.; Jiang, J.; Chen, Z.; Wu, C.; Zhang, Z.; Lin, H.; Shi, J. Iodinene Nanosheet-to-Iodine Molecule Allotropic Transformation for Antibiosis. *J. Am. Chem. Soc.* **2023**, *145*, 13249–13260. [CrossRef]
31. Liu, H.; Cao, S.; Chen, L.; Zhao, K.; Wang, C.; Li, M.; Shen, S.; Wang, W.; Ge, L. Electron acceptor design for 2D/2D iodinene/carbon nitride heterojunction boosting charge transfer and CO_2 photoreduction. *Chem. Eng. J.* **2022**, *433*, 133594. [CrossRef]
32. Wen, T.; Wang, Y.; Li, N.; Zhang, Q.; Zhao, Y.; Yang, W.; Zhao, Y.; Mao, H.-k. Pressure-Driven Reversible Switching between n- and p-Type Conduction in Chalcopyrite $CuFeS_2$. *J. Am. Chem. Soc.* **2019**, *141*, 505–510. [CrossRef]
33. Bu, K.; Luo, H.; Guo, S.; Li, M.; Wang, D.; Dong, H.; Ding, Y.; Yang, W.; Lü, X. Pressure-Regulated Dynamic Stereochemical Role of Lone-Pair Electrons in Layered Bi_2O_2S. *J. Phys. Chem. Lett.* **2020**, *11*, 9702–9707. [CrossRef] [PubMed]
34. Zhang, Q.; Liu, X.; Li, N.; Wang, B.; Huang, Q.; Wang, L.; Zhang, D.; Wang, Y.; Yang, W. Pressure-driven chemical lock-in structure and optical properties in Sillen compounds $PbBiO_2X$ (X = Cl, Br, and I). *J. Mater. Chem. A* **2020**, *8*, 13610–13618. [CrossRef]
35. Duan, D.; Jin, X.; Ma, Y.; Cui, T.; Liu, B.; Zou, G. Effect of nonhydrostatic pressure on superconductivity of monatomic iodine: An ab initio study. *Phys. Rev. B* **2009**, *79*, 064518. [CrossRef]
36. Sakai, N.; Takemura, K.-i.; Tsuji, K. Electrical properties of high-pressure metallic modification of iodine. *J. Phys. Soc. Japan* **1982**, *51*, 1811–1816. [CrossRef]
37. Kenichi, T.; Kyoko, S.; Hiroshi, F.; Mitsuko, O. Modulated structure of solid iodine during its molecular dissociation under high pressure. *Nature* **2003**, *423*, 971–974. [CrossRef]

38. San Miguel, A.; Libotte, H.; Gaspard, J.P.; Gauthier, M.; Itié, J.P.; Polian, A. Bromine metallization studied by X-ray absorption spectroscopy. *Eur. Phys. J. B Condens. Matter Complex Syst.* **2000**, *17*, 227–233. [CrossRef]
39. Li, Z.; Li, H.; Liu, N.; Du, M.; Jin, X.; Li, Q.; Du, Y.; Guo, L.; Liu, B. Pressure Engineering for Extending Spectral Response Range and Enhancing Photoelectric Properties of Iodine. *Adv. Opt. Mater.* **2021**, *9*, 2101163. [CrossRef]
40. Song, D.; Liu, K.; Li, J.; Zhu, H.; Sun, L.; Okulov, A. Mechanical tensile behavior-induced multi-level electronic transport of ultra-thin SiC NWs. *Mater. Today Commun.* **2023**, *36*, 106528. [CrossRef]
41. Liu, K.; Li, Y.; Liu, Q.; Song, D.; Xie, X.; Duan, Y.; Wang, Y.; Li, J. Superior electron transport of ultra-thin SiC nanowires with one impending tensile monatomic chain. *Vacuum* **2022**, *199*, 110950. [CrossRef]
42. Mlinar, V. Electronic and optical properties of nanostructured MoS$_2$ materials: Influence of reduced spatial dimensions and edge effects. *Phys. Chem. Chem. Phys.* **2017**, *19*, 15891–15902. [CrossRef]
43. Liu, K.; Li, J.; Liu, R.; Li, H.; Okulov, A. Non-equilibrium electronic properties of ultra-thin SiC NWs influenced by the tensile strain. *J. Mater. Res. Technol.* **2023**, *26*, 6955–6965. [CrossRef]
44. Li, J.; Li, T.; Zhou, Y.; Wu, W.; Zhang, L.; Li, H. Distinctive electron transport on pyridine-linked molecular junctions with narrow monolayer graphene nanoribbon electrodes compared with metal electrodes and graphene electrodes. *Phys. Chem. Chem. Phys.* **2016**, *18*, 28217–28226. [CrossRef]
45. Cao, J.; Dong, J.; Saglik, K.; Zhang, D.; Solco, S.F.D.; You, I.J.W.J.; Liu, H.; Zhu, Q.; Xu, J.; Wu, J.; et al. Non-equilibrium strategy for enhancing thermoelectric properties and improving stability of AgSbTe$_2$. *Nano Energy* **2023**, *107*, 108118. [CrossRef]
46. Brandbyge, M.; Mozos, J.-L.; Ordejón, P.; Taylor, J.; Stokbro, K. Density-functional method for nonequilibrium electron transport. *Phys. Rev. B* **2002**, *65*, 165401. [CrossRef]
47. Taylor, J.; Guo, H.; Wang, J. Ab initio modeling of quantum transport properties of molecular electronic devices. *Phys. Rev. B* **2001**, *63*, 245407. [CrossRef]
48. Zhang, L.; Yuan, J.; Shen, L.; Fletcher, C.; Li, H. Taper-shaped carbon-based spin filter. *Appl. Surf. Sci.* **2019**, *495*, 143501. [CrossRef]
49. Landauer, R. Electrical resistance of disordered one-dimensional lattices. *Philos. Mag.* **1970**, *21*, 863–867. [CrossRef]
50. Reed, M.A.; Zhou, C.; Muller, C.; Burgin, T.; Tour, J. Conductance of a molecular junction. *Science* **1997**, *278*, 252–254. [CrossRef]
51. Datta, S. *Electronic Transport in Mesoscopic Systems*; Cambridge University Press: Cambridge, UK, 1997.

Disclaimer/Publisher's Note: The statements, opinions and data contained in all publications are solely those of the individual author(s) and contributor(s) and not of MDPI and/or the editor(s). MDPI and/or the editor(s) disclaim responsibility for any injury to people or property resulting from any ideas, methods, instructions or products referred to in the content.

Article

Electron Transport Properties of Graphene/WS$_2$ Van Der Waals Heterojunctions

Junnan Guo [1], Xinyue Dai [2], Lishu Zhang [3] and Hui Li [1],*

[1] Key Laboratory for Liquid-Solid Structural Evolution and Processing of Materials, Ministry of Education, Shandong University, Jinan 250061, China; guojunnan113005@hotmail.com
[2] Materdicine Lab, School of Life Sciences, Shanghai University, Shanghai 200444, China; dxy1120@shu.edu.cn
[3] Peter Grünberg Institut (PGI-1) and Institute for Advanced Simulation (IAS-1), Forschungszentrum Jülich, Jülich 52428, Germany; lis.zhang@fz-juelich.de
* Correspondence: lihuilmy@sdu.edu.cn

Abstract: Van der Waals heterojunctions of two-dimensional atomic crystals are widely used to build functional devices due to their excellent optoelectronic properties, which are attracting more and more attention, and various methods have been developed to study their structure and properties. Here, density functional theory combined with the nonequilibrium Green's function technique has been used to calculate the transport properties of graphene/WS$_2$ heterojunctions. It is observed that the formation of heterojunctions does not lead to the opening of the Dirac point of graphene. Instead, the respective band structures of both graphene and WS$_2$ are preserved. Therefore, the heterojunction follows a unique Ohm's law at low bias voltages, despite the presence of a certain rotation angle between the two surfaces within the heterojunction. The transmission spectra, the density of states, and the transmission eigenstate are used to investigate the origin and mechanism of unique linear I–V characteristics. This study provides a theoretical framework for designing mixed-dimensional heterojunction nanoelectronic devices.

Keywords: graphene/WS$_2$ heterojunctions; electronic transport; first-principles calculation

Citation: Guo, J.; Dai, X.; Zhang, L.; Li, H. Electron Transport Properties of Graphene/WS$_2$ Van Der Waals Heterojunctions. *Molecules* 2023, 28, 6866. https://doi.org/10.3390/molecules28196866

Academic Editor: Bryan M. Wong

Received: 30 August 2023
Revised: 26 September 2023
Accepted: 27 September 2023
Published: 29 September 2023

Copyright: © 2023 by the authors. Licensee MDPI, Basel, Switzerland. This article is an open access article distributed under the terms and conditions of the Creative Commons Attribution (CC BY) license (https://creativecommons.org/licenses/by/4.0/).

1. Introduction

Two-dimensional (2D) layered materials have always been a cutting-edge field in condensed matter physics and materials research [1]. Various 2D layers can be combined using van der Waals (vdW) forces to build heterostructures with diverse functionalities [2]. These heterostructures exhibit a range of excellent properties and are applied to optoelectronic devices [3], providing unprecedented opportunities for the development of advanced nanoelectronics devices [4].

As one of the atomically thin 2D materials, graphene [5] has attracted worldwide attention due to its excellent optical [6], electrical [7], and mechanical properties [8], and it is expected to be used to build a new generation of miniaturized and intelligent electronic devices [9]. However, the absence of a band gap has limited the application of graphene, particularly in the semiconductor industry [10]. Significant efforts have been devoted to addressing this issue in the gap-opening of graphene, like functionalization [11], doping [12], and the construction of heterostructures [13]. Recently, many graphene-based vdW heterostructures have been investigated theoretically and experimentally [14]. For instance, Lan et al. transferred graphene grown on a copper foil to a sapphire substrate with Bi$_2$Te$_3$ crystals via low-pressure chemical vapor deposition (CVD). The crystallized Bi$_2$Te$_3$ was synthesized directly using spin-coated coring (SCCA). This procedure avoided any degradation of the nanoplates and significantly improved the quality of the heterojunction sample [15]. Hu et al. utilized a polymethyl methacrylate (PMMA)/polydimethylsiloxane (PDMS) blend to transfer metal-catalyzed CVD-fabricated graphene/SiNWS heterojunctions onto stretchable polytetrafluoroethylene (PTFE) substrates. The high preparation efficiency and outstanding quality were extremely encouraging for daily industrial production and life [16]. Ren et al. developed a novel flexible self-powered photodetector that transfers electrons through a solid electrolyte. The developed flexible WS_2/graphene photodetector displayed a quick photo response time and high photosensitivity [17]. Liu et al. fabricated Bi_2Se_3/graphene heterojunctions using molecular beam epitaxy and observed a spiral growth mechanism during the growth process [18]. By vertically stacking single-layer MoS_2/h-BN/graphene, Lee's team created random access memory with tunneling. It had excellent stretchability, long retention times, and highly dependable memory performance [19]. Additionally, Liu et al. investigated different conceivable atomic configurations of phosphorene/graphene in-plane heterojunctions and their effects on interfacial heat conductivity by using density functional theory calculations and molecular dynamics simulations [20]. Gao et al. simulated the heat transfer properties of graphene/MoS_2 heterojunctions using nonequilibrium molecular dynamics simulations and found that the degree of lattice matching of graphene and MoS_2 had an effect on phonon thermal transport [21]. However, the majority of these studies on graphene heterojunctions primarily focused on their electronic structures [22], preparation methods [23], and applications [24]. Little research has been conducted on their electron transport properties and intrinsic mechanisms.

In this context, constructing new graphene heterojunctions and studying their electron transport properties are essential if one wants to realize the practical application of graphene heterojunctions in nanoelectronic devices. With excellent electron mobility and a large direct band gap, monolayer WS_2 has a lot of potential uses in nanodevices [25]. In particular, in recent years, there have been significant breakthroughs in its synthesis and applications. For example, Prof. Feng's group produced monolayer triangular WS_2 single crystal wafers with excellent uniformity, large size, and high quality by controlling the nucleation density by changing the time of the introduction of the sulfur precursor and the distance between the tungsten source and the growth substrate [26]. Furthermore, some researchers have used chemical doping to significantly improve the optoelectronic performance of WS_2 field-effect transistors [27]. Inspired by these advancements, we selected monolayer WS_2 to create a series of graphene/WS_2 heterojunction models and design nanoelectronic devices. We systematically investigated their electronic structures and transport properties using first-principles methods based on the density functional theory (DFT) and nonequilibrium Green's function (NEGF) [28].

2. Results and Discussions

The hexagonal unit cell of WS_2 was the same as that of graphene. For graphene and WS_2, the optimized lattice parameters were 2.45 Å and 3.15 Å, respectively. The unit cell parameters we calculated closely matched experimental results [29,30].

To construct the graphene/WS_2 heterojunctions, we used a $3 \times 3 \times 1$ supercell of WS_2 and a $4 \times 4 \times 1$ supercell of graphene with 68 total atom numbers, and a $4 \times 4 \times 1$ supercell of WS_2 and a $5 \times 5 \times 1$ supercell of graphene with 109 total atom numbers. In this orientation, both components maintained their original hexagonal lattices without surface rotation and exhibited slight lattice mismatches of 3.1% and 2.4%, respectively. The interlayer spacings of the equilibrium geometries of these two heterojunctions were 3.41 Å and 3.46 Å, respectively, which are typical distances in graphene-based vdW heterostructures with weak interactions.

However, the devices built from the above two heterojunctions contained 366 and 603 atoms, separately. Due to the limitations of quantum-mechanics-based calculations used in this study, we continued to construct a series of heterojunctions with specific rotation angles between each surface to reduce the models' sizes. In these heterojunctions, the interatomic distances were consistently around 3.4 Å, indicating weak vdW interactions. At the same time, we could also analyze the electron transport properties of devices that had different rotation angles. The equilibrium geometries of heterojunctions and their related parameters are shown in Figure 1 and Table 1.

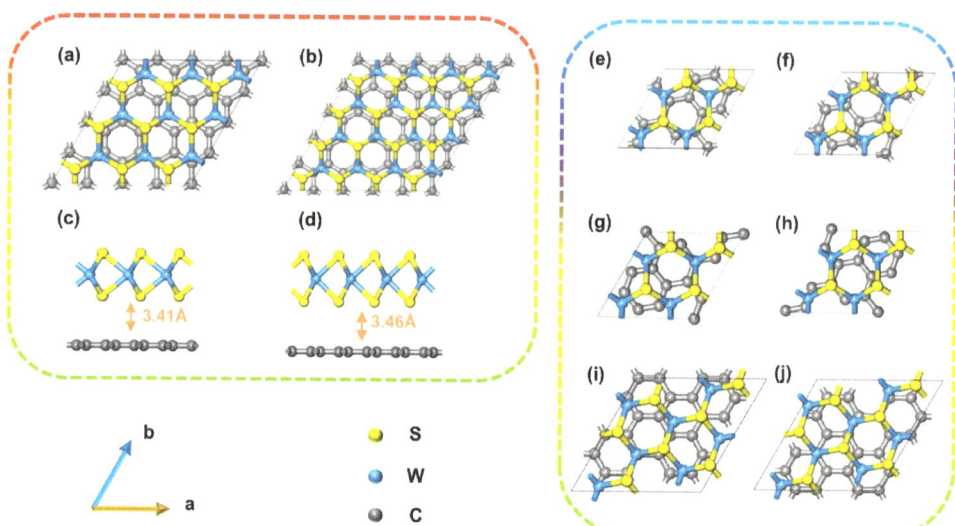

Figure 1. Top views of (**a**) Gr/WS$_2$-1, (**b**) Gr/WS$_2$-2, (**e**) Gr/WS$_2$-3, (**f**) Gr/WS$_2$-4, (**g**) Gr/WS$_2$-5, (**h**) Gr/WS$_2$-6, (**i**) Gr/WS$_2$-7, and (**j**) Gr/WS$_2$-8 ball-and-stick models. Side views of (**c**) Gr/WS$_2$-1 and (**d**) Gr/WS$_2$-2.

Table 1. The related parameters of heterojunctions.

Heterojunction	Lattice Parameters of Graphene (Å)	Rotation Angle of Graphene (°)	Lattice Parameters of WS$_2$ (Å)	Rotation Angle of WS$_2$ (°)	Lattice Mismatch (%)
Gr/WS$_2$-1	a = b = 9.8	0.0	a = b = 9.5	0.0	3.1
Gr/WS$_2$-2	a = b = 12.3	0.0	a = b = 12.6	0.0	2.4
Gr/WS$_2$-3	a = b = 6.5	21.8	a = b = 6.3	60.0	3.1
Gr/WS$_2$-4	a = b = 6.5	141.8	a = b = 6.3	60.0	3.1
Gr/WS$_2$-5	a = b = 6.5	21.8	a = b = 6.3	180.0	3.1
Gr/WS$_2$-6	a = b = 6.5	141.8	a = b = 6.3	180.0	3.1
Gr/WS$_2$-7	a = b = 8.5	0.0	a = b = 8.3	21.8	2.1
Gr/WS$_2$-8	a = b = 8.5	120.0	a = b = 8.3	21.8	2.1

In order to prove the thermodynamic stability of these heterojunctions, the binding energies of the graphene/WS$_2$ vdW heterojunctions were calculated to assess the system stability, as follows:

$$E_b = E(\text{heterojunction}) - E(\text{graphene}) - E(\text{WS}_2)$$

where E(heterojunction), E(graphene), and E(WS$_2$) represent the total energy of the heterojunctions, graphene layers, and WS$_2$ layers, respectively. The calculated binding energies are presented in Table 2. The negative binding energies in the table indicate the stability of these systems. Upon comparison, we observed that the most stable heterojunction was Gr/WS$_2$-1. Another regularity we found was that smaller heterojunctions were more stable when the two layers were not rotated. However, when there exist rotation angles between the two layers, the stability of the heterojunctions decreased and the larger heterojunctions were more stable.

Table 2. The binding energy of heterojunctions.

Heterojunction	Energy of Graphene (eV)	Energy of WS$_2$ (eV)	Energy of Heterojunction (eV)	Binding Energy (eV)
Gr/WS$_2$-1	−5038.3	−10,206.0	−15,246.9	−2.6
Gr/WS$_2$-2	−7874.7	−18,145.4	−26,022.2	−2.1
Gr/WS$_2$-3	−2204.4	−4563.7	−6741.6	−0.6
Gr/WS$_2$-4	−2204.4	−4563.7	−6741.6	−0.6
Gr/WS$_2$-5	−2204.4	−4563.7	−6741.6	−0.6
Gr/WS$_2$-6	−2204.4	−4563.7	−6741.6	−0.6
Gr/WS$_2$-7	−3779.3	−7939.2	−11,719.4	−0.9
Gr/WS$_2$-8	−3379.3	−7939.2	−11,719.4	−0.9

We initially investigated the electronic properties of Gr/WS$_2$-1 and Gr/WS$_2$-2 to determine whether they can be transported as electronic devices. As plotted in Figure 2a, graphene exhibits metallic properties with a zero bandgap semiconductor, where the top valence band and bottom conduction band intersect at the K point. In contrast, WS$_2$ is a semiconductor with a direct band gap of 1.95 eV, as shown in Figure 2b. It is worth noting that our calculations closely aligned with other theoretical predictions and were slightly lower than experimental values [31]. This discrepancy can be attributed to the inherent limitations of the GGA-PBE method, which tends to overestimate lattice constants and underestimate band gaps. Hybrid functionals, such as meta-GGA, HSE06, etc., are known to provide more accurate bandgap calculations [32]. However, the WS$_2$ bandgap calculated by meta-GGA was 2.13 eV, which was only slightly higher than the PBE value (1.95 eV). Thus, we believe that the GGA-PBE approach was accurate enough for our calculation and did not significantly impact other aspects of the analysis, such as energy band structure and electron transport.

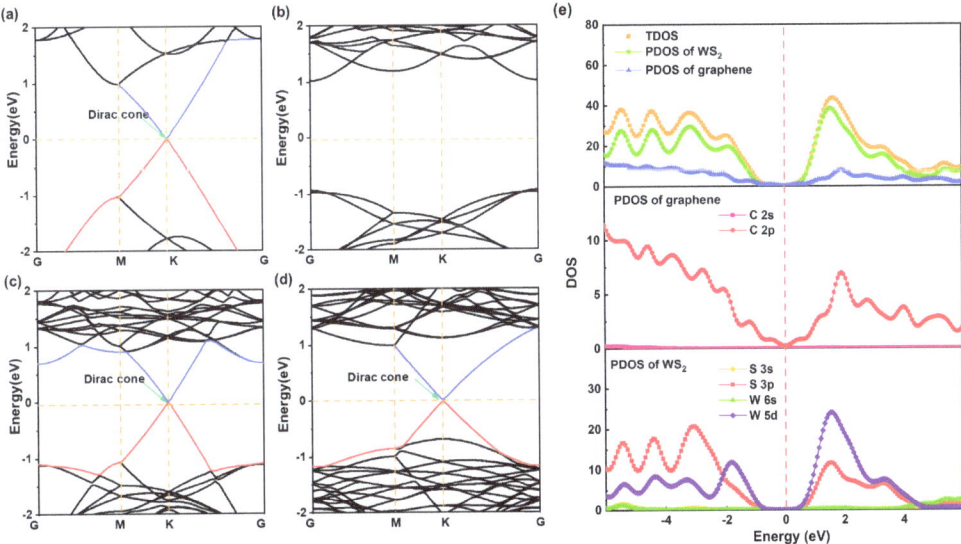

Figure 2. Band structures of the stand-alone (a) graphene and (b) WS$_2$; (c,d) are band structures of Gr/WS$_2$-1 and Gr/WS$_2$-2. The red and blue lines represent the top of the valence band and the bottom of the conduction band. (e) The PDOS and DOS of the graphene and WS$_2$ components in the vdW Gr/WS$_2$-1.

Figure 2c and d display the band structures of Gr/WS$_2$-1 and Gr/WS$_2$-2, which are simple superpositions of graphene and WS$_2$ and preserve their electronic systems. Notably, the valence band's top and the conduction band's bottom still intersected at the K point in the Brillouin zone, indicating that the Dirac point still exists in the heterojunction. Gr/WS$_2$-

1 behaved as an N-type semiconductor, with the E_c and E_v of WS_2 shifting downwards. Additionally, the Fermi energy level turned from near the top of the valence band to near the bottom of the conduction band. Conversely, Gr/WS_2-2 exhibited P-type semiconductor properties, with the Fermi energy level still close to the top of the valence band, but the conduction band bottom and valence band top shifted from the original G to the K point. This indicates that factors such as layer spacing, the degree of mismatch, and lattice parameters within the heterojunction influence its electronic energy band.

Next, we calculated the density of states (DOS) and the projected calculation density of states (PDOS). Due to the similarity in the calculation results, we present the results for Gr/WS_2-1 as an example. According to Figure 2e, near the Fermi level, the 2p orbital of the carbon atom in graphene plays a vital role in the density of states. The 5d orbital of the W atom also makes a contribution. Contributions from other valence electron orbitals can be disregarded. The absence of resonance peaks indicates that there was no bonding between WS_2 and C. Instead, weak van der Waals forces maintained the interlayer stability between the heterojunctions, corresponding to optimized interlayer spacing of around 3.4 Å. This weak hybridization between the graphene and WS_2 is another indication of why the graphene's Dirac points are still present in the heterojunctions.

As depicted in Figure 3, when there is a certain rotation angle between the two surfaces, no matter the change in the lattice constants or the rotation angle of graphene or WS_2, its effect on the energy band is little. But when the lattice parameter of heterojunctions is increased to around 8 Å, the Dirac cone of graphene shifts from K to G point due to the inequivalent K and K' points being folded and coupled into the same G-point (Figure 3e,f). However, the Dirac cone does not open. We predicted that these six heterojunctions had comparable electronic transport properties. Consequently, nanoelectronic devices could be built using heterostructures with rotation angles to reduce device size while maintaining their high transport properties.

With the Gr/WS_2-3 and Gr-1 (composed of graphene, with the same lattice parameter and rotation angle as Gr/WS_2-3), we built two devices, as depicted in Figure 4. As seen in the enlarged area, the rotation angle between graphene and WS_2 was still maintained. The poles of the device formed by themselves, the current transport direction was along the Z-axis, and the surface was perpendicular to the X-axis.

The I–V characteristics of the devices in a bias zone [0.0 V, 2.0 V] were calculated to explore the transport characteristics of these two devices, and the findings are shown in Figure 4. We can see from the current-voltage (I–V) characteristic curves (Figure 4c) that the heterojunction had comparable transport properties to graphene, unlike some typical heterojunction semiconductor devices. Interestingly, the Ohmic behavior of linear I–V curves was found in the 0–1.2 V bias voltage. After 1.2 V, the slope of the I–V curve gradually increased, leading to nonlinear transport properties. This was caused by a certain degree of rotation in the graphene and heterojunction, while the transport direction was primarily along the armchair direction of the graphene. Simultaneously, it became evident that the transport properties of both devices changed gradually as the voltage increased, signifying a weakened coupling between WS_2 and graphene. A nonlinear relationship only began to emerge at high bias voltages. Compared to other graphene-based heterojunctions, the transport current of graphene/WS_2 was nearly one order of magnitude higher than that of graphene/MoS_2 in-plane heterojunctions [33,34], graphene/BN heterojunctions [35], and so on. In addition, when compared to other WS_2-based heterojunctions, such heterojunctions could behave up to two orders of magnitude higher than that of WS_2/WSe_2 heterojunctions [36], with greater performance than that of MoS_2/WS_2 heterojunctions [37]. Thus, we can conclude that such heterojunctions can greatly enhance the transport current and decrease the contact resistance, which will be very important for achieving superior optoelectronic devices such as vertical field-effect transistors (FETs). Our calculations can reveal why graphene/WS_2 heterojunctions are widely used to build FETs and have superior behavioral properties [38–41]. In addition, the heterojunction used in our calculations not only maintained the perfect transport properties but also largely reduced the size of the electronic devices, which is very important in the post-Moore era.

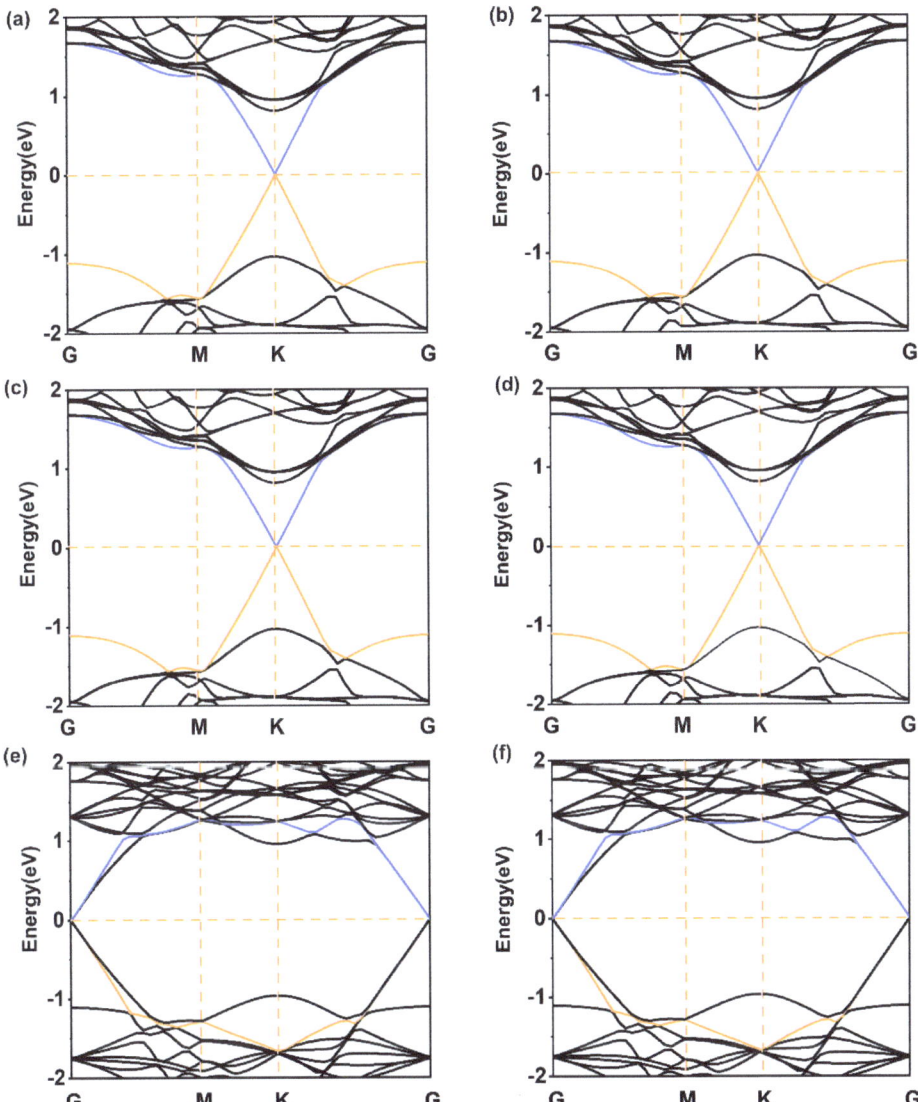

Figure 3. Band structures of (**a**) Gr/WS$_2$-3, (**b**) Gr/WS$_2$-4, (**c**) Gr/WS$_2$-5, (**d**) Gr/WS$_2$-6, (**e**) Gr/WS$_2$-7, and (**f**) Gr/WS$_2$-8. The orange and blue lines represent the top of the valence band and the bottom of the conduction band.

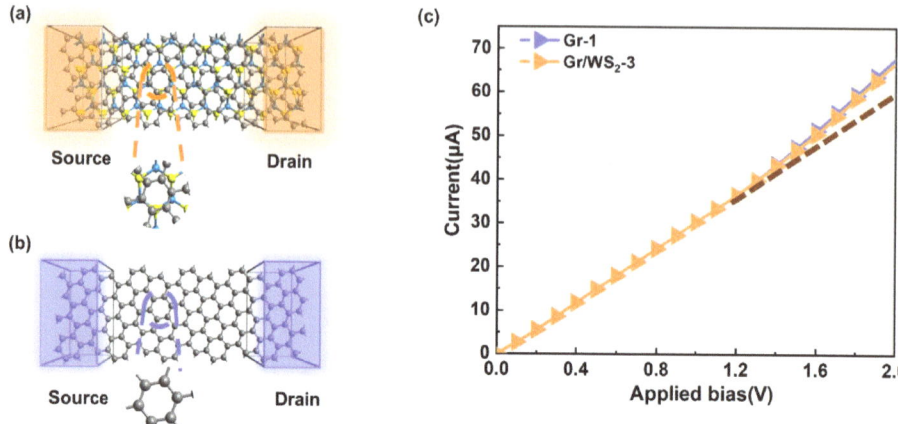

Figure 4. The device configuration with (**a**) Gr/WS$_2$-3 and (**b**) Gr-1. (**c**) I–V characteristics of devices.

Although the differences in transport properties between these two devices were slight, the transport mechanism exhibited different phenomena due to the weak vdW forces between the WS$_2$ and graphene. The most understandable depiction of the behavior of electron transport was the transmission spectrum $T(E)$, and the transmission coefficient of each energy point was determined by diagonalizing the transmission matrix from the eigenvalues of electron transmission. Therefore, we calculated the transmission spectra of the above devices to further study their transport properties.

Generally speaking, the magnitude of the transmission coefficient near the Fermi level represents the transport capability of the device, especially at the Fermi level. The larger the transmission coefficient at the Fermi level, the stronger the transport capability. As shown in Figure 5a,d, these two devices exhibited metallic properties, corresponding to the current–voltage curves. The electron transmission spectra of graphene devices and Gr/WS$_2$-3 devices displayed quantum steps between -1 eV and 1 eV, resembling the ideal one-dimensional nanowires. And the electron transmission probability at the Fermi energy level was almost zero, which shows a band gap feature between the conduction and valence bands, corresponding to a Dirac cone in the energy band structure. Although the system had almost no electrons passing through at this energy, at higher energies electrons could easily tunnel through the potential barrier, increasing their mobility and the step transmission coefficient, which indicates that there were several electron transmission channels in these devices. After the formation of the heterojunction, many spikes appeared away from the Fermi energy level, showing that the coupling between the graphene and WS$_2$ was weak. The band gap of graphene was not open, although it tends to be open, which does not have a significant influence on its transport properties.

To further shed light on the inherent mechanisms of these two devices, we discuss DOS around the Fermi level for these devices. Figure 5a,d illustrate that the DOS of the two devices were zero at the Fermi energy level, corresponding to their electron transmission spectrum. Before the construction of the heterojunction, the contribution of DOS near the Fermi energy level originated mainly from the 2p orbitals of the graphene carbon atoms. After the formation of the heterojunction, the contribution was mainly from the 2p orbital of the graphene carbon atom and the 5d of the W atom. We can see clearly that several peaks exceeded 100 in the Gr/WS$_2$-3, more than twice that of the Gr-1. The highest peaks in the valence band region were observed at -1.92 eV, while those in the conduction band region were found at 1.44 eV. These peaks serve to protect fewer delocalized states near the Fermi level.

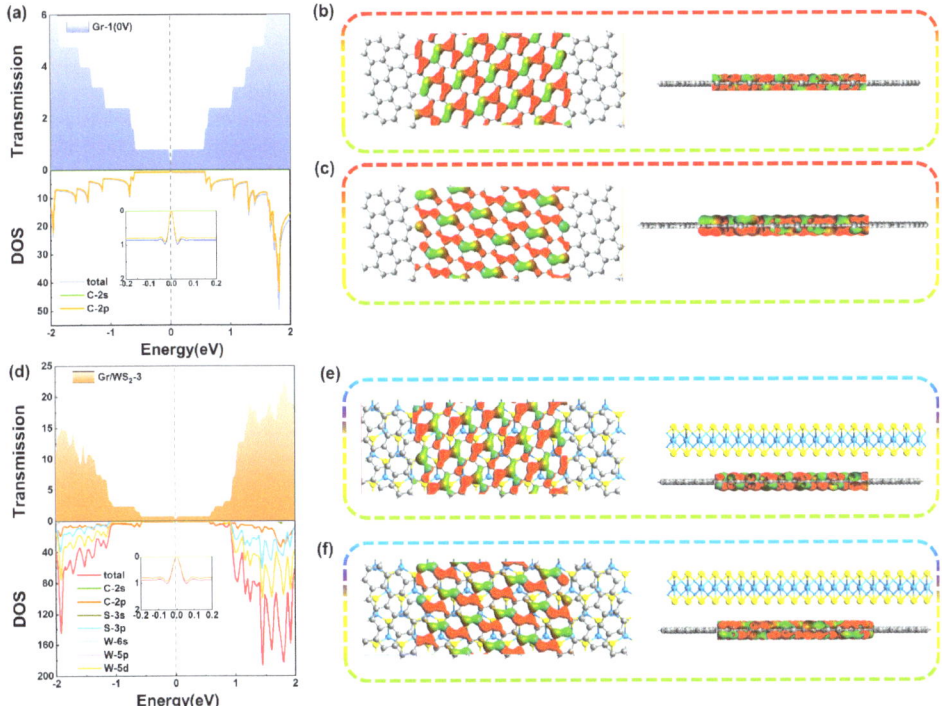

Figure 5. Transmission spectrum and DOS of (**a**) Gr-1 device and (**d**) Gr/WS$_2$-3 device at a free applied bias (0.0 V). Transmission eigenstate of Gr-1 (**b,c**) and Gr/WS$_2$-3 (**e,f**) around Fermi level with an isovalue of 0.21.

Here, the dominant transmission eigenstates near the Fermi energy level at equilibrium were calculated to explore the physical roots of their transport phenomena. The calculated results in Figure 5b,c,e, and f showed that the transmission of two devices around the Fermi level was provided by two major transport channels, both with transmission eigenvalues of nearly 1.000. The transmission eigenstates of both devices exhibited delocalization throughout the whole central region, resulting in significant transport capability near the Fermi energy level. We can see that the electronic states were evenly distributed in the diffusion region between the left and right electrodes, along the graphene armchair direction. This indicates that these states were all π-orbitals of the C atom of graphene, leading to their metallic characteristic. However, the contribution of WS$_2$ in Gr/WS$_2$-3 was almost negligible.

It is well known that the study of transmission spectra at non-zero bias voltages can provide useful information for the study of I–V characteristics. This is because the current is defined by the integrated area of the transmission curve within the bias window, as shown by the Landauer–Buttiker formula. As a result, we calculated the transmission spectra of the Gr-1 and Gr/WS$_2$-3 devices under 0.4, 0.8, 1.2, 1.6, and 2.0 to further reveal their transport phenomena (Figure 6a,b). The bias window's perimeter is represented by the colored parts. The effective integral area of the transmission curve within the bias window grew with increased bias, producing a linear I–V characteristic, as we can see from both devices. However, when the bias window increased to 1.2 V, the step transmission spectrum started to change shape and expand in an arc, so the I–V curve began to show non-linear features, and the slope subsequently increased. It is evident from the transmission spectrum that quantum steps are always present within the bias window at different bias voltages and that the steps shift as the bias window expands. The movement tendency of the steps in the conduction and valence band regions was indicated by the arrows, respectively. The number of wave valleys within the bias window in the Gr/WS$_2$-3 devices progressively increased. Spikes far from the Fermi energy level moved in the opposite direction and were unable to move inside the bias window, so the contribution of these spikes to the transport properties was almost negligible. Interestingly, the lowest transmission probability was always located at the boundary of the bias window, and as the bias increased from 0 V to 2 V, the gap shifted to the boundary of the bias window.

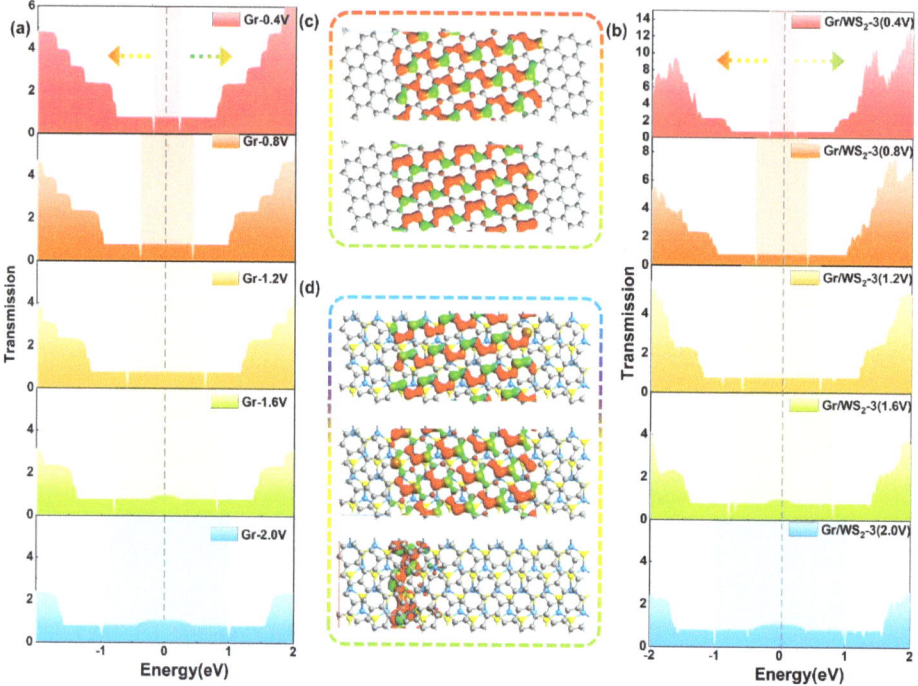

Figure 6. Transmission spectra of (**a**) Gr-1 device and (**b**) Gr/WS$_2$-3 device at nonzero bias voltage. Transmission eigenstates of Gr-1 (**c**) and Gr/WS$_2$-3 (**d**) at 2.0 V with an isovalue of 0.21. The colored arrows represent the movement of the steps in the conduction and valence band regions.

It is worth noting that the transmission eigenstates of electrons may change under external bias after forming the heterojunction. Therefore, the electronic transmission eigenstates of Gr-1 and Gr/WS$_2$-3 devices were calculated at different bias voltages. Before 2.0 V, the transmission eigenstates of both devices were mainly contributed by the two transmission channels of graphene. However, when the bias voltage increased to 2.0 V, the transmission channels at the Fermi level of Gr/WS$_2$-3 changed from two to multiple channels, as shown in Figure 6c,d. WS$_2$ started to participate in the transport, but its electronic state was localized at the left electrode and the transmission eigenvalue was so small that it can still be disregarded.

3. Computational Method

DFT implemented in the Atomistix ToolKit (ATK) package was used to optimize the geometry structures and calculate the electronic structures of graphene/WS$_2$ heterostructures. The exchange–correlation function is a generalized gradient approximation (GGA) [42] of Perdew–Burke–Ernzerhof (PBE) [43]. The selected valence electron configurations in our calculation were W $5d^4$ $6s^2$, S $3s^2$ $3p^4$, C $2s^2$ $2p^2$. In order to meet the computational precision, the linear combination of atomic orbitals (LCAO) basis was selected for all atoms.

Double-ζ plus polarization (DZP) basis sets were adopted for the local atomic numerical orbitals, and norm-conserving pseudo-potentials were employed. The Monkhorst–Pack k-points of $5 \times 5 \times 1$ were used to sample the Brillouin zone [44]. The cut-off energy for the density mesh and the electron temperature were set to 75 Ha and 300 K, accordingly.

The device's performances were studied with the DFT coupled with the NEGF method, using the ATK package. A 15 Å vacuum layer along the X-direction was used to avoid interactions between periodic images that were nearest neighbors. For self-consistent calculation, the k-points of $5 \times 5 \times 100$ were used for device models. The other parameters of the DFT calculation remained unchanged, and the energy convergence criterion was set to 10^{-4} eV. Before analyses, the devices were fully optimized by the quasi-Newton approach until all residual stresses on each atom were less than 0.05 eV. The devices' electronic properties were investigated by computing their currents, the density of states, and the transmission spectra, and the current I through the device was calculated using the Landauer–Buttiker equation [45]:

$$I = \frac{2e}{h} \int_{-\infty}^{+\infty} dE (T(E,V)(f_1(E) - f_2(E)))$$

The quantity $T(E)$ is the transmission function, which expresses the likelihood that electrons will go through the device from source to drain; $f_{1,2}(E)$ denote the Fermi functions of the source and drain electrodes; and e and h are the electron charge and Planck's constant, respectively.

4. Conclusions

In this work, we systematically studied the electronic transport properties and intrinsic mechanisms of graphene/WS$_2$ heterojunctions using first-principal calculations. Unique linear I–V characteristics were found among the devices. Even though there was an angle between the two surfaces, the heterojunction continued to exhibit this intriguing Ohm's law behavior. The transmission spectra, the density of states, and the transmission eigenstate were calculated to explain this phenomenon. After forming the heterojunctions, the quantum steps near the Fermi level approximated an ideal one-dimensional nanowire. The DOS shows that the vdW heterojunctions significantly increased the number of peaks and improved the maximum value of peaks, which protected less delocalized states near the Fermi level. The transmission eigenstates showed that the high transport properties came from the π orbitals of the C atoms in the graphene armchair direction. This study provides valuable insights into the transport properties of graphene heterojunctions and the potential fabrication of mixed-dimensional heterojunctions.

Author Contributions: Data curation, investigation, writing—original draft, writing—review & editing, J.G.; writing—review & editing, X.D., L.Z. and H.L. The manuscript was written through contributions of all authors. These authors contributed equally. All authors have read and agreed to the published version of the manuscript.

Funding: This research was funded by the National Natural Science Foundation of China (NNSFC) (Grant No. 52171038). This work was also supported by Special Funding in the Project of the Taishan Scholar Construction Engineering and the program of Jinan Science and Technology Bureau (2020GXRC019), as well as the new material demonstration platform construction project from the Ministry of Industry and Information Technology (2020-370104-34-03-043952-01-11) and the Key Research and Development Plan of Shandong Province (2021SFGC1001).

Institutional Review Board Statement: Not applicable.

Informed Consent Statement: Not applicable.

Data Availability Statement: Data will be made available on request.

Acknowledgments: The authors would like to thank Kangwei Wu and Jie Li for their help.

Conflicts of Interest: The authors declare no conflict of interest.

References

1. Coleman, J.N.; Lotya, M.; O'Neill, A.; Bergin, S.D.; King, P.J.; Khan, U.; Young, K.; Gaucher, A.; De, S.; Smith, R.J. Two-dimensional nanosheets produced by liquid exfoliation of layered materials. *Science* **2011**, *331*, 568–571. [CrossRef] [PubMed]
2. Britnell, L.; Ribeiro, R.M.; Eckmann, A.; Jalil, R.; Belle, B.D.; Mishchenko, A.; Kim, Y.-J.; Gorbachev, R.V.; Georgiou, T.; Morozov, S.V. Strong light-matter interactions in heterostructures of atomically thin films. *Science* **2013**, *340*, 1311–1314. [CrossRef]
3. Yu, W.J.; Liu, Y.; Zhou, H.; Yin, A.; Li, Z.; Huang, Y.; Duan, X. Highly efficient gate-tunable photocurrent generation in vertical heterostructures of layered materials. *Nat. Nanotechnol.* **2013**, *8*, 952–958. [CrossRef] [PubMed]
4. Chhowalla, M.; Jena, D.; Zhang, H. Two-dimensional semiconductors for transistors. *Nat. Rev. Mater.* **2016**, *1*, 16052. [CrossRef]
5. Novoselov, K.S.; Colombo, L.; Gellert, P.; Schwab, M.; Kim, K. A roadmap for graphene. *Nature* **2012**, *490*, 192–200. [CrossRef]
6. Sun, Z.; Chang, H. Graphene and graphene-like two-dimensional materials in photodetection: Mechanisms and methodology. *ACS Nano* **2014**, *8*, 4133–4156. [CrossRef]
7. Kandpal, H.; Anand, S.; Vaishya, J. Experimental observation of the phenomenon of spectral switching for a class of partially coherent light. *IEEE J. Quantum Electron.* **2002**, *38*, 336–339. [CrossRef]
8. Lee, C.; Wei, X.; Kysar, J.W.; Hone, J. Measurement of the elastic properties and intrinsic strength of monolayer graphene. *Science* **2008**, *321*, 385–388. [CrossRef]
9. Yang, J.; Tang, L.; Luo, W.; Feng, S.; Leng, C.; Shi, H.; Wei, X. Interface engineering of a silicon/graphene heterojunction photodetector via a diamond-like carbon interlayer. *ACS Appl. Mater. Interfaces* **2021**, *13*, 4692–4702. [CrossRef]
10. Xia, F.; Wang, H.; Jia, Y. Rediscovering black phosphorus as an anisotropic layered material for optoelectronics and electronics. *Nat. Commun.* **2014**, *5*, 4458. [CrossRef]
11. Haberer, D.; Vyalikh, D.; Taioli, S.; Dora, B.; Farjam, M.; Fink, J.; Marchenko, D.; Pichler, T.; Ziegler, K.; Simonucci, S. Tunable band gap in hydrogenated quasi-free-standing graphene. *Nano Lett.* **2010**, *10*, 3360–3366. [CrossRef] [PubMed]
12. Kaplan, D.; Swaminathan, V.; Recine, G.; Balu, R.; Karna, S. Bandgap tuning of mono-and bilayer graphene doped with group IV elements. *J. Appl. Phys.* **2013**, *113*, 183701. [CrossRef]
13. Geim, A.K.; Grigorieva, I.V. Van der Waals heterostructures. *Nature* **2013**, *499*, 419–425. [CrossRef] [PubMed]
14. Shin, Y.; Kwon, J.; Jeong, Y.; Watanabe, K.; Taniguchi, T.; Im, S.; Lee, G.H. Graphene Via Contact Architecture for Vertical Integration of vdW Heterostructure Devices. *Small* **2022**, *18*, 2200882. [CrossRef]
15. Lan, J.-C.; Qiao, J.; Sung, W.-H.; Chen, C.-H.; Jhang, R.-H.; Lin, S.-H.; Ng, L.-R.; Liang, G.; Wu, M.-Y.; Tu, L.-W. Role of carrier-transfer in the optical nonlinearity of graphene/Bi$_2$Te$_3$ heterojunctions. *Nanoscale* **2020**, *12*, 16956–16966. [CrossRef]
16. Hu, J.; Li, L.; Wang, R.; Chen, H.; Xu, Y.; Zang, Y.; Li, Z.; Feng, S.; Lei, Q.; Xia, C. Fabrication and photoelectric properties of a graphene-silicon nanowire heterojunction on a flexible polytetrafluoroethylene substrate. *Mater. Lett.* **2020**, *281*, 128599. [CrossRef]
17. Ren, X.; Wang, B.; Huang, Z.; Qiao, H.; Duan, C.; Zhou, Y.; Zhong, J.; Wang, Z.; Qi, X. Flexible self-powered photoelectrochemical-type photodetector based on 2D WS$_2$-graphene heterojunction. *FlatChem* **2021**, *25*, 100215. [CrossRef]
18. Liu, Y.; Weinert, M.; Li, L. Spiral growth without dislocations: Molecular beam epitaxy of the topological insulator Bi$_2$Se$_3$ on epitaxial graphene/SiC (0001). *Phys. Rev. Lett.* **2012**, *108*, 115501. [CrossRef]
19. Vu, Q.A.; Shin, Y.S.; Kim, Y.R.; Nguyen, V.L.; Kang, W.T.; Kim, H.; Luong, D.H.; Lee, I.M.; Lee, K.; Ko, D.-S. Two-terminal floating-gate memory with van der Waals heterostructures for ultrahigh on/off ratio. *Nat. Commun.* **2016**, *7*, 12725. [CrossRef]
20. Liu, X.; Gao, J.; Zhang, G.; Zhang, Y.-W. Design of phosphorene/graphene heterojunctions for high and tunable interfacial thermal conductance. *Nanoscale* **2018**, *10*, 19854–19862. [CrossRef]
21. Gao, Y.; Liu, Q.; Xu, B. Lattice mismatch dominant yet mechanically tunable thermal conductivity in bilayer heterostructures. *ACS Nano* **2016**, *10*, 5431–5439. [CrossRef]
22. Sun, X.; Li, X.; Zeng, Y.; Meng, L. Improving the stability of perovskite by covering graphene on FAPbI3 surface. *Int. J. Energy Res.* **2021**, *45*, 10808–10820. [CrossRef]
23. Wang, X.; Long, R. Rapid charge separation boosts solar hydrogen generation at the graphene–MoS2 Junction: Time-domain Ab initio analysis. *J. Phys. Chem. Lett.* **2021**, *12*, 2763–2769. [CrossRef] [PubMed]
24. Wang, H.; Gao, S.; Zhang, F.; Meng, F.; Guo, Z.; Cao, R.; Zeng, Y.; Zhao, J.; Chen, S.; Hu, H. Repression of interlayer recombination by graphene generates a sensitive nanostructured 2D vdW heterostructure based photodetector. *Adv. Sci.* **2021**, *8*, 2100503. [CrossRef] [PubMed]
25. Cong, C.; Shang, J.; Wang, Y.; Yu, T. Optical properties of 2D semiconductor WS$_2$. *Adv. Opt. Mater.* **2018**, *6*, 1700767. [CrossRef]
26. Yue, Y.; Chen, J.; Zhang, Y.; Ding, S.; Zhao, F.; Wang, Y.; Zhang, D.; Li, R.; Dong, H.; Hu, W. Two-dimensional high-quality monolayered triangular WS$_2$ flakes for field-effect transistors. *ACS Appl. Mater. Interfaces* **2018**, *10*, 22435–22444. [CrossRef] [PubMed]
27. Iqbal, M.W.; Iqbal, M.Z.; Khan, M.F.; Kamran, M.A.; Majid, A.; Alharbi, T.; Eom, J. Tailoring the electrical and photo-electrical properties of a WS$_2$ field effect transistor by selective n-type chemical doping. *RSC Adv.* **2016**, *6*, 24675–24682. [CrossRef]
28. Brandbyge, M.; Mozos, J.-L.; Ordejón, P.; Taylor, J.; Stokbro, K. Density-functional method for nonequilibrium electron transport. *Phys. Rev. B* **2002**, *65*, 165401. [CrossRef]
29. Jiang, J.-W. Graphene versus MoS2: A short review. *Front. Phys.* **2015**, *10*, 287–302. [CrossRef]
30. Georgiou, T.; Yang, H.; Jalil, R.; Chapman, J.; Novoselov, K.S.; Mishchenko, A. Electrical and optical characterization of atomically thin WS$_2$. *Dalton Trans.* **2014**, *43*, 10388–10391. [CrossRef]

31. Ding, Y.; Wang, Y.; Ni, J.; Shi, L.; Shi, S.; Tang, W. First principles study of structural, vibrational and electronic properties of graphene-like MX2 (M=Mo, Nb, W, Ta; X=S, Se, Te) monolayers. *Phys. B Phys. Condens. Matter* **2011**, *406*, 2254–2260. [CrossRef]
32. Heyd, J.; Scuseria, G.E.; Ernzerhof, M. Hybrid functionals based on a screened Coulomb potential. *J. Chem. Phys.* **2003**, *118*, 8207, Erratum in *J. Chem. Phys.* **2006**, *124*, 219906. [CrossRef]
33. Li, W.; Wei, J.; Bian, B.; Liao, B.; Wang, G. The effect of different covalent bond connections and doping on transport properties of planar graphene/MoS2/graphene heterojunctions. *Phys. Chem. Chem. Phys.* **2021**, *23*, 6871–6879. [CrossRef] [PubMed]
34. Zhou, Y.; Yang, Y.; Guo, Y.; Wang, Q.; Yan, X. Influence of length and interface structure on electron transport properties of graphene-MoS_2 in-plane heterojunction. *Appl. Surf. Sci.* **2019**, *497*, 143764. [CrossRef]
35. Dong, J.C.; Li, H. Monoatomic Layer Electronics Constructed by Graphene and Boron Nitride Nanoribbons. *J. Phys. Chem. C* **2012**, *116*, 17259–17267. [CrossRef]
36. Kim, H.; Kim, J.; Uddin, I.; Phan, N.A.N.; Whang, D.; Kim, G.-H. Dual-Channel WS_2/WSe_2 Heterostructure with Tunable Graphene Electrodes. *ACS Appl. Electron. Mater.* **2023**, *5*, 913–919. [CrossRef]
37. Zhou, Y.; Dong, J.C.; Li, H. Electronic transport properties of in-plane heterostructures constructed by MoS_2 and WS_2 nanoribbons. *RSC Adv.* **2015**, *5*, 66852–66860. [CrossRef]
38. Zheng, J.; Li, E.; Ma, D.; Cui, Z.; Wang, X. Effect on Schottky Barrier of Graphene/WS_2 Heterostructure with Vertical Electric Field and Biaxial Strain. *Phys. Status Solidi (b)* **2019**, *256*, 1900161. [CrossRef]
39. Georgiou, T.; Jalil, R.; Belle, B.D.; Britnell, L.; Gorbachev, R.V.; Morozov, S.V.; Kim, Y.J.; Gholinia, A.; Haigh, S.J.; Makarovsky, O. Vertical field-effect transistor based on graphene-WS2 heterostructures for flexible and transparent electronics. *Nat. Nanotechnol.* **2013**, *8*, 100–103. [CrossRef]
40. Bai, Z.; Xiao, Y.; Luo, Q.; Li, M.; Peng, G.; Zhu, Z.; Luo, F.; Zhu, M.; Qin, S.; Novoselov, K. Highly Tunable Carrier Tunneling in Vertical Graphene-WS_2-Graphene van der Waals Heterostructures. *ACS Nano* **2022**, *16*, 7880–7889. [CrossRef]
41. Xia, C.; Xiong, W.; Xiao, W.; Du, J.; Jia, Y. Enhanced Carrier Concentration and Electronic Transport by Inserting Graphene into van der Waals Heterostructures of Transition-Metal Dichalcogenides. *Phys. Rev. Appl.* **2018**, *10*, 024028. [CrossRef]
42. Perdew, J.P.; Burke, K.; Wang, Y. Generalized gradient approximation for the exchange-correlation hole of a many-electron system. *Phys. Rev. B* **1996**, *54*, 16533. [CrossRef]
43. Perdew, J.P.; Burke, K.; Ernzerhof, M. Generalized gradient approximation made simple. *Phys. Rev. Lett.* **1996**, *77*, 3865. [CrossRef] [PubMed]
44. Monkhorst, H.J.; Pack, J.D. Special points for Brillouin-zone integrations. *Phys. Rev. B* **1976**, *13*, 5188. [CrossRef]
45. Pastawski, H.M. Classical and quantum transport from generalized Landauer-Büttiker equations. *Phys. Rev. B* **1991**, *44*, 6329. [CrossRef] [PubMed]

Disclaimer/Publisher's Note: The statements, opinions and data contained in all publications are solely those of the individual author(s) and contributor(s) and not of MDPI and/or the editor(s). MDPI and/or the editor(s) disclaim responsibility for any injury to people or property resulting from any ideas, methods, instructions or products referred to in the content.

Article

Evaluation of the Toxicity Potential of the Metabolites of Di-Isononyl Phthalate and of Their Interactions with Members of Family 1 of Sulfotransferases—A Computational Study

Silvana Ceauranu [1,2], Alecu Ciorsac [3], Vasile Ostafe [1,2] and Adriana Isvoran [1,2,*]

1. Department of Biology Chemistry, West University of Timisoara, 16 Pestalozzi, 300115 Timisoara, Romania; silvana.ceauranu96@e-uvt.ro (S.C.); vasile.ostafe@e-uvt.ro (V.O.)
2. Advanced Environmental Research Laboratories, West University of Timisoara, 4 Oituz, 300086 Timisoara, Romania
3. Department of Physical Education and Sport, University Politehnica Timisoara, 2. Piata Victoriei, 300006 Timisoara, Romania; alecu.ciorsac@upt.ro
* Correspondence: adriana.isvoran@e-uvt.ro

Citation: Ceauranu, S.; Ciorsac, A.; Ostafe, V.; Isvoran, A. Evaluation of the Toxicity Potential of the Metabolites of Di-Isononyl Phthalate and of Their Interactions with Members of Family 1 of Sulfotransferases—A Computational Study. *Molecules* 2023, 28, 6748. https://doi.org/10.3390/molecules28186748

Academic Editors: Shiling Yuan and Heng Zhang

Received: 30 August 2023
Revised: 19 September 2023
Accepted: 20 September 2023
Published: 21 September 2023

Copyright: © 2023 by the authors. Licensee MDPI, Basel, Switzerland. This article is an open access article distributed under the terms and conditions of the Creative Commons Attribution (CC BY) license (https://creativecommons.org/licenses/by/4.0/).

Abstract: Di-isononyl phthalates are chemicals that are widely used as plasticizers. Humans are extensively exposed to these compounds by dietary intake, through inhalation and skin absorption. Sulfotransferases (SULTs) are enzymes responsible for the detoxification and elimination of numerous endogenous and exogenous molecules from the body. Consequently, SULTs are involved in regulating the biological activity of various hormones and neurotransmitters. The present study considers a computational approach to predict the toxicological potential of the metabolites of di-isononyl phthalate. Furthermore, molecular docking was considered to evaluate the inhibitory potential of these metabolites against the members of family 1 of SULTs. The metabolites of di-isononyl phthalate reveal a potency to cause liver damage and to inhibit receptors activated by peroxisome proliferators. These metabolites are also usually able to inhibit the activity of the members of family 1 of SULTs, except for SULT1A3 and SULT1B1. The outcomes of this study are important for an enhanced understanding of the risk of human exposure to di-isononyl phthalates.

Keywords: metabolism; molecular docking; di-isononyl phthalate monoesters; toxicological effects

1. Introduction

Xenobiotics are chemical compounds that do not occur naturally in the human body and, if not metabolized, they could reach toxic concentrations. Humans are exposed daily to numerous xenobiotics: drugs, food additives, cosmetic ingredients, pesticides, environmental pollutants, etc. The metabolism of xenobiotics usually occurs through enzymatic reactions, and the enzymes involved are classified into phase I, phase II, and transporter enzymes. Phase I enzymes (such as cytochromes P450) convert lipophilic xenobiotics into more polar compounds for easier excretion and provide sites for conjugation reactions. Phase II enzymes (such as glutathione transferases and sulfotransferases) carry out the conjugation reactions and can interact directly with xenobiotics, but most of the time they interact with the metabolites obtained in the phase I metabolism [1].

This study focuses on family 1 of sulfotransferases (SULT1), containing enzymes that are known to catalyze the sulfate conjugation of pharmacologically important endobiotics and xenobiotics by transferring a sulfate group from the donor 3′-phosphoadenosine 5′-phosphosulfate (PAPS) to the hydroxyl group of an acceptor [2]. In humans, there are four subfamilies of SULT1: SULT1A (SULT1A1, SULT1A2, SULT1A3), SULT1B1, SULT1C (SULT1C1/C2, SULT1C3, SULT1C4) and SULT1E1 [3]. SULT1A1 and SULT1A2 have specificity for small phenolic compounds and SULT1A3 preferably sulfonates monoamines. SULT1B1 and SULT1C1 are involved in the sulfonation of thyroid hormones, whereas SULT1E1 sulfonates steroids, preferentially estrogens.

Several polymorphisms have been identified for the SULT1 family of enzymes that proved to alter enzyme activities and presented differences in the metabolism of endogenous compounds, drugs and other xenobiotics [4,5]. The frequently identified SULT1A1 polymorphic variants are SULT1A1*1 (the wild type, WT), SULT1A1*2 (amino acids substitution R213H) and SULT1A1*3 (amino acids substitution M223V) [4]. Significant differences in the sulfation activities of SULT1A1*1, SULT1A1*2 and SULT1A1*3 have been identified, with SULT1A1*2 revealing lower catalytic activity and thermostability [5].

Among other xenobiotics, humans are exposed to phthalates because these compounds are widely used as solvents, additives and/or plasticizers in numerous consumer products. People can be exposed to phthalates by ingesting them from food or drink, by inhaling phthalate-contaminated air and dust resulting from paintings or furniture, by using pharmaceuticals and cosmetics, and by skin contact with products and/or clothing that contain them [6,7]. There are specific studies confirming the contamination of humans with phthalates as they reveal the presence of phthalate metabolites in human urine, saliva, breast milk and some serum samples [8–10]. Frequently observed harmful effects of phthalates on humans concern the reproductive, neurological and developmental systems [11]. A computational study conducted on 25 of the most used phthalates emphasized that they have the potential to interact with numerous molecular targets in the human organism (cytochromes, transporters, transcription factors, membrane receptors, kinases, phosphatases) [12]. These interactions may lead to harmful effects: skin and eye irritation, endocrine disruption, non-genotoxic carcinogenicity and toxicity of the respiratory and gastrointestinal tracts [12].

In humans, the metabolism of phthalates occurs in at least two steps: (i) a phase I hydrolysis when the diester phthalate is hydrolyzed into the monoester phthalate and (ii) a phase II conjugation [13]. Some in vitro and in vivo animal studies exposed that the monoester phthalates resulting from the metabolism of the diester phthalates are often more bioactive compounds [14]. Literature data reveal that benzylbutyl phthalate in high concentrations inhibits SULT1A1, whereas dibutyl phthalate and benzylbutyl phthalate inhibit SULT1E1 [15]. Furthermore, numerous phthalate metabolites revealed strong inhibition potential towards members of SULT1, especially SULT1A1, SULT1B1 and SULT1E1 [16]. SULT1A1 activity was strongly inhibited by mono-hexyl phthalate, monobenzyl phthalate, mono-octyl phthalate and mono-ethylhexyl phthalate. Mono hexyl phthalate, mono-ethylhexyl phthalate, mono-octyl phthalate, mono-butyl phthalate and mono-cyclohexyl phthalate inhibited the activity of SULT1B1. Mono-hexyl phthalate, mono-ethylhexyl phthalate and mono-octyl phthalate inhibited the activity of SULT1E1 [16].

Di-isononyl phthalates (DiNP) are a group of isomeric forms of dinonyl phthalates widely used as plasticizers and found in various products such as cables, foil, PVC flooring, toys, shoe materials, tablecloths, etc. [17]. It was already assessed that upon ingestion, at least 50% of DiNP is expected to be absorbed through the gastrointestinal tract, with absorption decreasing as the dose increases (https://echa.europa.eu/documents/10162/31b4067e-de40-4044-93e8-9c9ff1960715, accessed on 15 June 2023). Investigations conducted on rats and on human adult volunteers have shown that DiNP has been metabolized to a number of secondary metabolites before being excreted in the urine: monoisononyl phthalate (MiNP), mono-hydroxy-isononyl phthalate (MHiNP), mono-carboxy-isooctyl phthalate (MCiOP) and mono-oxo-isononyl phthalate (MOiNP) [18–20].

This study aims to predict the possible toxicological effects on humans of the metabolites of DiNP and to assess the interactions of these xenobiotics with members of the SULT1 family, also taking into account the most frequently identified polymorphic variants of SULT1A1 enzymes. To obtain this information, a computational study was implemented.

Sometimes the reliability of the results obtained by using calculation methods is called into question. Considering both the cost of laboratory studies and the ethical concerns of using animals for testing, the role of computational studies is becoming well recognized. There has been an explosive growth in both the scale and diversity of data in the natural sciences, leading to the creation of freely accessible databases and enabling scientists to build accurate computational models for toxicology assessments. Consequently, the Organization for Economic Development and Cooperation (OECD) created guidelines for quantitative structure activity relationship (QSAR) models in 2004. The principles for building QSAR models and calculation methods have been described in detail since 2007 [21]. The Registration, Evaluation and Authorization of Chemicals (REACH) regulation has recognized QSAR techniques for studying the toxicological profile of chemicals [22]. Last but not least, molecular modeling methods offer the possibility to assess the risk to human health due to exposure to different chemicals as they can simulate possible modes of action. All these virtual screening tools have been useful for prioritizing bioassay requirements, saving resources and limiting the use of laboratory animals for testing. In this particular study, the computational approach was considered for a better understanding of the risk of human exposure to phthalates.

2. Results

2.1. Physicochemical and Structural Properties of the Metabolites of Di-Isononyl Phthalate and of the Ligands That Are Present in the Crystallographic Structures of SULT1 Enzymes

The physicochemical and structural properties of the DiNP metabolites and of the ligands that are present in the crystallographic structures of SULT1 enzymes are presented in Table 1.

Table 1. Physicochemical and structural properties of the metabolites of di-isonyl phthalate and of the ligands that are present in the crystallographic structures of SULT1 enzymes: MW—molecular weight, logP—partition coefficient, HBD—hydrogen bonds donors, HBA—hydrogen bonds acceptors, RB—rotatable bonds, tPSA—topological polar surface area.

Compound	MW (g/mol)	LogP	HBD	HBA	RB	tPSA ($Å^2$)	Area ($Å^2$)	Volume ($Å^3$)
mono-carboxy-isooctyl phthalate	322.40	3.4	2	6	10	101.0	285.1	287.3
mono-hydroxy-isononyl phthalate	308.40	3.9	2	5	10	83.8	293.5	284.6
mono-isononyl phthalate	292.40	5.6	1	4	10	63.6	288.8	268.6
mono-oxo-isononyl phthalate	306.40	2.8	1	5	10	80.7	278.8	288.0
3,3′,5,5′-tetrachloro-4,4′-biphenyldiol	324.00	5.4	2	2	1	40.5	265.5	252.3
p-nitrophenol	139.11	1.9	1	3	0	66.0	128.6	106.1
dopamine	153.18	−1.0	3	3	2	66.5	156.8	132.4
resveratrol	228.24	3.1	3	3	2	60.7	217.7	191.0

Figure 1 reveals the molecular lipophilicity potential (MLP) on the molecular surface for the DiNP metabolites and for the ligands that are present in the crystallographic structures of SULT1 enzymes. In this figure, the hydrophobic parts of the surface are colored by violet and blue and the hydrophilic ones by orange and red.

Data presented in Table 1 and Figure 1 reveal that DiNP metabolites have physicochemical and structural properties that better resemble those of the 3,3′,5,5′-tetrachloro-4,4′-biphenyldiol, the SULT1E1 inhibitor. DiNP metabolites poses a higher number of rotatable bonds and consequently reveals a higher flexibility compared to that of 3,3′,5,5′-tetrachloro-4,4′-biphenyldiol. DiNP metabolites are bigger than dopamine, resveratrol and p-nitrophenol, which are the ligands that are present in the structural complexes of some of the SULT1 enzymes. It is important to notice that there are two molecules of p-nitrophenol in the structural file of the complex of SULT1A1*2 with this substrate.

2.2. Toxicological Effects of the Di-Isononyl Phthalate and of Its Metabolites

Considering the fact that DiNP metabolites were observed in urine from human subjects, predictions about their toxicity on humans have been also obtained by using the ADMETLab2.0 computational tool [23,24]. This tool uses quantitative structure activity relationship regression or classification models for predicting the absorption, distribution, metabolism, excretion and toxicity (ADMET) profile of chemicals. For the ADMET properties that are predicted by the regression models, concrete predictive values are delivered, whereas for ADMET properties predicted by the classification models, the prediction probability values are provided. A positive value for the probability reveals that the studied molecule is more likely to produce the biological activity under investigation. Predicted probabilities with values higher than 0.700 are considered as reliable predictions [24].

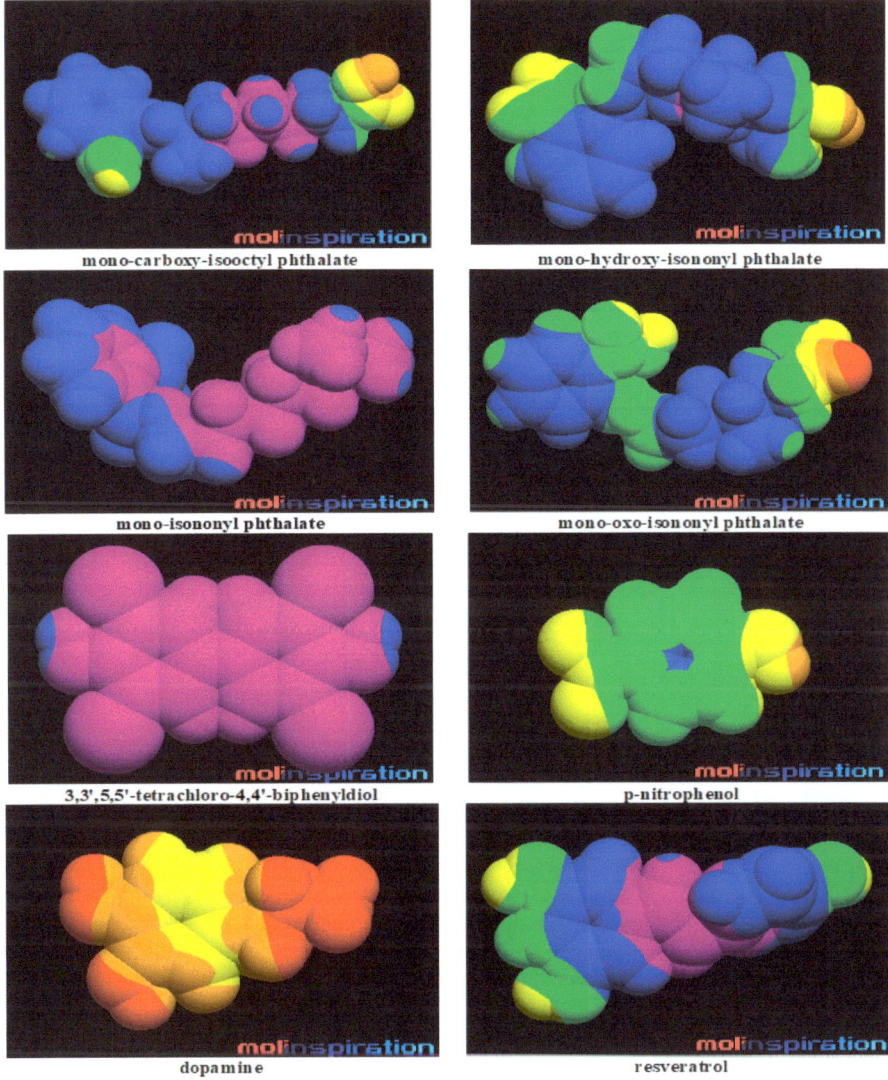

Figure 1. Molecular lipophilicity potential on the molecular surface for the metabolites of di-isononyl phthalate and for the ligands that are present in the crystallographic structures of SULT1 enzymes. The hydrophobic parts of the surface are colored by violet and blue and the hydrophilic ones by orange and red.

The predicted toxicological effects of DiNP and of its metabolites on the human organism are illustrated in Table 2. In this table, there are delivered predicted concrete values for clearance (CL) and plasma protein binding (PPB) using regression models. Based on classification models, predicted probabilities are also computed for the ability of DiNP metabolites to penetrate the blood–brain barrier (BBB), to be inhibitors or substrates of P-glycoprotein (PgpI/PgpS), to produce cardiotoxicity by inhibiting the h-ERG channel (hERG), to produce drug liver injuries (DILI), to produce mutagenicity (Ames mutagenicity) and carcinogenicity, respectively, and to affect the nuclear receptors (NR). In the case of nuclear receptors, the probabilities that DiNP metabolites bind to the active sites of the androgen receptor (NR-AR-LBD), the aryl hydrocarbon receptor (NR-AhR), the estrogen receptor (NR-ER-LBD), the receptors activated by peroxisome proliferators (NR-PPAR-gamma) and, respectively, the probability to produce endocrine disruption (NR-aromatase) are predicted.

Table 2. Probabilities regarding the possible toxicological effects of DiNP metabolites: CL—clearance, BBB—blood–brain barrier penetration, PPB—plasma protein binding, PgpI—inhibitor of P-glycoprotein, PgpS—substrate of P-glycoproteine, hERG—human cardiotoxicity, DILI—induced liver damage, NR-AR-LBD—binding domain of androgen receptor ligands, NR-AhR—aryl hydrocarbon receptor, NR-aromatase—endocrine disrupting chemicals, NR-ER-LBD—estrogen receptor ligand binding domain, NR-PPAR-gamma—activated receptors of peroxisome proliferators. Cells colored in orange reveal mean toxicological effects.

Compound/Toxicity	Di-Isononyl Phthalate	Mono-Carboxy-Isooctyl Phthalate	Mono-Hydroxy-Isononyl Phthalate	Mono-Isononyl Phthalate	Mono-Oxo-Isononyl phthalate
CL (mL/min)	8.329	1.369	7.210	3.032	5.149
BBB	0.011	0.109	0.379	0.106	0.336
PPB (%)	97.84	95.97	90.69	97.34	92.87
PgpI	0.974	0.004	0.003	0.006	0.026
PgpS	0.000	0.001	0.026	0.000	0.004
hERG	0.190	0.053	0.209	0.260	0.198
DILI	0.358	0.795	0.743	0.852	0.770
Ames mutagenicity	0.002	0.007	0.004	0.004	0.005
Carcinogenicity	0.302	0.019	0.019	0.300	0.017
NR-AR-LBD	0.002	0.060	0.007	0.003	0.007
NR-AhR	0.071	0.012	0.015	0.057	0.042
NR-aromatase	0.086	0.005	0.009	0.011	0.008
NR-ER-LBD	0.230	0.064	0.036	0.010	0.012
NR-PPAR-gamma	0.060	0.744	0.890	0.872	0.785

The following can be observed from Table 2: (i) MiNP and MCiOP show a low elimination rate from human organisms as the computed values for clearance are low, 3.032 mL/min for MiNP and 1.369 mL/min for MCiOP; (ii) DiNp may be an inhibitor of P-glycoprotein as the computed probability for this effect is very high, 0.974; (iii) DiNP and its metabolites may strongly bind to plasma proteins, with all predicted values for binding being higher than 90%; (iv) all DiNP metabolites reveal the capacity to cause liver damage (the computed probabilities for producing this effect being higher than 0.700), with MiNP enlightening the higher probability to produce liver injuries; (v) all metabolites of DiNP reveal the potential to inhibit the receptors activated by peroxisome proliferators (computed probabilities to produce these effects are higher than 0.700); (vi) none of the listed metabolites show cardiotoxicity, mutagenicity, carcinogenicity, do not inhibit androgenic, estrogenic, endocrine and aryl hydrocarbon receptors, with the predicted probabilities corresponding to these toxicological effects being low.

2.3. Evaluation of the Interactions of DiNP Metabolites with the Family 1 of Human Sulfotransferases

The outcomes of the molecular docking are revealed in Figures 2–7 and in Table 3. MCiOP and MiNP are able to bind to the active site of SULT1A1*1 (Figure 2), whereas MHiNP and MOiNP are not able to bind to the active site of this enzyme (Supplementary Material, Figure S1).

Table 3. Binding energies of DiNP metabolites to the active sites of SULT1 enzymes.

Enzyme	ΔG (kcal/mol)			
	Mono-Carboxy-Isooctyl Phthalate	Mono-Hydroxy-Isononyl Phthalate	Mono-Isononyl Phthalate	Mono-Oxo-Isononyl Phthalate
SULT1A1*1	−8.06	-	−6.79	-
SULT1A1*2	-	-	−7.51	−8.30
SULT1A1*3	−8.36	−11.36	−8.46	−8.29
SULT1A2	−7.3	-	−7.75	−6.98
SULT1C1	−8.04	−9.99	−7.60	−7.76
SULT1E1	−7.95	−10.91	−7.54	−7.83

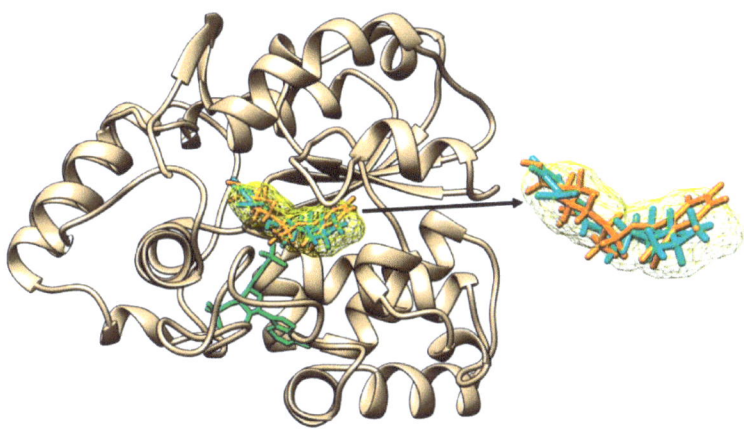

Figure 2. Binding of mono-carboxy-isooctyl phthalate (orange sticks) and of mono-isononyl phthalate (cyan sticks) to the active site of SULT1A1*1. The binding poses correspond to the binding position of p-nitrophenol (yellow mesh surface) to SULT1A1*2 and superimposed on the structure of SULT1A1*1. The inactive cofactor adenosine-3′-5′-diphosphate that is present in the structural file of SULT1A1*1 is revealed in green sticks.

In the case of the allelic variant SULT1A1*2, MiNP and MOiNP are the DiNP metabolites that are able to bind to the active site of the enzyme (Figure 3).

As a control, the molecular docking of p-nitrophenol (the substrate that is present in the crystallographic structure) to SULT1A1*2 emphasizes that the binding mode of p-nitrophenol obtained through molecular docking strongly corresponds to the position of one of the p-nitrophenol molecules in the crystallographic structure, the binding energy being −6.05 kcal/mol (Supplementary Material, Figure S2). The binding modes of the other two metabolites, MCiOP and MHiNP, that do not correspond to the active site of SULT1A1*2 are revealed in Supplementary Material, Figure S3.

All the DiNP metabolites are able to bind to the active site of SULT1A1*3 (Figure 4).

In the case of SULT1A2, MCiOP, MiNP and MOiNP are able to bind to the active site of SULT1A2 (Figure 5). MHiNP does not bind to the active site of this enzyme (Supplementary Material, Figure S4). MHiNP has the higher topological polar surface area compared to the other DiNP metabolites; this property may be responsible for the fact that it does not bind to the active site of SULT1A2.

None of the DiNP metabolites bind to SULT1A3 and SULT1B1 active sites (Supplementary Material, Figures S5 and S6).

In order to assess the binding of DiNP metabolites to SULT1C1, first the molecular docking of iodothyronine (the typical substrate of SULT1C1) to the enzyme has been completed. The binding to the active site of SULT1C1 has been emphasized by superposition of the binding modes of iodothyronine to SULT1C1 with the structures of SULT1A1*2 and SULT1B1 (Supplementary Figure S7). Further, the binding modes of DiNP metabolites to SULT1C1 were compared to the position of iodothyronine.

All the DiNP metabolites are able to bind to the active site of SULT1C1 (Figure 6) and the binding poses correspond to the position of iodothyronine.

All the DiNP metabolites are also able to bind to the active site of SULT1E1 (Figure 7), with the binding poses matching the binding position of the inhibitor 3,3′,5,5′-tetrachloro-4,4′-biphenyldiol.

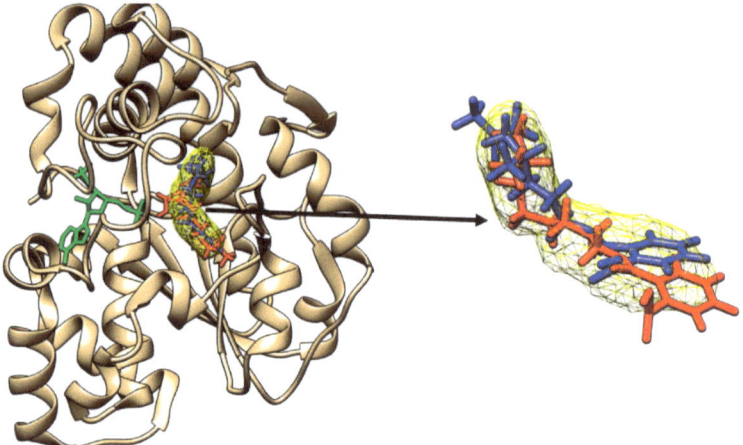

Figure 3. Binding of mono-isononyl phthalate (red sticks) and of mono-oxo-isononyl phthalate (blue sticks) to the active site of SULT1A1*2. The binding poses correspond to the binding position of p-nitrophenol (yellow mesh surface). The inactive cofactor adenosine-3′-5′-diphosphate is revealed in green sticks.

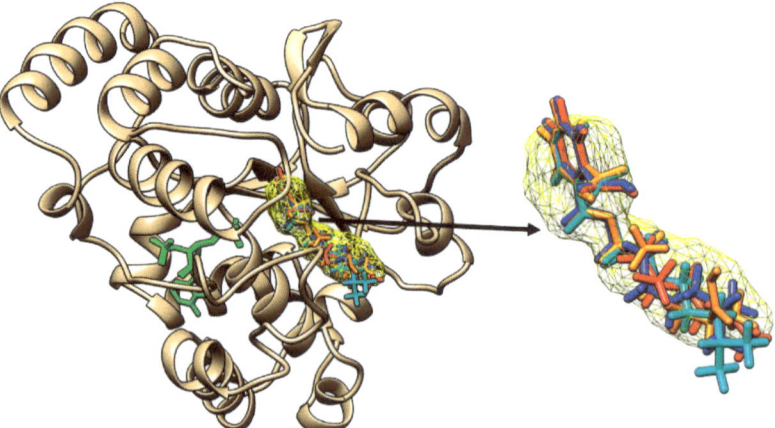

Figure 4. Binding of mono-carboxy-isononyl phthalate (orange sticks), of mono-hidroxy-isononyl phthalate (cyan sticks), of mono-isononyl phthalate (blue sticks) and of mono-oxo-isononyl phthalate (red sticks) to the active site of SULT1A1*3. All the binding poses correspond to the binding position of p-nitrophenol (yellow mesh surface). The inactive cofactor adenosine-3′-5′-diphosphate is revealed in green sticks.

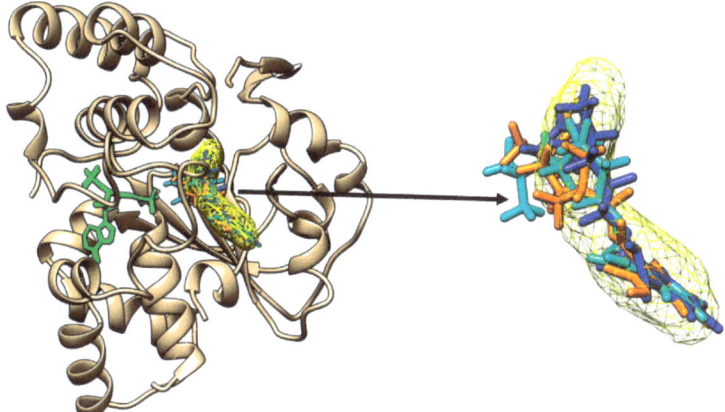

Figure 5. Binding of mono-carboxy-isononyl phthalate (orange sticks), of mono-isononyl phthalate (cyan sticks) and of mono-oxo-isononyl phthalate (blue sticks) to the active site of SULT1A2. All the binding poses correspond to the binding position of p-nitrophenol (yellow mesh surface) to SULT1A1*2. The inactive cofactor adenosine-3′-5′-diphosphate is revealed in green sticks.

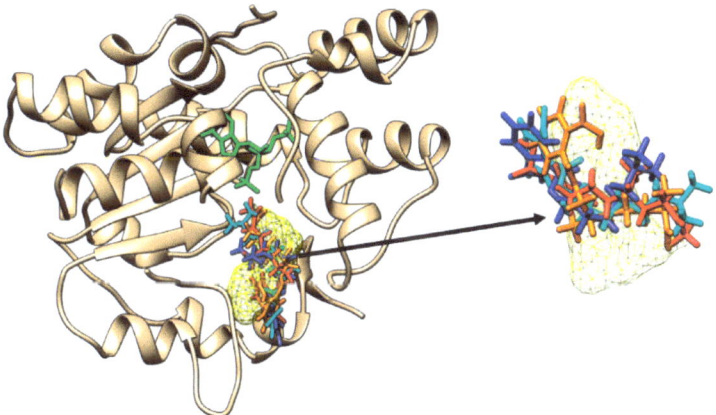

Figure 6. Binding of mono-carboxy-isononyl phthalate (orange sticks), of mono-hidroxy-isononyl phthalate (cyan sticks), of mono-isononyl phthalate (blue sticks) and of mono-oxo-isononyl phthalate (red sticks) to the active site of SULT1C1. All the binding poses correspond to the binding position of iodothyronine (yellow mesh surface). The inactive cofactor adenosine-3′-5′-diphosphate is revealed in green sticks.

The energies resulting from the molecular docking study and corresponding to the binding modes matching the active sites of SULT1 enzymes are presented in Table 3.

Data presented in Table 3 emphasize the inhibition of SULT1A1*1, SULT1A1*2, SULT1A1*3, SULT1A2, SULT1C1 and SULT1E1 by the DiNP metabolites. Furthermore, MiNP is the metabolite that is able to inhibit almost all SULT1 enzymes, except SULT1A3 and SULT1B1, with the highest binding energy corresponding to SULT1A1*3.

The results obtained using PLIP 2021 software regarding the identification of the residues of the SULT1 enzymes that are involved in the noncovalent interactions with DiNP metabolites are presented in Table 4.

Table 4. Amino acids belonging to SULT1 enzymes and involved in the noncovalent interactions with their substrates/inhibitors and the DiNP metabolites: MCiOP—monocarboxy-izooctyl phthalate, MHiNP—monohydroxy-isononyl phthalate, MiNP—mono-isononyl phthalate, MOiNP—monooxo-isononyl phthalate, PNP—p-nitrophenol, TBD—3,5,3′,5′-tetrachloro-biphenyl-4,4′-diol.

Complex	Hydrophobic Contacts	Hydrogen Bonds	π Staking	π Cations	Salt Bridges
SULT1A1*1—MCiOP	PHE76, PHE81, PHE84, ILE89, PHE142	TYR139, TYR140	PHE84	-	LYS106, HIS149
SULT1A1*1—MiNP	ILE21, PRO47, PHE76, PHE84, PHE142, TYR169, TYR240, VAL243, PHE214, PHE255	-	-	-	LYS106
SULT1A1*2—PNP in the crystallographic structure	PHE76, PHE84, ILE89, VAL243, PHE247	-	PHE76	-	-
SULT1A1*2—PNP as a result of docking	PHE76, PHE84, ILE89, VAL243, PHE247	-	PHE76	-	-
SULT1A1*2—MiNP	PRO47, PHE76, PHE81, PHE84, ILE89, PHE142, TYR240,	LYS48	PHE142	-	LYS48, LYS106
SULT1A1*2—MOiNP	ILE21, PHE24, PRO47, PHE76, PHE81, PHE84, ILE89, TYR139, ALA146, VAL148, TYR169	TYR240	PHE84	-	LYS106
SULT1A1*3—MCiOP	ILE21, PHE24, PRO47, PHE76, PHE84, ILE89, PHE142, TYR169, TYR240, VAL243	-	-	-	LYS106
SULT1A1*3—MHiNP	PRO47, PHE76, PHE84, ILE89, PHE142, VAL243, PHE247	TYR240	-	-	LYS48, LYS106 (3)
SULT1A1*3—MiNP	ILE21, PHE76, PHE81, PHE84, ILE89, PHE142, TYR169, TYR240, VAL243	-	-	-	LYS106
SULT1A1*3—MOiNP	ILE21, PHE24, PHE76, PHE81, PHE84, ILE89, VAL148, TYR149	-	PHE84	-	LYS106
SULT1A2—MCiOP	PHE76, PHE81, PHE84, ILE89, THR95	TYR240 (2)	-	-	LYS106
SULT1A2—MiNP	LYS48, PHE76, PHE81, PHE84, ILE89, PHE142, PHE255	-	PHE84	-	LYS106
SULT1A2—MOiNP	ILE21, PHE24, PHE76, PHE81, PHE84, ILE89, VAL148, TYR149	-	PHE84	-	LYS106
SULT1C1—iodothyronine	PHE82, MET149, TRP 170	GLN22	-	-	LYS96
SULT1C1—MCiOP	TRP145, PHE143, LEU150	LYS49, ASN147	-	-	LYS49
SULT1C1—MHiNP	PHE143, MET149	PHE256	-	-	LYS49 (3)

Table 4. *Cont.*

Complex	Hydrophobic Contacts	Hydrogen Bonds	π Staking	π Cations	Salt Bridges
SULT1C1—MiNP	GLN22, ALA24, ARG87, PHE143, TRP170	THR25, TRP85, ALA86	-	-	-
SULT1C1—MOiNP	GLN22, PHE82, PHE143, TRP170	GLN22, TRP85	-	-	-
SULT1E1—TBD in crystallographic structure	TYR20, PHE25, VAL145	LYS105, HIS107	PHE141	HIS148	-
SULT1E1—TBD as a result of docking	TYR20, ASP22, PHE23, PHE141, VAL145, ALA146, TYR239, ILE246	ASP22, LYS105, HIS107	PHE80, PHE141	HIS148	
SULT1E1—MCiOP	PHE23, PRO46, LYS47, LYS85PHE141, ILE246	TYR239			LYS105
SULT1E1—MHiNP	TYR20, PRO46, LYS47, PHE80, PHE141, VAL145	LYS47, TYR239			LYS47, LYS105 (3)
SULT1E1—MiNP	PHE23, PRO46, LYS47, PHE141, VAL175, PHE254	-			LYS105
SULT1E1—MOiNP	TYR20, PRO46, LYS47, PHE80, PHE141, VAL145, ILE246	LYS47, THR50			LYS47, LYS105 (2)

Information presented in Table 4 confirms the good correlation between the complexes of SULT1 enzymes with their substrates resulted from docking and the complexes that are present in the crystallographic structures.

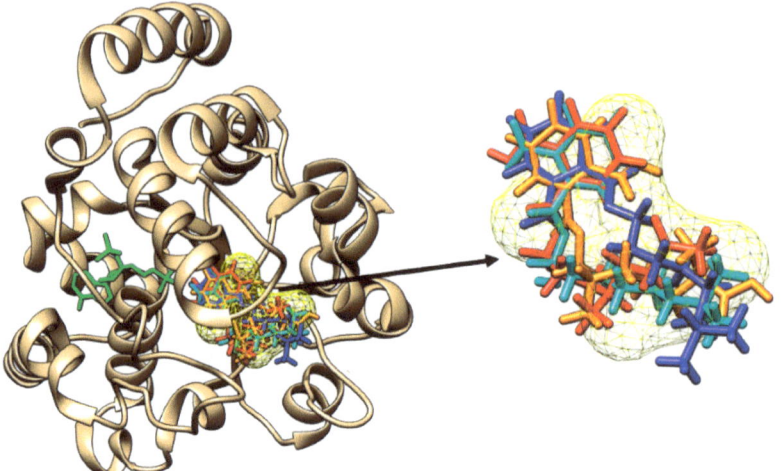

Figure 7. Binding of mono-carboxy-isononyl phthalate (orange sticks), of mono-hidroxy-isononyl phthalate (cyan sticks), of mono-isononyl phthalate (blue sticks) and of mono-oxo-isononyl phthalate (red sticks) to the active site of SULT1A1E1. All the binding poses correspond to the binding position of 3,3′,5,5′-tetrachloro-4,4′-biphenyldiol (yellow mesh surface). The inactive cofactor adenosine-3′-5′-diphosphate is revealed in green sticks.

3. Discussion

3.1. Toxicological Effects of the Di-Isononyl Phthalate and of Its Metabolites

To the best of our knowledge, this is the first study dealing with the possible toxicity of DiNP and its metabolites. Predictions obtained using the ADMETLab2.0 tool revealed that DiNp may be an inhibitor of P-glycoprotein, underlining that DiNP may potentially affect the absorption, distribution, metabolism and elimination of the substrates of this protein. The high degree of binding to plasma proteins of the DiNP and its metabolites may lead to long acting xenobiotics as their bound fractions are not available for metabolism or excretion.

The present study emphasizes that the DiNP metabolites may produce liver damage and this outcome is in good correlation with literature data revealing that several phthalate metabolites produced liver injuries [25]. The potential of the DiNP metabolites to inhibit the receptors activated by peroxisome proliferators predicted in this study is also in good agreement with published data revealing the interactions of several monoester phthalates (including MiNP) with receptors activated by peroxisome proliferators [26].

None of the DiNP metabolites are considered to produce cardiotoxicity, mutagenicity or carcinogenicity. These predictions correlate with no carcinogenic [27] and mutagenic [28] effects noticed for some other phthalate metabolites. This study reveals no effects produced by the DiNP metabolites against the androgenic, estrogenic, endocrine and aryl hydrocarbon receptors. This finding is also in good correlation with specific literature data revealing no estrogenic and androgenic activities of mono-(2-ethylhexyl) phthalate [29].

Several toxicological effects on humans have been predicted for numerous other xenobiotics [30–33] and all these results should be used to guide the decision making regarding their use if human contamination is expected.

3.2. Evaluation of the Interactions of DiNP Metabolites with the Family 1 of Human Sulfotransferases

A molecular docking study reveals that MCiOP and MiNP bind to the active site of SULT1A1*1, but MHiNP and MOiNP do not bind to the active site of this enzyme. This distinct behavior of MCiOP and MiNP that binds to SULT1A1*1 may be associated with their higher molecular lipophilicity potential (as presented in Figure 1); it is known that the distribution of lipophilicity in the molecular surface is important for the biological activity of a chemical compound. Literature data reveal that other monoester phthalates are strong inhibitors of SULT1A1*1: mono-hexyl phthalate, monobenzyl phthalate, mono-octyl phthalate and mono-ethylhexyl phthalate [16].

MiNP and MOiNP are the only DiNP metabolites that bind to the active site of the allelic variant SULT1A1*2. The binding energies for MiNP and MOiNP to the active site of SULT1A1*2 are higher (Table 3) than the binding energy of the substrate p-nitrophenol, underlining the strong inhibitory effects of these monoesters against SULT1A1*2.

The distinct interactions of the DiNP metabolites with SULT1A1*2 compared to SULT1A1*1 may be due to the decreased flexibility of the active site of SULT1A1*2 [34], with MCiOP and MhiNP being higher than the other two metabolites. This outcome is also in good correlation with literature data revealing that an R213H mutation corresponding to SULT1A1*2 induces a local conformational change and results in lower catalytic activity of the enzyme [35].

In the case of SULT1A1*3, all DiNP metabolites bind to the active site of the enzyme. It was already shown that an M223V mutation corresponding to the SULT1A1*3 allelic variant produced an increase in the local flexibility of the active site of this enzyme [35]. This increased flexibility may add to the ability of this variant to accommodate all the DiNP metabolites, when compared to the SULT1A1*1 and SULT1A1*2 variants. This outcome is also in good agreement with information resulting from a molecular dynamics study and emphasizing that an M223V mutation resulted in the loss of a hydrophobic contact between M223 and M60 and led to increased local flexibility and the ability to accommodate larger compounds [5].

DiNP metabolites do not bind to the active sites of SULT1A3 and SULT1B1, respectively. As substrates, SULT1A3 has monoamines and biogenic catecholamines [36] that are hydrophilic compounds. Consequently, it is not an unexpected result that the hydrophobic DiNP metabolites are not able to bind to the active site of SULT1A3. The output of the molecular docking study regarding the binding of dopamine (the known substrate of SULT1A3) to SULT1A3 revealed that the binding mode obtained by docking corresponds to the position of dopamine in the crystallographic structure (Supplementary Material Figure S8), with the binding energy being -9.41 kcal/mol and underlying the correct predictions obtained through docking.

In the case of SULT1B1, the crystal structure that was used for molecular docking contains the enzyme in the complex with A3P and the substrate resveratrol. Table 1 and Figure 1 reveal dissimilar molecular properties and molecular lipophilic potential for resveratrol, compared to those of the DiNP metabolites. Consequently, the physicochemical properties of DiNP metabolites do not fit those of the active site of the enzyme. Another molecular docking study revealed that several other monoester phthalates (mono-hexyl phthalate, mono-octyl phthalate, mono-ethylhexyl phthalate, mono-butyl phthalate, mono-cyclohexyl phthalate) were able to inhibit SULT1B1 [16]. Furthermore, it was shown that phthalate monoesters, having shorter chains, significantly inhibited SULT1B1 but not the other SULT1 enzymes [16]. The compounds analyzed in the study of Huang et al. (2022) [16] are usually smaller than the DiNP metabolites and their physicochemical properties are more similar to those of resveratrol (Supplementary Material, Table S1). The output of the molecular docking study regarding the binding of resveratrol (the known substrate of SULT1B1) to the SULT1B1 structure highlighted that the binding mode obtained by docking matches the position of resveratrol in the crystallographic structure (Supplementary Material Figure S9), with the binding energy being -7.16 kcal/mol. It underlines the correct predictions obtained through docking.

All DiNP metabolites bind to the active sites of SULT1C1 and of SULT1E1, respectively. The binding of DiNP metabolites to SULT1C1 can be explained by the resemblance of their properties with those of the known SULT1C1 substrates: thyroid hormones and phenolic drugs [37]. The thyroid hormones are hydrophobic compounds and have a higher molecular weight (and consequently dimensions) compared to DiNP metabolites. In the case of SULT1E1, the crystal structure that was used in the molecular docking study contains the inhibitor 3,5,3′,5′-tetrachloro-biphenyl-4,4′-diol, a molecule that has similar physicochemical and geometric properties to those of the DiNP metabolites (see Table 1 and Figure 1). Moreover, the output of the molecular docking study regarding the binding of the inhibitor 3,5,3′,5′-tetrachloro-biphenyl-4,4′-diol to the SULT1E1 structure emphasized that the binding mode obtained by docking matches the position of the 3,5,3′,5′-tetrachloro-biphenyl-4,4′-diol in the crystallographic structure (Supplementary Material Figure S10) and underlies the correct predictions obtained through docking. The binding energy of the inhibitor 3,5,3′,5′-tetrachloro-biphenyl-4,4′-diol to SULT1E1 is −6.91 kcal/mol, lower than the binding energies of the DiNP metabolites to the active site of this enzyme (Table 3). It sustains the inhibitory potential of DiNP metabolites against SULT1E1. Another study revealed that mono-hexyl phthalate, mono-octyl phthalate and mono-ethylhexyl phthalate were potent inhibitors of the activity of SULT1E1 [16].

SULT1A1*1, SULT1A1*2, SULT1A1*3, SULT1A2, SULT1C1 and SULT1E1 were inhibited by some or all of the DiNP metabolites. MiNP is the metabolite that inhibits almost all of these members of the SULT1 family, with the highest binding energy corresponding to SULT1A1*3. The stronger interaction of MiNp with SULT1A1*3 may be explained by the fact that, among the DiNP metabolites, MiNP has the lowest molecular weight and the highest hydrophobicity. It was revealed that anM223V mutation corresponding to the SULT1A1*3 allelic variant produced an increase in the local hydrophobicity of the catalytic site [34].

Data obtained by using the PLIP 2021 software emphasize the amino acids that are involved in the noncovalent interactions of the DiNP metabolites and SULT1 enzymes. This outcome allows a full description of the inhibition of SULT1 by these compounds. Taking into account the hydrophobic character of the DiNP metabolites, the predominance of hydrophobic contacts was expected.

The inhibition of phase I or/and phase 2 of metabolism may result in toxic concentrations of both endo- and xenobiotics. Furthermore, for numerous xenobiotics, the enzymes responsible for metabolizing them or/and the molecular targets of xenobiotics in humans are unknown. Predicting the molecular targets and what enzymes are required for metabolism allows for a better understanding of how humans handle a xenobiotic based on a given exposure.

4. Materials and Method

4.1. Materials

Several metabolites of di-isononyl phthalate (IUPAC name bis(7-methyloctyl) benzene-1,2-dicarboxylate) are considered in the present study (Figure 8): mono-carboxy-isooctyl phthalate (MCiOP, IUPAC name 2-(8-carboxyoctan-2-yloxycarbonyl)benzoic acid), mono-hydroxy-isononyl phthalate (MHiNP, IUPAC name 2-(1-hydroxy-7-methyloctoxy)carbonylbenzoate), mono-isononyl phthalate (MiNP, IUPAC name 2-(7-methyloctoxycarbonyl)benzoic acid) and mono-oxo-isononyl phthalate (MOiNP, IUPAC name 2-[[(4-methyl-7-oxooctyl)oxy]carbonyl}benzoic acid). Their structures have been extracted from PubChem database [38] and visualized using Chimera 1.16 software [39]. In order to implement the molecular docking study (see further), structures of SULT1 enzymes and of their allelic variants in complex with cofactor and ligands (if available) are also needed. These structures have been extracted from Protein Data Bank (PDB, [40]) and are presented in Table 5. All structures have been prepared for molecular docking using UCSF Chimera 1.16 software.

Table 5. Structural files available in Protein Data Bank (PDB) corresponding to SULT1 enzymes and their allelic variants in complex with cofactor and substrates/ligands. PAP—the inactive cofactor adenosine-3′-5′-diphosphate.

PDB ID of the Structural File	Description	References
4GRA	Crystal structure of SULT1A1*1 (wild type) complexed with PAP	[41]
1LS6	Crystal structure of SULT1A1*2 complexed with PAP and the substrate p-pitrophenol	[42]
1Z28	Crystal structure of SULT1A1*3 complexed with PAP	[43]
1Z29	Crystal structure of SULT1A2 complexed with PAP, acetic acid and Ca^{2+}	[43]
2A3R	Crystal structure of SULT1A3 in complex with PAP and the substrate dopamine	[44]
3CKL	Crystal structure SULT1B1 in complex with PAP and the substrate resveratrol	[45]
3BFX	Crystal structure SULT1C1 in complex with PAP	[46]
1G3M	Crystal structure SULT1E1 in complex with PAP and the inhibitor 3,5,3′,5′-tetrachloro-biphenyl-4,4′-diol	[47]

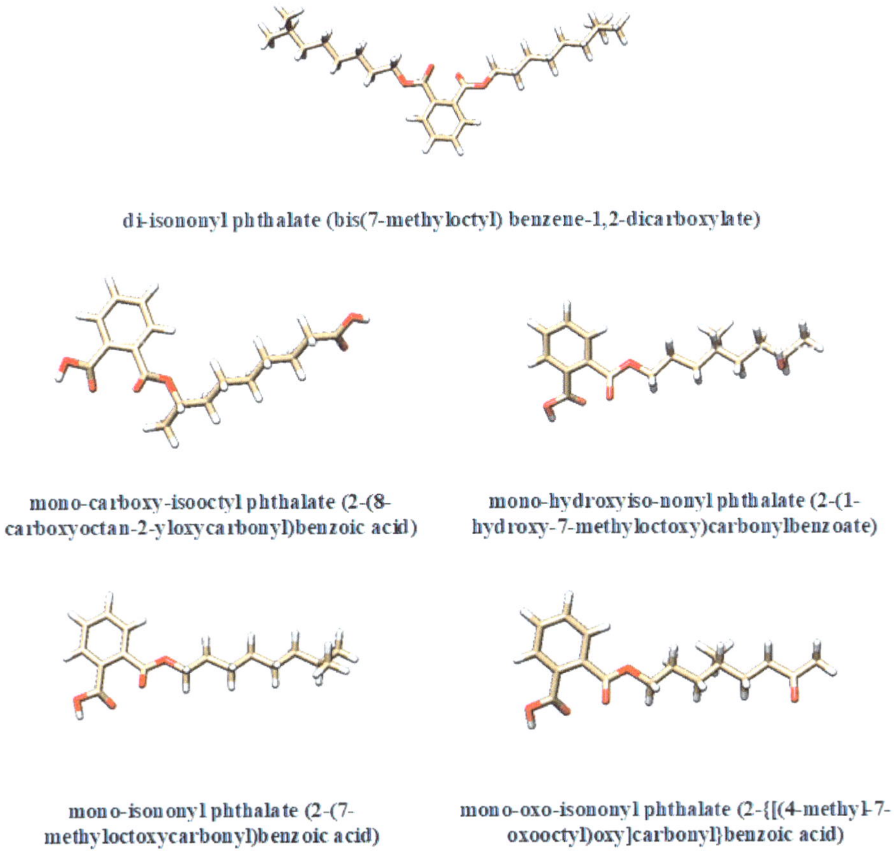

di-isononyl phthalate (bis(7-methyloctyl) benzene-1,2-dicarboxylate)

mono-carboxy-isooctyl phthalate (2-(8-carboxyoctan-2-yloxycarbonyl)benzoic acid)

mono-hydroxyiso-nonyl phthalate (2-(1-hydroxy-7-methyloctoxy)carbonylbenzoate)

mono-isononyl phthalate (2-(7-methyloctoxycarbonyl)benzoic acid)

mono-oxo-isononyl phthalate (2-{[(4-methyl-7-oxooctyl)oxy]carbonyl}benzoic acid)

Figure 8. Three-dimensional structures of the di-isononyl phthalate and of its metabolites. The atoms are coloured like this: carbon—brown, oxygen—red, hydrogen—white.

4.2. Extraction of the Physicochemical and Computation of the Structural Properties of the Metabolites of Di-Isononyl Phthalate and of the Ligands That Are Present in the Crystallographic Structures of SULT1 Enzymes

Physicochemical properties and three-dimensional structures of the metabolites of DiNP have been extracted from PubChem database [38]. Surface area and volume of these compounds have been computed using Chimera 1.16 software and their molecular lipophilicity potential has been obtained using Molinspiration Galaxy 3D Structure Generator v2021.01 (https://www.molinspiration.com/cgi/galaxy, accessed on 4 June 2023). For comparison purposes, the physicochemical and structural properties of ligands that are present in the crystallographic structures of SULT1 enzymes have been also obtained.

4.3. Prediction of the Toxicological Effects of Di-Isononyl Phthalate and of Its Metabolites

ADMETLab2.0 server has been used in order to obtain information regarding the possible human organ toxicological effects of DiNP and of its metabolites. The classification models used for predicting the toxicity endpoints have a minimum accuracy of 80%; regarding the regression models, most of them achieved an $R^2 > 0.72$. There are numerous other computational tools that can be used to assess the ADMET properties, with ADMETLab2.0 tool being used in this study because it is based on well-performed prediction models and has been used by more than 50,0c millions of compounds screened [23,24]. The SMILES (Simplified Molecular Input Line Entry System) formulas of the metabolites of DiNP were used as the entry data, and the pharmacokinetics and toxicological endpoints under investigation were: clearance, penetration of the blood–brain barrier, possibility to be substrate or inhibitors of P-glycoprotein, plasma protein binding, hepatotoxicity, nephrotoxicity, cardiotoxicity, mutagenicity, carcinogenicity and endocrine effects.

4.4. Evaluation of the Interactions of DiNP Metabolites with the Family 1 of Human Sulfotransferases

The molecular docking approach has been considered for assessing the possible interactions of the DiNP metabolites with the SULT1 family. In order to implement the molecular docking study, the structures of enzymes and those of the metabolites of di-isononyl phthalates are necessary. Structures of DiNP metabolites were extracted from PubChem database. Structures of SULT1 enzymes have been extracted from Protein Data Bank and are described in Table 5. All structures have been prepared for molecular docking using UCSF Chimera 1.16 software [38]. Molecular docking has been implemented using SwissDock web service [48] that is based on EADock algorithm [49]. An accurate and blind docking has been selected.

It is known that both SULT1A1 and SULT1A2 have p-nitrophenol as a specific substrate; in order to interpret the docking outputs, structures of the SULT1A2 and of the variants SULT1A1*1 and SULT1A1*3 (that do not have a ligand in the structural file) have been superimposed to the structure of SULT1A1*2 that corresponds to the complex formed by the SULT1A1*2 variant with the inactive cofactor adenosine-3′-5′-diphosphate (A3P) and two molecules of p-nitrophenol (Supplementary Material, Figure S11). The superimposition reveals a high similarity between the structures and emphasized the most probable position of p-nitrophenol when binding to SULT1A1*1, SULT1A1*3 and SULT1A2, respectively. Consequently, the binding modes for the DiNP metabolites that resulted from molecular docking were compared to the position occupied by the two p-nitrophenol molecules that are present in the crystallographic structure of their complex with SULT1A1*2.

In the case of SULT1C1, in order to locate the catalytic cavity and to more easily interpret the molecular docking outcomes, its crystal structure has been superposed with both that of SULT1A1*2 containing the inactive cofactor A3P and two molecules of p-nitrophenol, and that of SULT1B1 that contains the substrate resveratrol, with the superposition also reflecting a high similarity between the three structures (Supplementary Material, Figure S7). Furthermore, as a control, molecular docking has been implemented for every enzyme with its substrate/inhibitor that is present in the crystallographic structure, when available. In the case of SULT1C1 as a control, molecular docking has been implemented for assessing its interaction with its known substrate, iodothyronine [37].

The noncovalent interactions between the DiNP metabolites and the SULT1 enzymes have been analyzed using the Protein–Ligand Interaction Profiler tool [50].

5. Conclusions

The results obtained in the present study reveal several human health risks of the metabolites of di-isononyl phthalate. These compounds revealed the capacity to cause liver damage, to inhibit receptors activated by peroxisome proliferators and are usually able to inhibit the activity of family 1 of sulfotransferases, except SULT1A3 and SULT1B1. Taking into account that humans are extensively exposed to phthalates and the important role of sulfotransferases for the detoxification and elimination of numerous endogenous and exogenous molecules in the body, information regarding the biological effects of the metabolites of phthalates and their possible inhibitory effects against the activities of sulfotransferases contributes to better understanding the risk of human exposure to phthalates.

Supplementary Materials: The following supporting information can be downloaded at: https://www.mdpi.com/article/10.3390/molecules28186748/s1, Figure S1. Binding of mono-hydroxy-isononyl phthalate (a) and mono-oxo-isononyl phthalate (b) to SULT1A1*1; Figure S2. Outcome of the molecular docking study for docking p-nitrophenol to SULT1A1*2 compared with the crystallographic structure of the complex between SULT1A1*2a and p-nitrophenol; Figure S3. Binding of mono-carboxy-iso-octyl phthalate (a) and mono-hydroxy-isononyl phthalate (b) to SULT1A1*2; Figure S4. Binding of mono-hydroxy-isononyl phthalate to SULT1A2; Figure S5. Binding of mono-carboxy-iso-octyl phthalate (a), mono-hydroxy-isononyl phthalate (b), mono-isononyl phthalate (c) and mono-oxo-isononyl phthalate (d) to SULT1A3; Figure S6. Binding of mono-carboxy-iso-octyl phthalate (a), mono-hydroxy-isononyl phthalate (b), mono-isononyl phthalate (c) and mono-oxo-isononyl phthalate (d) to SULT1B1; Figure S7. Superimposition of the outcome of the binding mode resulting from molecular docking of iodothyronine to SULT1C1 to the crystallographic structure of the complex between SULT1B1 and resveratrol and to the structure of SULT1A1*2 in complex with two molecules of p-nitrophenol; Figure S8. Outcome of the molecular docking study for docking dopamine to SULT1A3 compared with the crystalographic structure of the complex between SULT1A3 and dopamine; Figure S9. Outcome of the molecular docking study for docking resveratrol to SULT1B1 compared with the crystalographic structure of the complex between SULT1B1 and resveratrol; Figure S10. Outcome of the molecular docking study for docking the inhibitor 3,5,3',5'-tetrachloro-biphenyl-4,4'-diol to SULT1E1 compared with the crystallographic structure of the complex between SULT1E1 and 3,5,3',5'-tetrachloro-biphenyl-4,4'-diol; Figure S11. Superimposition of the structure of SULT1A1*2 with structures of SULT1A1*, SULT1A1* and SULT1A2; Figure S12. Superimposition of the structure of SULT1C1 (with structure of SULT1B1 and with structure of SULT1A1*2; Table S1. Physicochemical and structural properties of the metabolites of di-isononyl phthalate and of the ligands that are present in the crystallographic structures of SULT1 enzymes.

Author Contributions: Conceptualization, A.I. and V.O.; methodology, A.I.; formal analysis, S.C. and A.C.; investigation, S.C. and A.C.; data curation, S.C.; writing—original draft preparation A.I. and A.C.; writing—review and editing, S.C. and A.I.; visualization, V.O.; supervision, A.I. All authors have read and agreed to the published version of the manuscript.

Funding: This research received no external funding.

Institutional Review Board Statement: Not applicable.

Informed Consent Statement: Not applicable.

Data Availability Statement: All the data are available in the article and the Supplementary Materials.

Conflicts of Interest: The authors declare no conflict of interest.

Sample Availability: Not applicable.

References

1. Croom, E. Metabolism of xenobiotics of human environments. *Prog. Mol. Biol. Transl. Sci.* **2012**, *112*, 31–88. [PubMed]

2. Negishi, M.; Pedersen, L.G.; Petrotchenko, E.; Shevtsov, S.; Gorokhov, A.; Kakuta, Y.; Pedersen, L.C. Structure and function of sulfotransferases. *Arch. Biochem. Biophys.* **2001**, *390*, 149–157. [CrossRef] [PubMed]
3. Gamage, N.; Barnett, A.; Hempel, N.; Duggleby, R.G.; Windmill, K.F.; Martin, J.L.; McManus, M.E. Human Sulfotransferases and Their Role in Chemical Metabolism. *Toxicol. Sci.* **2006**, *90*, 5–22. [CrossRef] [PubMed]
4. Kurogi, K.; Rasool, M.I.; Alherz, F.A.; El Daibani, A.A.; Bairam, A.F.; Abunnaja, M.S.; Yasuda, S.; Wilson, L.J.; Hui, Y.; Liu, M.C. SULT genetic polymorphisms: Physiological, pharmacological and clinical implications. *Expert Opin. Drug Metab. Toxicol.* **2021**, *17*, 767–784. [CrossRef] [PubMed]
5. Isvoran, A.; Peng, Y.; Ceauranu, S.; Schmidt, L.; Nicot, A.B.; Miteva, M.A. Pharmacogenetics of human sulfotransferases and impact of amino acid exchange on Phase II drug metabolism. *Drug Discov. Today* **2022**, *27*, 103349. [CrossRef]
6. Wittassek, M.; Koch, H.M.; Angerer, J.; Brüning, T. Assessing exposure to phthalates—The human biomonitoring approach. *Mol. Nutr. Food Res.* **2011**, *55*, 7–31. [CrossRef]
7. Gennings, C.; Hausser, R.; Koch, H.M.; Kortenkamp, A.; Lioy, P.J.; Mirkes, P.E.; Schwetz, B.A. Report to the U.S. Consumer Product Safety Commission by the Chronic Hazard Advisory Panel on Phthalates and Phthalate Alternatives. 2014. Available online: https://www.cpsc.gov/chap (accessed on 18 May 2023).
8. Hines, E.P.; Calafat, A.M.; Silva, M.J.; Mendola, P.; Fenton, S.E. Concentrations of phthalate metabolites in milk, urine, saliva, and Serum of lactating North Carolina women. *Environ. Health Perspect.* **2009**, *117*, 86–92. [CrossRef]
9. Onipede, O.J.; Adewuyi, G.O.; Ayede, I.A.; Olayemi, O.; Bello, F.A.; Osamor, O. Phthalate Esters in Blood, Urine and Breast-milk Samples of Transfused Mothers in Some Hospitals in Ibadan Metropolis Southwestern Nigeria. *Chem. Sci. J.* **2018**, *9*, 1000188.
10. Yalçin, S.S.; Erdal, İ.; Oğuz, B.; Duzova, A. Association of urine phthalate metabolites, bisphenol A levels and serum electrolytes with 24-h blood pressure profile in adolescents. *BMC Nephrol.* **2022**, *23*, 141. [CrossRef]
11. Wang, Y.; Qian, H. Phthalates and Their Impacts on Human Health. *Healthcare* **2021**, *9*, 603. [CrossRef]
12. Craciun, D.; Dascalu, D.; Isvoran, A. Computational assessment of the ADME-Tox profiles and harmful effects of the most common used phthalates on the human health. *Stud. Univ. Babes-Bolyai Chem.* **2019**, *64*, 71–92. [CrossRef]
13. Frederiksen, H.; Skakkebaek, N.E.; Andersson, A.M. Metabolism of phthalates in humans. *Mol. Nutr. Food. Res.* **2007**, *51*, 899–911. [CrossRef]
14. Heindel, J.J.; Powell, C.J. Phthalate ester effects on rat Sertoli cell function in vitro: Effects of phthalate side chain and age of animal. *Toxicol. Appl. Pharmacol.* **1992**, *115*, 116–123. [CrossRef]
15. Harris, R.M.; Waring, R.H. Sulfotransferase inhibition: Potential impact of diet and environmental chemicals on steroid metabolism and drug detoxification. *Curr. Drug Metab.* **2008**, *9*, 269–275. [CrossRef] [PubMed]
16. Huang, H.; Lan, B.D.; Zhang, Y.J.; Fan, X.J.; Hu, M.C.; Qin, G.Q.; Wang, F.G.; Wu, Y.; Zheng, T.; Liu, J.H. Inhibition of Human Sulfotransferases by Phthalate Monoesters. *Front. Endocrinol.* **2022**, *13*, 868105. [CrossRef]
17. Saeidnia, S. Phthalates. In *Encyclopedia of Toxicology*; Academic Press: Cambridge, MA, USA, 2014; pp. 928–933.
18. Silva, M.J.; Kato, K.; Wolf, C.; Samandar, E.; Silva, S.S.; Gray, E.L.; Needham, L.L.; Calafat, A.M. Urinary biomarkers of di-isononyl phthalate in rats. *Toxicology* **2006**, *223*, 101–112. [CrossRef] [PubMed]
19. Silva, M.J.; Reidy, J.A.; Preau, J.L., Jr.; Needham, L.L.; Calafat, A.M. Oxidative metabolites of diisononyl phthalate as biomarkers for human exposure assessment. *Environ. Health Perspect.* **2006**, *114*, 1158–1161. [CrossRef] [PubMed]
20. Koch, H.M.; Muller, J.; Angerer, J. Determination of secondary, oxidised di-iso-nonylphthalate (DINP) metabolites in human urine representative for the exposure to commercial DINP plasticizers. *J. Chromatogr. B Analyt. Technol. Biomed. Life Sci.* **2007**, *847*, 114–125. [CrossRef]
21. Joint Research Centre; Institute for Health and Consumer Protection; Worth, A.; Lo Piparo, E. Review of QSAR Models and Software Tools for Predicting Developmental and Reproductive Toxicity, Publications Office. 2010. Available online: https://data.europa.eu/doi/10.2788/9628 (accessed on 19 September 2023).
22. Kleandrova, V.V.; Speck-Planche, A. Regulatory issues in management of chemicals in OECD member countries. *Front. Biosci.* **2013**, *1*, 375–398. [CrossRef] [PubMed]
23. Dong, J.; Wang, N.N.; Yao, Z.J.; Zhang, L.; Cheng, Y.; Ouyang, D.; Lu, A.P.; Cao, D.S. ADMETlab: A platform for systematic ADMET evaluation based on a comprehensively collected ADMET database. *J. Cheminform.* **2018**, *10*, 29. [CrossRef]
24. Xiong, G.; Wu, Z.; Yi, J.; Fu, L.; Yang, Z.; Hsieh, Y.C.M.; Zeng, X.; Wu, C.; Lu, A.; Chen, X.; et al. ADMETlab 2.0: An integrated online platform for accurate and comprehensive predictions of ADMET properties. *Nucleic Acids Res.* **2021**, *49*, W5–W14. [CrossRef]
25. Yu, L.; Yang, M.; Cheng, M.; Fan, L.; Wang, X.; Xu, T.; Wang, B.; Chen, W. Associations between urinary phthalate metabolite concentrations and markers of liver injury in the US adult population. *Environ. Int.* **2021**, *155*, 106608. [CrossRef] [PubMed]
26. Meling, D.D.; De La Torre, K.M.; Arango, A.S.; Gonsioroski, A.; Deviney, A.R.K.; Neff, A.M.; Laws, M.J.; Warner, G.R.; Tajkhorshid, E.; Flaws, J.A. Phthalate monoesters act through peroxisome proliferator-activated receptors in the mouse ovary. *Reprod. Toxicol.* **2022**, *110*, 113–123. [CrossRef]
27. Liu, G.; Cai, W.; Liu, H.; Jiang, H.; Bi, Y.; Wang, H. The Association of Bisphenol A and Phthalates with Risk of Breast Cancer: A Meta-Analysis. *Int. J. Environ. Res. Public Health* **2021**, *18*, 2375. [CrossRef] [PubMed]
28. Dirven, H.A.; Theuws, J.L.; Jongeneelen, F.J.; Bos, R.P. Non-mutagenicity of 4 metabolites of di(2-ethylhexyl) phthalate (DEHP) and 3 structurally related derivatives of di(2-ethylhexyl) adipate (DEHA) in the Salmonella mutagenicity assay. *Mutat. Res.* **1991**, *260*, 121–130. [CrossRef] [PubMed]

29. Kim, D.H.; Park, C.G.; Kim, S.H.; Kim, Y.J. The Effects of Mono-(2-Ethylhexyl) Phthalate (MEHP) on Human Estrogen Receptor (hER) and Androgen Receptor (hAR) by YES/YAS in Vitro Assay. *Molecules* **2019**, *24*, 1558. [CrossRef]
30. Roman, D.L.; Isvoran, A.; Filip, M.; Ostafe, V.; Zinn, M. In silico assessment of pharmacological profile of low molecular weight oligo-hydroxyalkanoates. *Front. Bioeng. Biotechnol.* **2020**, *8*, 584010.
31. Roman, D.L.; Roman, M.; Som, C.; Schmutz, M.; Hernandez, E.; Wick, P.; Casalini, T.; Perale, G.; Ostafe, V.; Isvoran, A. Computational Assessment of the Pharmacological Profiles of Degradation Products of Chitosan. *Front. Bioeng. Biotechnol.* **2019**, *7*, 214. [CrossRef]
32. Gridan, I.M.; Ciorsac, A.A.; Isvoran, A. Prediction of ADME-Tox properties and toxicological endpoints of triazole fungicides used for cereals protection. *ADMET DMPK* **2019**, *7*, 161–173. [CrossRef]
33. Alves, V.M.; Muratov, E.N.; Zakharov, A.; Muratov, N.N.; Andrade, C.H.; Tropsha, A. Chemical toxicity prediction for major classes of industrial chemicals: Is it possible to develop universal models covering cosmetics, drugs, and pesticides? *Food Chem. Toxicol.* **2018**, *112*, 526–534. [CrossRef]
34. Ceauranu, S.; Ostafe, V.; Isvoran, A. Impaired local hydrophobicity, structural stability and conformational flexibility due to point mutations in sult1 family of enzymes. *J. Serb. Chem. Soc.* **2023**, *88*, 841–857. [CrossRef]
35. Dash, R.; Ali, M.C.; Dash, N.; Azad, M.A.K.; Hosen, S.M.Z.; Hannan, M.A.; Moon, I.S. Structural and dynamic characterizations highlight the deleterious role of SULT1A1 R213H polymorphism in substrate binding. *Int. J. Mol. Sci.* **2019**, *20*, 6256. [CrossRef] [PubMed]
36. Reiter, C.; Mwaluko, G.; Dunnette, J.; Van Loon, J.; Weinshilboum, R. Thermolabile and thermostable human platelet phenol sulfotransferase. Substrate specificity and physical separation. *Naunyn-Schmiedeberg's. Archiv. Pharmacol.* **1983**, *324*, 140–147. [CrossRef] [PubMed]
37. Allali-Hassani, A.; Pan, P.W.; Dombrovski, L.; Najmanovich, R.; Tempel, W.; Dong, A.; Arrowsmith, C.H. Structural and chemical profiling of the human cytosolic sulfotransferases. *PLoS Biol.* **2007**, *5*, e97.
38. Kim, S.; Chen, J.; Cheng, T.; Gindulyte, A.; He, J.; He, S.; Li, Q.; Shoemaker, B.A.; Thiessen, P.A.; Yu, B.; et al. PubChem 2023 update. *Nucleic Acids Res.* **2023**, *51*, D1373–D1380. [CrossRef]
39. Pettersen, E.F.; Goddard, T.D.; Huang, C.C.; Couch, G.S.; Greenblatt, D.M.; Meng, E.C.; Ferrin, T.E. UCSF Chimera a visualization system for exploratory research and analysis. *J. Comput. Chem.* **2004**, *25*, 1605–1612. [CrossRef]
40. Berman, H.M.; Westbrook, J.; Feng, Z.; Gilliland, G.; Bhat, T.N.; Weissig, H.; Shindyalov, I.N.; Bourne, P.E. The Protein Data Bank. *Nucleic Acids Res.* **2000**, *28*, 235–242. [CrossRef]
41. Cook, I.; Wang, T.; Almo, S.C.; Kim, J.; Falany, C.N.; Leyh, T.S. The Gate That Governs Sulfotransferase Selectivity. *Biochemistry* **2012**, *52*, 415–424. [CrossRef]
42. Gamage, N.U.; Duggleby, R.G.; Barnett, A.C.; Tresillian, M.; Latham, C.F.; Liyou, N.E.; Martin, J.L. Structure of a Human Carcinogen-converting Enzyme, SULT1A1. *J. Biol. Chem.* **2002**, *278*, 7655–7662. [CrossRef]
43. Lu, J.; Li, H.; Zhang, J.; Li, M.; Liu, M.Y.; An, X.; Chang, W. Crystal structures of SULT1A2 and SULT1A1*3: Insights into the substrate inhibition and the role of Tyr149 in SULT1A2. *Biochem. Biophys. Res. Commun.* **2010**, *396*, 429–434. [CrossRef]
44. Lu, J.H.; Li, H.T.; Liu, M.C.; Zhang, J.P.; Li, M.; An, X.-M.; Chang, W.-R. Crystal structure of human sulfotransferase SULT1A3 in complex with dopamine and 3′-phosphoadenosine 5′-phosphate. *Biochem. Biophys. Res. Commun.* **2005**, *335*, 417–423. [CrossRef] [PubMed]
45. Pan, P.W.; Tempel, W.; Dong, A.; Loppnau, P.; Kozieradzki, I.; Edwards, A.M.; Arrowsmith, C.H.; Weigelt, J.; Bountra, C.; Bochkarev, A.; et al. Crystal structure of human cytosolic sulfotransferase SULT1B1 in complex with PAP and resveratrol. **2008**, *to be published*.
46. Dombrovski, L.; Dong, A.; Bochkarev, A.; Plotnikov, A.N. Crystal structures of human sulfotransferases SULT1B1 and SULT1C1 complexed with the cofactor product adenosine-3′-5′-diphosphate (PAP). *Proteins Struct. Funct. Bioinform.* **2006**, *64*, 1091–1094. [CrossRef] [PubMed]
47. Shevtsov, S.; Petrotchenko, E.V.; Pedersen, L.C.; Negishi, M. Crystallographic analysis of a hydroxylated polychlorinated biphenyl (OH-PCB) bound to the catalytic estrogen binding site of human estrogen sulfotransferase. *Environ. Health Perspect.* **2003**, *111*, 884–888. [CrossRef] [PubMed]
48. Grosdidier, A.; Zoete, V.; Michielin, O. SwissDock, a protein-small molecule docking web service based on EADock DSS. *Nucleic Acids Res.* **2011**, *39*, W270–W277. [CrossRef]
49. Grosdidier, A.; Zoete, V.; Michielin, O. EADock: Docking of small molecules into protein active sites with a multiobjective evolutionary optimization. *Proteins* **2007**, *67*, 1010–1025. [CrossRef]
50. Adasme, M.F.; Linnemann, K.L.; Bolz, S.N.; Kaiser, F.; Salentin, S.; Haupt, V.J.; Schroeder, M. PLIP 2021: Expanding the scope of the protein-ligand interaction profiler to DNA and RNA. *Nucleic Acids Res.* **2021**, *49*, W530–W534. [CrossRef]

Disclaimer/Publisher's Note: The statements, opinions and data contained in all publications are solely those of the individual author(s) and contributor(s) and not of MDPI and/or the editor(s). MDPI and/or the editor(s) disclaim responsibility for any injury to people or property resulting from any ideas, methods, instructions or products referred to in the content.

Article

Molecular Dynamics Simulations on the Adsorbed Monolayers of N-Dodecyl Betaine at the Air–Water Interface

Chengfeng Zhang [†], Lulu Cao [†], Yongkang Jiang, Zhiyao Huang, Guokui Liu, Yaoyao Wei [*] and Qiying Xia [*]

School of Chemistry and Chemical Engineering, Linyi University, Linyi 276000, China; zcf100660@163.com (C.Z.); 19560890625@163.com (L.C.); justy838@163.com (Y.J.); yaoshuu@163.com (Z.H.); liuguokui@lyu.edu.cn (G.L.)
* Correspondence: weiyaoyao@lyu.edu.cn (Y.W.); xiaqiying@lyu.edu.cn (Q.X.)
[†] These authors contributed equally to this work.

Abstract: Betaine is a kind of zwitterionic surfactant with both positive and negative charge groups on the polar head, showing good surface activity and aggregation behaviors. The interfacial adsorption, structures and properties of *n*-dodecyl betaine (NDB) at different surface coverages at the air–water interface are studied through molecular dynamics (MD) simulations. Interactions between the polar heads and water molecules, the distribution of water molecules around polar heads, the tilt angle of the NDB molecule, polar head and tail chain with respect to the surface normal, the conformations and lengths of the tail chain, and the interfacial thickness of the NDB monolayer are analyzed. The change of surface coverage hardly affects the locations and spatial distributions of the water molecules around the polar heads. As more NDB molecules are adsorbed at the air–water interface, the number of hydrogen bonds between polar heads and water molecules slightly decreases, while the lifetimes of hydrogen bonds become larger. With the increase in surface coverage, less gauche defects along the alkyl chain and longer NDB chain are obtained. The thickness of the NDB monolayer also increases. At large surface coverages, tilted angles of the polar head, tail chain and whole NDB molecule show little change with the increase in surface area. Surface coverages can change the tendency of polar heads and the tail chain for the surface normal.

Keywords: betaine; molecular dynamics simulations; air-water interface; monolayer structure

Citation: Zhang, C.; Cao, L.; Jiang, Y.; Huang, Z.; Liu, G.; Wei, Y.; Xia, Q. Molecular Dynamics Simulations on the Adsorbed Monolayers of N-Dodecyl Betaine at the Air-Water Interface. *Molecules* **2023**, *28*, 5580. https://doi.org/10.3390/molecules28145580

Academic Editor: Fabio Ganazzoli

Received: 28 June 2023
Revised: 14 July 2023
Accepted: 21 July 2023
Published: 22 July 2023

Copyright: © 2023 by the authors. Licensee MDPI, Basel, Switzerland. This article is an open access article distributed under the terms and conditions of the Creative Commons Attribution (CC BY) license (https:// creativecommons.org/licenses/by/ 4.0/).

1. Introduction

Betaine widely exists in animals and plants [1,2]. It is a kind of important osmotic adjustment substance. For plants, betaine contributes to enhance stress resistance, such as salt resistance and drought resistance. When the tail chain becomes longer, betaine possesses interfacial activity. These betaines with long tail chains are classified to be zwitterionic surfactants, which are important kinds of surfactants and have important applications in industry and life [3].

Due to the wide application of betaine, it has attracted much scientific attention. Liu et al. used surface tension and small angle neutral-scattering methods to study the interaction between four kinds of betaines and the lipopeptide surfactin. They found that the synergistic effect between them was related to the structure and molar ratio of betaine [4]. Hines et al. studied the aggregation structure of n-dodecyl-N,N-dimethylamino acetate amphoteric surfactants at the gas–liquid interface using the neutral reflection method, and they proposed the aggregation model of related systems at the gas–liquid interface [5]. Through surface tension and thermokinetic analysis, Erfani et al. studied the effect of cocamidopropyl betaine on the adsorption of bovine serum albumin at the gas–liquid interface. The results showed that betaine could effectively prevent the adsorption of protein at the interface [6]. With small angle neutral scaling and visibility measurements, McCoy et al. [7] studied the effect of organic additives on the micelles formed by oleyl amidopropyl betaine. When nonpolar organic additives are added, wormlike micelles could transform into microemulsions, while polar additives only affected fluid rheology, not changing micelles into microemulsions. Gao et al. [8] compared the interfacial adsorption of alkylbetaine with the one of phenyl-containing betaine, showing that betaine with a benzene ring has better interfacial activity. Hussain et al. [9] studied the effect of different ethoxylation on the properties of betaine so as to study the betaine that can be dissolved in aqueous solution containing salt.

In addition to experimental research, molecular dynamics (MD) simulation also plays an important role in studying surfactant systems [10–15]. Liu et al. used the MD simulation method to study the structural characteristics of several betaines and carboxylic acid mixtures at the n-decane–water interface. They found that polar head tends to be parallel to the interface, while the tail tends to be in the oil phase. The benzene ring on an alkyl chain would affect the distribution of the tail chain [16]. Furthermore, Cai et al. studied the microstructure characteristics of two betaines and three ionic surfactants mixtures at the n-decane–water interface by MD simulation, focusing on the changes of microstructure characteristics with different mixture systems [17]. Combining experimental and simulation methods, Su et al. proposed that the benzene ring has a great influence on the synergistic effect for betaine and anionic surfactants mixture systems [18]. Ergin et al. studied the aggregation behavior of betaine and sodium dodecyl sulfate (SDS) surfactant mixtures at the gas–liquid interface from the perspectives of the polar head, tail chain and water structure through MD simulation [19]. In the previous study, we focused on the change of the configuration entropy of betaine molecules at the interface, bulk phase and very dilute solution [20]. The distribution of local configuration entropy in betaine micelle was used to study the state of the micelle core [21]. In this paper, we focus on the effect of surface adsorption areas of betaine (n-dodecyl betaine, NDB) molecules on its microstructure and properties at the gas–liquid interface.

2. Results and Discussion

2.1. Distribution of Water Molecules around Polar Heads

In order to study the distribution of water molecules around the polar heads, we calculate the radial distribution functions (RDFs) of water molecules around the N (N1) and carboxyl-C (C16) atoms of polar heads. As shown in Figure 1, the curve shapes and the locations of peaks hardly change with the increase in surface area. However, as the interfacial area per surfactant increases from 0.44 to 64 nm^2, the heights of different peaks gradually increase. This means that the local densities of water molecules around peak locations increase, which can also be confirmed by the analysis of the number of hydrogen bonds in the later section. The first peaks of all RDFs around the N1 and C16 atoms are around 0.44 nm and 0.35 nm, respectively. This difference is reasonable because the CH$_2$ and CH$_3$ groups are connected with the N atom to cause some steric hindrance, and the C16 atom has closer contact with the water phase than the N atom in the view of the NDB structure. The location of the first peak for C16-OW RDF indicates the existence of hydrogen bonds between carboxyl groups and water molecules. From Figure 1, both the N1-OW and C16-OW RDF curves have two peaks. The difference is that there is an acromion on the right side of the first peak for N1 atom, which is located at about 0.58 nm. One interesting finding is that the location of the first peak for N1-OW RDF is the one of the first valley for C16-OW RDF, and the location of the second peak for N-OW RDF is the one of the second valley for C16-OW RDF.

RDF analysis provides the averaged distribution of other atoms or molecules around the specifically functional group, while the spatial distribution function (SDF) provides a more detailed three-dimensional distribution. The SDFs of the first shell water molecules around the N1 and C16 atoms for all the studied systems are plotted in Figure 2. The graph shows the similar distributions around all polar heads, specifically around the oxygen (O) atom on the outer side of the C=O bond on the polar head. For the carboxyl group, water molecules are mainly located around two O atoms, presenting a circular distribution. There is basically no water distribution in the direction of the extension line of the C=O bond. Water molecules exhibit a relatively broad distribution around nitrogen (N) atoms. As the interfacial area per surfactant increases from 0.44 to 64 nm^2, the distributions of water molecules around the N1 and C16 atoms become more fragmented, especially for those around N1 atoms. This change indicates a higher level of freedom for the surrounding water molecules. It should be noted that the N1 atom already has four covalent bonds, which prevents the N1 atom from forming hydrogen bonds with water molecules. The surrounding water molecules are restrained by weak interactions such as van der Waals forces.

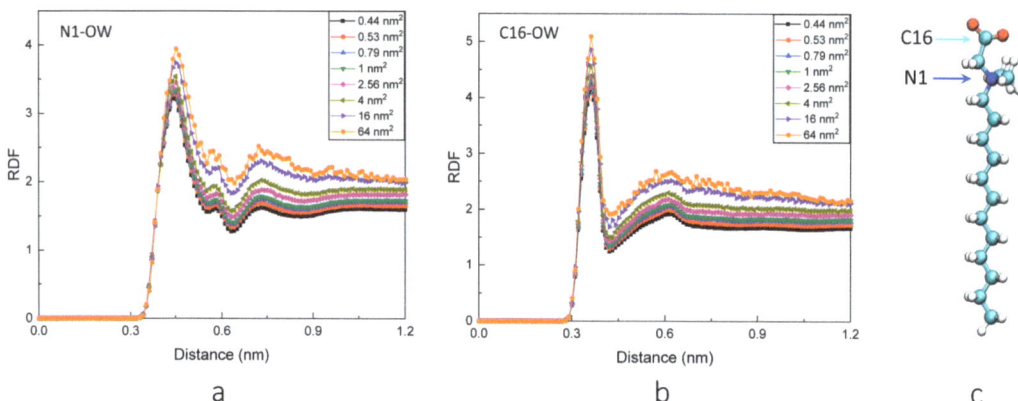

Figure 1. The RDFs of water molecules around N1 and C16 atoms. (**a**,**b**) represent the RDF between N1 and water molecules and the RDF between C16 and water molecules, respectively. (**c**) shows the structure of the NDB molecule and labels the calculated N1 and C16 atoms.

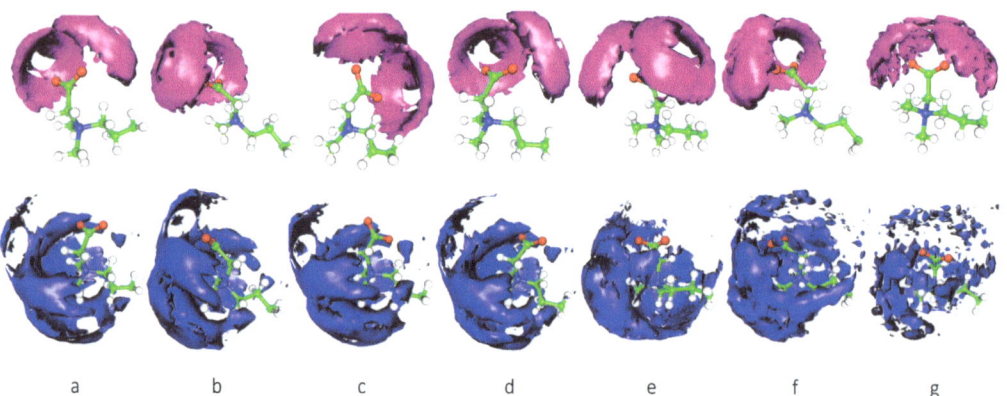

Figure 2. The SDFs of water molecules around C16 and N1 atoms. The top seven graphs present those between C16 and water molecules, and the bottom seven graphs present those between N1 and water molecules. The (**a**–**g**) symbols represent 0.44 nm^2, 0.53 nm^2, 0.79 nm^2, 1 nm^2, 2.56 nm^2, 4 nm^2, 16 nm^2, and 64 nm^2 systems, respectively. In order to maintain clarity, only a portion of the tail chain is displayed.

2.2. Hydrogen Bonds between Water Molecules and Polar Heads

The hydrogen bond is a type of intermolecularly chemical bond, which typically refers to the interactions between the hydrogen atom and the more electronegative atom (such as nitrogen, oxygen, fluorine) in another molecule. The hydrogen bond is often stronger than other non-covalent bonds and can have important effects on many biochemical and physical processes. The hydrogen bonds are determined using a geometric criterion, where the distance between the chosen donor–acceptor pairs is within 3.5 Å and the H-O-H angle is less than 120°. Based on this definition, the hydrogen bond numbers between water molecules and polar heads are calculated. Table 1 shows the averaged hydrogen bond number of one polar head. From the table, the number of hydrogen bonds increases from 4.14 to 4.98 as the interfacial area per surfactant increases from 0.44 to 64 nm^2. At a small surface area per surfactant, the interfacial concentration is large, and surfactant molecules tend to aggregate at the interface. Thus, some water molecules around the interface are displaced, resulting in a small decrease for the hydrogen bond numbers. To give more detailed information for the water molecules around polar heads, the coordination number is calculated. The coordination number is calculated by integrating RDFs from zero to the first location of the valley, which counts the coordination numbers of water molecules in the first hydration layer. Table 1 lists all the counted coordination numbers. As the interfacial area per surfactant decreases from 64 to 0.44 nm^2, the coordination number also decreases, which is consistent with previous discussions of the hydrogen bonds between water molecules and polar heads. Differently, all coordination numbers of the same interfacial area per surfactant are larger than the hydrogen bond numbers, meaning that not all coordinated water molecules around the polar heads can form hydrogen bonds with polar heads. This phenomenon is reasonable because the hydrogen bonds are defined with rigorously geometric criteria. Some water molecules are restrained around the polar heads but cannot satisfy the criteria to form hydrogen bonds with the polar heads.

Table 1. The hydrogen bond numbers, coordination numbers and lifetimes between water molecules and polar heads.

System	0.44 nm^2	0.53 nm^2	0.79 nm^2	1 nm^2	2.56 nm^2	4 nm^2	16 nm^2	64 nm^2
Hydrogen bond number	4.14 ± 0.08	4.21 ± 0.09	4.31 ± 0.11	4.36 ± 0.13	4.47 ± 0.19	4.61 ± 0.24	4.85 ± 0.45	4.98 ± 0.87
Coordination number	5.31 ± 0.11	5.47 ± 0.13	5.62 ± 0.16	5.70 ± 0.19	5.89 ± 0.29	6.12 ± 0.36	6.60 ± 0.64	6.89 ± 1.21
Lifetime	14.84 ps	14.17 ps	12.94 ps	12.31 ps	10.78 ps	9.75 ps	8.63 ps	8.10 ps

To further study the interactions between water molecules and polar heads, we calculate the time correlation function C(t) of hydrogen bonds for all systems as shown in Figure 3. With the increase in surface area, the relaxation curves decline more rapidly. This indicates that the water molecules around the NDB head group in systems with smaller surface concentrations are freer to move, and the formed hydrogen bonds are weaker. In addition, the larger the surface concentration of NDB systems, the slower their relaxation curves decline. This suggests that the water molecules around the polar heads in systems with larger surface concentrations are more restricted and less likely to escape. To verify these phenomena, the lifetimes of hydrogen bonds are calculated by integrating the C(t), which is fit through the multiple exponential function of $C(t) = \sum a_i e^{(-t/b_i)}$ [22,23]. As shown in Table 1, the lifetime of hydrogen bonds decreases from 14.84 to 8.10 ps with the increase in surface area from 0.44 to 64 nm^2. This decrease is consistent with the analysis of C(t) curves in different surface coverages. A longer lifetime indicates the stronger hydrogen bond interactions at large surface coverage.

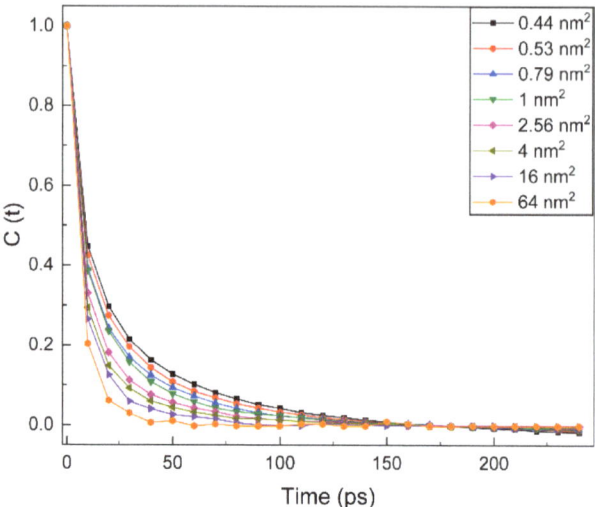

Figure 3. Time correlation function C(t) for the hydrogen bonds formed between polar heads and water molecules.

2.3. Tilted Angle at the Interface

A very important analysis index of surfactant adsorption is the tilted angle of the surfactant molecule at the air–water interface. In this study, we choose several types of tilted angles along the Z-axis (surface normal) to analyze. As shown in Figure 4, the tilted angle between the tail chain and Z-direction (A_Tail), tilted angle between the whole NDB molecule and the Z-direction (A_Whole), and the tilted angle between the polar head and the Z-direction (A_Polar) are considered. All calculated results are averaged. From Figure 4, A_Tail is defined by the angle between the $C_4 \rightarrow C_{15}$ vector and Z-direction away from the water phase, A_Whole is defined by the angle between the vector $C_{16} \rightarrow C_{15}$ and Z-direction away from the water phase, and A_Polar is defined by the angle between the vector $C_{16} \rightarrow C_4$ and Z-direction away from the water phase. Through these definitions, the 0° and 180° angles mean the perpendicular orientation of selected vectors to the vapor–water interface. Differently, the 0° angle denotes the preference to vapor phase, while the 180° angle denotes the preference to water phase. The 90° angle represents that the selected vector is parallel to the vapor–water interface. When 90° is reached, the NDB molecule or tail chain or polar head lies on the interface.

From Figure 4, we can see that A_Tail and A_Whole increase as away from the water phase surface area increases. This is reasonable because fewer NDB molecules are absorbed at the vapor–water interface at a large surface area. With the increasement of surface area, a larger extending space among NDB molecules exists and the preference of away from the water phase NDB molecule and alkyl chain to the interface also increases. A singly absorbed NDB molecule at the interface (64 nm²) can have the largest extending space, making whole molecule and tail chain incline to the interface. In comparison, the angle change is small (<5°) when the surface area is small. This is similar to experimental results that indicate the tilted angle hardly changes with the change of surface concentration [5]. At a very large surface area (corresponding to extremely dilute solution), the angle rises ca. 10°. On the contrary, A_Polar decreases with the increase in surface area, showing an opposite trend with A_Tail and A_Whole. This opposite trend may be related to the more alkyl chain lying on the interface at a large surface area. The deflection of the tail chain will also deflect with the polar head, which makes the angle between the polar head and Z-direction smaller. Since the arrangement of the polar head and tail chain has mutual affect, the angle distribution of the overall molecule is slightly lower than that of the tail chain. In comparison, A_Polar angles are larger than A_Tail and A_Whole angles at small surface area (<16 nm^2), indicating that the tail chain has a larger tendency to surface normal than the polar head. From 16 nm^2, the A_Tail angle begins to be larger than the A_Polar angle. When the surface coverage is low, the NDB polar head prefers to surface normal more than the tail chain. The change of surface coverage has different effects on the inclination of NDB polar heads and alkyl chain at the air–water interface.

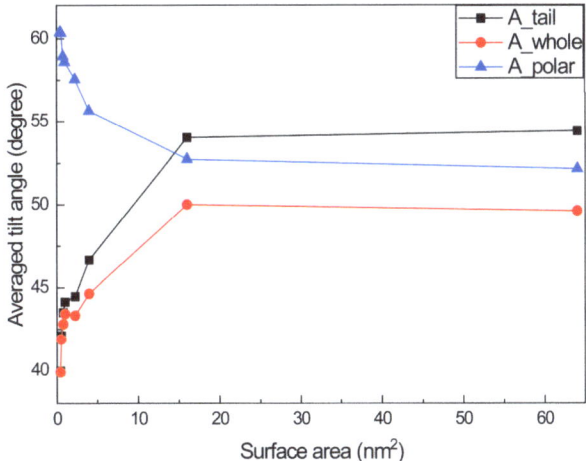

Figure 4. Tilted angles of polar head, tail chain and whole NDB molecule at different surface area.

2.4. Gauche Defects of NDB Molecule

The distribution of dihedral conformations can be well described by the probability of gauche defects. A dihedral angle that deviates from 180° exceeding 60° is defined as a gauche angle. Twelve dihedral angles are defined and calculated. As shown in the mark at the top of Figure 5, mark 1 is defined as a CCNC dihedral angle, and gradually, mark 12 is a CCCC dihedral angle at the end of the tail chain. All curves in Figure 5 have similar distribution trends. From dihedral 1 to 12, the gauche probabilities begin from ca. 0.5 to the highest value at dihedral 2, and the probability rapidly decreases to the lowest value (<0.1) at dihedral 3. For dihedral 4, it again increases. Then, the probability shows a U-shape distribution from dihedral 5 to dihedral 12 at surface areas less than 16 nm^2. When the surface area is large for 16 and 64 nm^2, the gauche probability reaches the approximate platform from dihedral 5. Until the last one (dihedral 12), it slightly increases. In detail, the gauche probabilities of dihedral 1 of all systems are around 0.5, which means that gauche and trans conformations account for half, respectively. For dihedral angle 2, gauche defects reach the highest value, showing that the proportion of gauche conformation is high and this chain section bends to a large extent. Dihedrals 1 and 2 are two dihedral angles including the polar head, and their gauche distributions slightly decrease with the increase in surface area. For dihedral 3, gauche defects decrease to about 0.08, which means that the trans conformation occupies a very high proportion. Interestingly, the probability of dihedral 3 hardly changes with the increase in surface area. Contrary to dihedrals related to polar head, the gauche defects of dihedrals 4–12 along the alkyl chain increase with the increase in surface area. For all systems, the gauche probabilities related to the polar head are larger than those with alkyl chains.

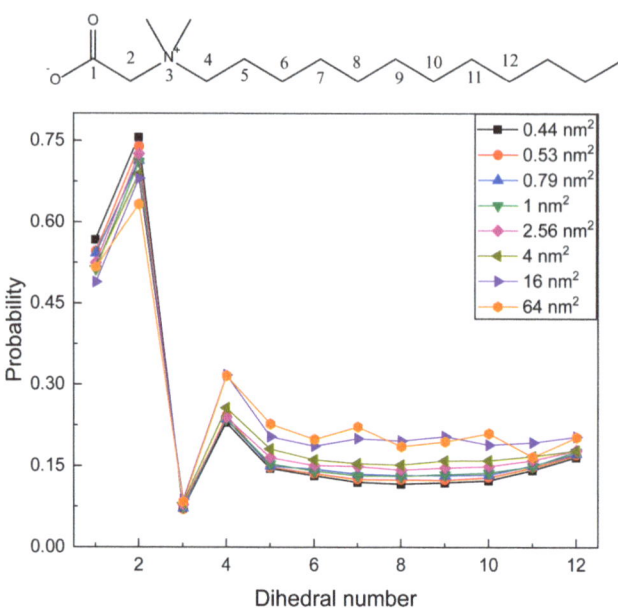

Figure 5. Gauche defects of NDB molecule of different studied systems.

Gauche defects along the tail chain are related to the length of the alkyl chain. The higher the gauche defects, the shorter the chain length. The length distribution of the NDB chain is shown in Figure 6. With the increase in surface area, the distribution peak shifts to the decreasing direction of the tail length. This trend is in good agreement with the one of chain gauche distribution. When the NDB molecule has a small surface area, the interfacial monolayer is compact. The whole NDB molecule and alkyl chain have less of a chance of spreading on the air–water interface and will extend more uprightly, causing more trans-conformations and a longer alkyl chain.

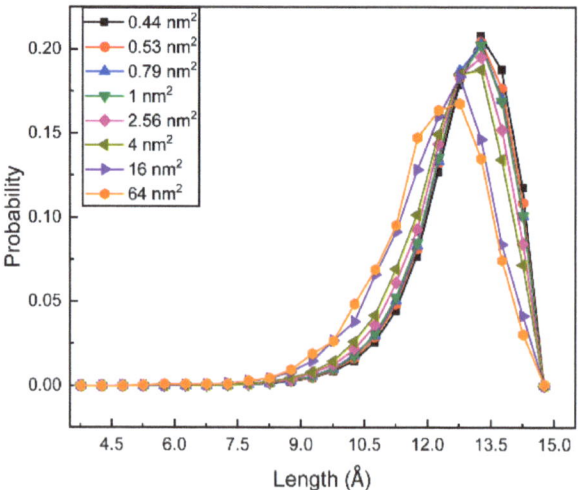

Figure 6. Length distributions of NDB tail chain for different studied systems.

2.5. The Thickness of Different NDB Monolayers

The thickness of the interfacial monolayer can be calculated with the "10–90" rule [24,25]. The length between 10% and 90% of the bulk value on the number density profiles of water is defined as the thickness of the interfacial monolayer. The calculated monolayer thicknesses are listed in Table 2. As the interfacial area increases, the thickness decreases and eventually reaches around 0.6 nm. This phenomenon is caused by two factors discussed earlier: firstly, in extremely dilute conditions, alkyl chains bend more severely with more gauche defects and become shorter; secondly, in extremely dilute conditions, the tilt of the NDB molecule is the largest, and the NDB molecule is more likely to be parallel to the interface rather than perpendicular.

Table 2. The interfacial thickness of different NDB monolayers.

			Interfacial	Thickness	(nm)			
System	$0.44\ nm^2$	$0.53\ nm^2$	$0.79\ nm^2$	$1\ nm^2$	$2.25\ nm^2$	$4\ nm^2$	$16\ nm^2$	$64\ nm^2$
Thickness	1.50	1.48	1.40	1.33	0.97	0.74	0.62	0.60

2.6. Weak Interaction Analysis

The interactions between different components of the NDB molecule and the interactions between NDB and water molecules are analyzed with the method [26] proposed by Johnson et al. The reduced density gradient function (RDG) is defined as $\text{RDG}(r) = \frac{1}{2(3\pi^2)^{1/3}} \frac{|\nabla \rho(r)|}{\rho(r)^{4/3}}$ [27]. As shown in Figure 7, red, green, and blue colors in the color bar represent the repulsion, van der Waals, and attractive interactions, respectively. The sign of λ_2 is utilized to distinguish bonded ($\lambda_2 < 0$) and nonbonded ($\lambda_2 > 0$) interactions. The quantity $\text{sign}(\lambda_2)\rho$ represents the product of ρ and the sign of λ_2, where λ_2 is the second largest eigenvalue of the Hessian matrix of ρ. One or more spikes that appear in low-density and low-gradient regions indicate the existence of weak interactions. Figure 7 is generated with the Multiwfn 3.8 program [28] and VMD 1.9.3 [29].

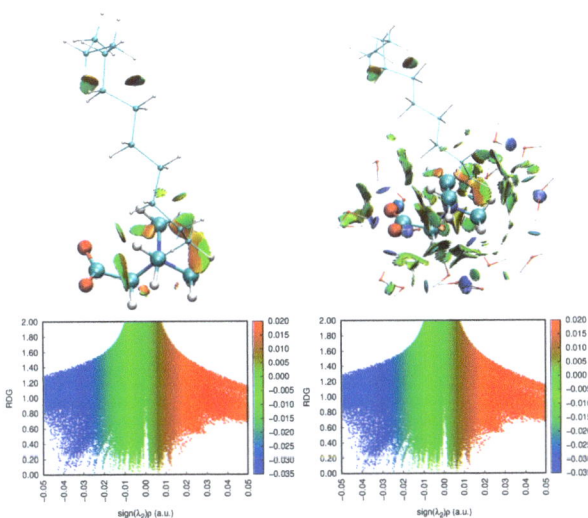

Figure 7. The reduced density gradient (RDG) vs. $\text{sign}(\lambda_2) \times \rho(r)$ for NDB and water molecules. The red, green, and blue colors in the color bar represent the repulsion, van der Waals, and attractive interactions, respectively.

For one single molecule, the weak interaction isosurface will appear in the area where the dihedral angle bends more severely. From Figure 7, green and red parts in the isosurface represent van der Waals interactions and spatially potential resistance interactions, respectively. The gauche defects of the dihedral angles on the polar head are higher; thus, green and red fragments are obvious near the polar head. Furthermore, some green and red distributions exist at terminal tail parts, which also correspond to the increased distribution of gauche defects. In order to study the interactions between NDB and water molecules, we select all water molecules in the first hydration layer around NDB molecules to calculate. In regions around polar heads, more green and red parts occur for NDB and the water system. Moreover, some blue fragments appear, which correspond to the hydrogen bonds formed between O atoms of the polar head and water molecules or between water and water molecules around the polar head. It should be noted that N atoms cannot form hydrogen bonds with water molecules due to the four bonds connected by N atoms.

3. Materials and Methods

The optimized NDB structure was obtained by Gaussian 09 software. All MD simulations were performed using GROMACS 2019 software [30]. The general AMBER force field (gaff2) is used as the force field. The atom-centered point charges are obtained with the restrained electrostatic potential (RESP) methods [31] at HF/6-31G* level. TIP3P water models [32] were applied. To build initial models, the $8 \times 8 \times 24$ nm^3 rectangular box with the Z-axis being perpendicular to the interface was used. In the center of the box, a water slab of $8 \times 8 \times 8$ nm^3 was placed to construct the water–vapor interface with the *gmx editconf* command. Equal amounts of NDB molecules were respectively placed at two sides of the water slab to build surfactant layers with the *gmx solvate* command. All trans alkyl chains of NDB molecules extended away from the water phase and were perpendicular to the interface. For each system, corresponding NDB molecules were added to obtain interfacial systems from extremely dilute surface concentration to saturated concentration. Detailed information of all simulated systems is provided in Table 3. In the table, the change of surface concentration was characterized by the change of surface adsorption area of one NDB molecule. For example, for a system with an area of 64 nm^2, one molecule was placed on each of two interfaces, resulting in a total of two NDB molecules in the system. The surface area of each interface is 8×8 nm, which amounts to 64 nm^2; hence, it is labeled as 64 nm^2. Other systems are labeled similarly. It should be noted that the decreasing interface area occupied by each molecule from 64 to 0.44 nm^2 means that more NDB molecules are gathering together. After constructing initial models, the steepest descent method was applied to minimize systems. Then, the NVT simulation with 200 ns time-length was carried out to equilibrate the system and obtain MD trajectories. To keep the temperature of 298 K, the v-rescale temperature coupling method was used [33]. During the simulation, the LINCS algorithm was used to constrain the bond length [34]. To calculate electrostatic interactions, the particle mesh Ewald (PME) summation method was used [35]. The 1.0 nm cut-off length was set for van der Waals interactions.

Table 3. Detailed composition of studied systems.

	0.44 nm^2	0.53 nm^2	0.79 nm^2	1 nm^2	2.56 nm^2	4 nm^2	16 nm^2	64 nm^2
NDB	288	242	162	128	50	32	8	2
H$_2$O	16,586	16,667	16,824	16,904	17,042	17,074	17,115	17,128

4. Conclusions

At the air–water interface, the structures and properties of NDB monolayers from extremely dilute surface coverage to saturated surface coverage are analyzed with MD simulations. Interactions between NDB monolayers and water molecules are analyzed by the RDF, SDF and hydrogen bond analyses. As the surface area increases, the RDF locations and SDF graphs are similar. For the polar carboxyl group, water molecules are mainly located around oxygen atoms with two circular distributions. Although the number of hydrogen bonds between polar heads and water molecules and coordination numbers of water molecules around polar heads increase with the increase in surface area, the time correlation function decays faster, and the lifetime of hydrogen bonds decreases. Hydrogen bonds formed at small surface areas are stronger than those at large surface areas. An analysis of weak interactions also shows the hydrogen bonds' interactions between polar heads and water molecules. As more NDB molecules are adsorbed at the surface, gauche defects including the polar head increase, but gauche defects along the alky chain decrease. The NDB alkyl chains become longer at large surface coverage. Moreover, gauche defects related to the polar head are larger than those with alkyl chains for each NDB monolayer, which is consistent with the results of the weak interaction iso-surface. Both the tail chain and NDB molecule tilt further away from the surface angle with the increase in surface area, while the polar head shows the opposite change. When surface coverages are large, the change of tilted angle is little. The tail chain and whole molecules rises ca. 10°, but the polar head reduces ca. 7° under extremely dilute conditions. Contrary to intuitive thoughts, the tilted angle shows that polar heads are more inclined toward the interfacial plane at large surface coverages, while tail chains prefer to tilt toward the interface at low surface coverages. Based on the MD results, typical NDB monolayers at saturation adsorption and extremely dilute condition are proposed as shown in Figure 8. The MD models of NDB monolayers give a molecule-level presentation of different arrangements for NDB molecules at the air–water interface, monolayer thickness, and the inclination of NDB fragments at the air–water interface. This study provides the theoretical reference for further exploration and application of NDB adsorption at the interface and for other zwitterionic surfactant micelle systems.

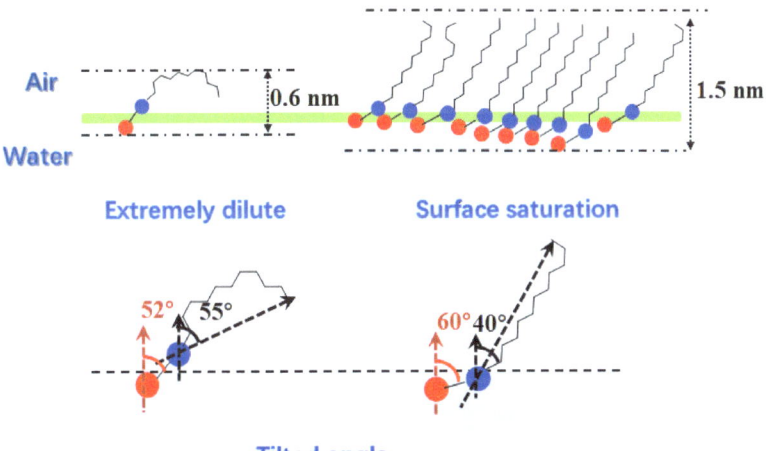

Figure 8. The schematic diagram of typical NDB monolayers at saturation adsorption and extremely dilute solution. Red and blue balls in the figure represent the negative and positive charge center, respectively.

Author Contributions: Conceptualization, C.Z., L.C., Y.W. and Q.X.; methodology, C.Z. and L.C.; software, C.Z. and L.C.; validation, G.L., Y.W. and Q.X.; formal analysis, C.Z., L.C., Y.J. and Z.H.; investigation, C.Z., L.C., Y.J., Z.H., G.L., Y.W. and Q.X.; data curation, G.L., Y.W. and Q.X.; writing—original draft preparation, C.Z. and L.C.; writing—review and editing, G.L., Y.W. and Q.X.; supervision, G.L., Y.W. and Q.X.; funding acquisition, G.L. and Y.W. All authors have read and agreed to the published version of the manuscript.

Funding: This research was funded by the Natural Science Foundation of Shandong Province, grant number ZR2021QB153 and ZR2022QB043, and Shandong Students' innovation and entrepreneurship training program.

Institutional Review Board Statement: Not applicable.

Informed Consent Statement: Not applicable.

Data Availability Statement: The data presented will be made available on request by the corresponding authors.

Conflicts of Interest: The authors declare no conflict of interest.

Sample Availability: Not applicable.

References

1. Clendennen, S.K.; Boaz, N.W. Chapter 14—Betaine amphoteric surfactants—Synthesis, properties, and applications. In *Biobased Surfactants*, 2nd ed.; Hayes, D.G., Solaiman, D.K.Y., Ashby, R.D., Eds.; AOCS Press: London, UK, 2019; pp. 447–469.
2. Di Gioacchino, M.; Bruni, F.; Ricci, M.A. Aqueous solution of betaine: Hydration and aggregation. *J. Mol. Liq.* **2020**, *318*, 114253. [CrossRef]
3. Kelleppan, V.T.; King, J.P.; Butler, C.S.G.; Williams, A.P.; Tuck, K.L.; Tabor, R.F. Heads or tails? The synthesis, self-assembly, properties and uses of betaine and betaine-like surfactants. *Adv. Colloid Interface Sci.* **2021**, *297*, 102528. [CrossRef] [PubMed]
4. Liu, F.; Xiao, J.; Garamus, V.M.; Almásy, L.; Willumeit, R.; Mu, B.; Zou, A. Interaction of the biosurfactant, surfactin with betaines in aqueous solution. *Langmuir* **2013**, *29*, 10648–10657. [CrossRef] [PubMed]
5. Hines, J.D.; Garrett, P.R.; Rennie, G.K.; Thomas, R.K.; Penfold, J. Structure of an adsorbed layer of n-dodecyl-n,n-dimethylamino acetate at the air/solution interface as determined by neutron reflection. *J. Phys. Chem. B* **1997**, *101*, 7121–7126. [CrossRef]
6. Erfani, A.; Khosharay, S.; Flynn, N.H.; Ramsey, J.D.; Aichele, C.P. Effect of zwitterionic betaine surfactant on interfacial behavior of bovine serum albumin (bsa). *J. Mol. Liq.* **2020**, *318*, 114067. [CrossRef]
7. McCoy, T.M.; King, J.P.; Moore, J.E.; Kelleppan, V.T.; Sokolova, A.V.; de Campo, L.; Manohar, M.; Darwish, T.A.; Tabor, R.F. The effects of small molecule organic additives on the self-assembly and rheology of betaine wormlike micellar fluids. *J. Colloid Interface Sci.* **2019**, *534*, 518–532. [CrossRef]
8. Gao, S.; Song, Z.; Lan, F.; Li, P.; Zhang, A.; Hu, J.; Jiang, Q. Studies on physicochemical properties and aggregation behavior of two pairs of betaine surfactants. *Ind. Eng. Chem. Res* **2019**, *58*, 22260–22272. [CrossRef]
9. Hussain, S.M.S.; Fogang, L.T.; Kamal, M.S. Synthesis and performance evaluation of betaine type zwitterionic surfactants containing different degrees of ethoxylation. *J. Mol. Struct.* **2018**, *1173*, 983–989. [CrossRef]
10. Chandrakar, A.; Bhargava, B.L. Aqueous solutions of hydroxyl-functionalized ionic liquids: Molecular dynamics studies. *J. Mol. Graph. Model.* **2020**, *101*, 107721. [CrossRef]
11. Dash, S.; Chowdhury, U.D.; Bhargava, B.L. The effect of external salts on the aggregation of the multiheaded surfactants: All-atom molecular dynamics studies. *J. Mol. Graph. Model.* **2022**, *111*, 108110. [CrossRef]
12. Nadimi, H.; Housaindokht, M.R.; Moosavi, F. The effect of anion on aggregation of amino acid ionic liquid: Atomistic simulation. *J. Mol. Graph. Model.* **2020**, *101*, 107733. [CrossRef]
13. Peredo-Mancilla, D.; Dominguez, H. Adsorption of phenol molecules by sodium dodecyl sulfate (sds) surfactants deposited on solid surfaces: A computer simulation study. *J. Mol. Graph. Model.* **2016**, *65*, 108–112. [CrossRef]
14. Gao, S.; Bao, X.; Yu, L.; Wang, H.; Li, J.; Chen, X. Molecular dynamics study of "quasi-gemini" surfactant at n-decane/water interface: The synergistic effect of hydrophilic headgroups and hydrophobic tails of surfactants on the interface properties. *Colloids Surf. A* **2022**, *634*, 127899. [CrossRef]
15. Wei, Y.; Wang, X.; Dong, L.; Liu, G.; Xia, Q.; Yuan, S. Molecular dynamics study on the effect of surfactant mixture on their packing states in mixed micelles. *Colloids Surf. A* **2021**, *631*, 127714. [CrossRef]
16. Liu, Z.-Y.; Xu, Z.; Zhou, H.; Wang, Y.; Liao, Q.; Zhang, L.; Zhao, S. Interfacial behaviors of betaine and binary betaine/carboxylic acid mixtures in molecular dynamics simulation. *J. Mol. Liq.* **2017**, *240*, 412–419. [CrossRef]
17. Cai, H.-Y.; Zhang, Y.; Liu, Z.-Y.; Li, J.-G.; Gong, Q.-T.; Liao, Q.; Zhang, L.; Zhao, S. Molecular dynamics simulation of binary betaine and anionic surfactant mixtures at decane—Water interface. *J. Mol. Liq.* **2018**, *266*, 82–89. [CrossRef]
18. Su, L.; Sun, J.; Ding, F.; Gao, X.; Zheng, L. Effect of molecular structure on synergism in mixed zwitterionic/anionic surfactant system: An experimental and simulation study. *J. Mol. Liq.* **2021**, *322*, 114933. [CrossRef]
19. Ergin, G.; Lbadaoui-Darvas, M.; Takahama, S. Molecular structure inhibiting synergism in charged surfactant mixtures: An atomistic molecular dynamics simulation study. *Langmuir* **2017**, *33*, 14093–14104. [CrossRef] [PubMed]
20. Wei, Y.; Liu, G.; Wang, H.; Xia, Q.; Yuan, S. Exploring relationship of the state of n-dodecyl betaine in the solution monomer, at the interface and in the micelle via configurational entropy. *Colloids Surf. A* **2020**, *600*, 124975. [CrossRef]
21. Wei, Y.; Gao, F.; Wang, H.; Liu, G.; Xia, Q.; Yuan, S. A molecular dynamics study combining with entropy calculation on the packing state of hydrophobic chains in micelle interior. *J. Mol. Liq.* **2019**, *283*, 860–866. [CrossRef]

22. van der Spoel, D.; van Maaren, P.J.; Larsson, P.; Tîmneanu, N. Thermodynamics of hydrogen bonding in hydrophilic and hydrophobic media. *J. Phys. Chem. B* **2006**, *110*, 4393–4398. [CrossRef] [PubMed]
23. Zhang, Y.; Maginn, E.J. Direct correlation between ionic liquid transport properties and ion pair lifetimes: A molecular dynamics study. *J. Phys. Chem. Lett.* **2015**, *6*, 700–705. [CrossRef] [PubMed]
24. Rivera, J.L.; McCabe, C.; Cummings, P.T. Molecular simulations of liquid-liquid interfacial properties: Water--n-alkane and water-methanol-n-alkane systems. *Phys. Rev. E* **2003**, *67*, 011603. [CrossRef] [PubMed]
25. Wei, Y.; Wang, H.; Xia, Q.; Yuan, S. A molecular dynamics study on the dependence of phase behaviors and structural properties of two-dimensional interfacial monolayer on surface area. *Appl. Surf. Sci.* **2018**, *459*, 741–748. [CrossRef]
26. Johnson, E.R.; Keinan, S.; Mori-Sánchez, P.; Contreras-García, J.; Cohen, A.J.; Yang, W. Revealing noncovalent interactions. *J. Am. Chem. Soc.* **2010**, *132*, 6498–6506. [CrossRef]
27. Perdew, J.P.; Yue, W. Accurate and simple density functional for the electronic exchange energy: Generalized gradient approximation. *Phys. Rev. B* **1986**, *33*, 8800–8802. [CrossRef]
28. Lu, T.; Chen, F. Multiwfn: A multifunctional wavefunction analyzer. *J. Comput. Chem.* **2012**, *33*, 580–592. [CrossRef]
29. Humphrey, W.; Dalke, A.; Schulten, K. Vmd: Visual molecular dynamics. *J. Mol. Graph.* **1996**, *14*, 33–38. [CrossRef]
30. Lindahl, E.; Abraham, M.; Hess, B.; van der Spoel, D. Gromacs 2019 Source Code. Available online: https://doi.org/10.5281/zenodo.2424362 (accessed on 21 July 2023).
31. Bayly, C.I.; Cieplak, P.; Cornell, W.; Kollman, P.A. A well-behaved electrostatic potential based method using charge restraints for deriving atomic charges: The resp model. *J. Phys. Chem.* **1993**, *97*, 10269–10280. [CrossRef]
32. Jorgensen, W.L.; Chandrasekhar, J.; Madura, J.D.; Impey, R.W.; Klein, M.L. Comparison of simple potential functions for simulating liquid water. *J. Chem. Phys.* **1983**, *79*, 926–935. [CrossRef]
33. Bussi, G.; Donadio, D.; Parrinello, M. Canonical sampling through velocity rescaling. *J. Chem. Phys.* **2007**, *126*, 014101. [CrossRef] [PubMed]
34. Hess, B.; Bekker, H.; Berendsen, H.J.C.; Fraaije, J.G.E.M. Lincs: A linear constraint solver for molecular simulations. *J. Comput. Chem.* **1997**, *18*, 1463–1472. [CrossRef]
35. Essmann, U.; Perera, L.; Berkowitz, M.L.; Darden, T.; Lee, H.; Pedersen, L.G. A smooth particle mesh ewald method. *J. Chem. Phys.* **1995**, *103*, 8577–8593. [CrossRef]

Disclaimer/Publisher's Note: The statements, opinions and data contained in all publications are solely those of the individual author(s) and contributor(s) and not of MDPI and/or the editor(s). MDPI and/or the editor(s) disclaim responsibility for any injury to people or property resulting from any ideas, methods, instructions or products referred to in the content.

Article

Detachment of Dodecane from Silica Surfaces with Variable Surface Chemistry Studied Using Molecular Dynamics Simulation

Binbin Jiang [1], Huan Hou [2], Qian Liu [2,*], Hongyuan Wang [2], Yang Li [2], Boyu Yang [2], Chen Su [1] and Min Wu [1]

[1] State Key Laboratory of Water Resource Protection and Utilization in Coal Mining, China Energy Investment Group, Beijing 102211, China; 20029907@chnenergy.com.cn (B.J.)
[2] School of Chemical and Environmental Engineering, China University of Mining and Technology (Beijing), Beijing 100083, China
* Correspondence: liuqian331@163.com

Abstract: The adsorption and detachment processes of n-dodecane ($C_{12}H_{26}$) molecules were studied on silica surfaces with variable surface chemistry (Q^2, Q^3, Q^4 environments), using molecular dynamics simulations. The area density of the silanol groups varied from 9.4 to 0 per nm^2. The shrinking of the oil–water–solid contact line was a key step for the oil detachment, due to water diffusion on the three-phase contact line. The simulation results showed that oil detachment was easier and faster on a perfect Q^3 silica surface which had ($\equiv$Si(OH))-type silanol groups, due to the H-bond formation between the water and silanol groups. When the surfaces contained more Q^2 crystalline type which had ($\equiv$Si(OH)$_2$)-type silanol groups, less oil detached, due to the formations of H-bonds among the silanol groups. There were no silanol groups on the Si-OH 0 surface. Water cannot diffuse on the water–oil–silica contact line, and oil cannot detach from the Q^4 surface. The detachment efficiency of oil from the silica surface not only depended on the area density, but also on the types of silanol groups. The density and type of silanol groups depend on the crystal cleavage plane, particle size, roughness, and humidity.

Keywords: n-dodecane; SiO$_2$ surface; adsorption; detachment; MD simulation

Citation: Jiang, B.; Hou, H.; Liu, Q.; Wang, H.; Li, Y.; Yang, B.; Su, C.; Wu, M. Detachment of Dodecane from Silica Surfaces with Variable Surface Chemistry Studied Using Molecular Dynamics Simulation. *Molecules* **2023**, *28*, 4765. https://doi.org/10.3390/molecules28124765

Academic Editors: Shiling Yuan and Heng Zhang

Received: 4 May 2023
Revised: 10 June 2023
Accepted: 13 June 2023
Published: 14 June 2023

Copyright: © 2023 by the authors. Licensee MDPI, Basel, Switzerland. This article is an open access article distributed under the terms and conditions of the Creative Commons Attribution (CC BY) license (https://creativecommons.org/licenses/by/4.0/).

1. Introduction

The displacement mechanism of oil from solid surfaces plays a crucial role in a variety of technological applications such as enhanced oil recovery, self-cleaning materials, and flotation. Although the operating parameters and methods are different during the technological processes, two common essential and fundamental steps are involved, which are the liberation of oil from reservoir solids or host rocks, and the separation of oil and water [1,2]. The interface properties of the solid–oil–water system determine the oil detachment from the rock surfaces [3–7].

Due to the different intrinsic composition and structure of reservoir rocks, the reservoir rocks exhibit different characteristic properties and wettability. In general, the reservoir rocks can be divided into the hydrophobic type (carbonate rocks) and the hydrophilic type (silicate rocks). Experimental studies have mostly focused on hydrophilic mineral surfaces, specifically adsorption and wettability studies for crude oil extraction, mineral flotation, removal of contaminants, etc. [8–10]. The hydrophobicity of mineral surfaces is an important factor which affects the adsorption and detachment of crude oil, and the altering of the solid wettability can affect the oil recovery rate.

Nowadays, molecular dynamics (MD) simulations have been used to evaluate complex interactions from an atomistic point of view, and provide comprehensive information about the static and dynamic properties of the system at the molecular level [11]. For example,

MD simulations were used to study the aggregation mechanism of asphaltenes on the oil–water interface and in different types of solvents, the effect of different surfactant structure, the force field on the interfacial properties of water and oil, and the wettability variation of reservoir rocks at the molecular scale [12].

Liu et al. [13] firstly evaluated intermolecular interactions in oil–water surfactant hydrophilic silica systems, using MD simulations. Their simulation results demonstrated the mechanism of oil detachment from silica surfaces in the presence of dodecyl trimethyl ammonium bromide (DTAB). A three-stage model was revealed, which was composed of the formation of water channels, the diffusion of water on the solid surface, and the solubilization of alkane molecules in surfactant micelles. Surfactant molecules can lower interfacial tension between the oil and water interface, accompany water molecules via their tail or head groups into a water channel, and facilitate the process of oil detachment from a rock surface. Due to the H-bonding and electrostatic interaction, water diffused on the oil interface, and formed a gel layer on the hydrophilic silica surface. Oil molecules were completely removed and solubilized into the surfactant micelles.

The adsorption of surfactant molecules altered the wetting properties of the solid surface, which was important for an enhanced oil recovery. Moncayo-Riascos et al. [14–16] investigated how the ester groups of surfactants could interact with the rock surface and change its wettability. The contact angles of water were measured on the amorphous silica coated by six different organosilicon surfactants with different chain lengths. In addition, they evaluated the process of wettability alteration on glass surfaces in the presence of surfactants. The adsorption of surfactants onto the glass surfaces could change the surface wettability. Zhang et al. [17] evaluated the effect of ethoxylate group numbers of a non-anionic surfactant on the wettability alteration of a lignite surface. Based on their simulation results, it was found that when the surfactants had more ethoxylate groups, lignite surfaces became more hydrophobic after the adsorption of non-anionic surfactants.

In contact with crude oil, the rock surface changed into a hydrophobic one, due to the adsorption of polar components. Liu et al. [18] investigated whether the {1 0 − 1 4} surface of calcite in carbonate reservoirs was indeed water-wet under initial conditions. The contact angle of the calcite surface was 68.47 ± 3.6 deg. However, when the reservoirs came into contact with the crude oil, the components containing multiple hydroxyl functional groups, such as glycerol (GLYC), were adsorbed onto the water-wet calcite surface by hydrogen bonding and Coulomb interaction, and the alcite surfaces became hydrophobic. Mohammadail et al. [19] used MD simulation to evaluate the wettability alteration under a steam injection process for bitumen and heavy oil recovery. The simulation results showed that when there were more asphaltenes, the adsorption energy was higher between the bitumen/heavy oil and quartz surfaces, due to the Coulomb interactions. Additionally, quartz surfaces became more oil-wet when temperatures were above the water boiling temperature. They were extremely water-wet at ambient conditions.

The wettability alteration process on different types of reservoir rocks was studied by Mohammed and Gadikota through MD simulations [20]. Three rock surfaces, including illite, calcite and quartz, were constructed to evaluate the process of wettability alteration in contact with asphaltene molecules. The illite was the most hydrophobic among these three surfaces. Time-dependent self-diffusion of asphaltene monomers was realized, and played a crucial role in the wettability alteration process. Li et al. [21] studied the oil detachment process from four different types of rock surfaces, including quartz, siderite, calcite and dolomite. The wettability of dolomite rock surfaces was changed by all types of surfactants. However, the non-anionic one took the longest time among the other surfactants. The water channel formed quicker in the case of the anionic surfactant than with other types of surfactants.

The wetting characteristics of water were investigated on three typical inorganic minerals (calcite, quartz and montmorillonite) by Yang et al. [22]. The water–mineral interaction properties were obtained, including the interaction energy and adhesion work of the water–mineral interface, mineral wettability, and the structural and diffusion properties of water molecules near the surface. Their simulation results revealed that the diffusion properties of water molecules on mineral surfaces played an important role in the wetting process.

Most studies focused on rocks with different chemical compositions. However, surface characteristics (such as cleavage plane, particle size and porosity), pH, and ionic strength were shown to determine the adsorption and self-assembly of complex molecules. Zhu et al. [23] studied the effect of rock surface roughness on the oil detachment process from a quartz surface in the presence of β-cyclodextrins (βCD). Fateme, S. et al. [6], obtained a silica surface model database for the full range of variable surface chemistry and pH (Q^2, Q^3, Q^4 environments with adjustable degree of ionization), and explained the mechanistic details of the molecular adsorption of water vapor.

The number and arrangement of hydroxyl groups exposed on the mineral surface may also have an effect on the adsorption and detachment of oil molecules. Silica exists in different crystalline polymorphs and amorphous phases. The surface properties of different types of silica were related to the cleavage crystal plane, syntheses methods, roughness and thermal pretreatment. In this work, MD simulations were used to study the dynamic process of the aggregation and detachment of n-dodecane from silica surfaces with variable surface chemistry. Silica surfaces were obtained by mixing different crystalline silica surfaces (Q^2, Q^3, Q^4 environments) and non-ionizing. The adsorption process and configurations of $C_{12}H_{26}$ molecules were simulated on different silica surfaces. Then, the detachment processes of $C_{12}H_{26}$ molecules from different silica surfaces were studied in the presence of sodium dodecyl sulphate (SDS) solutions.

2. Results

2.1. Adsorption of $C_{12}H_{26}$ on Different Silica Surfaces

The arrangement of oil molecules on the solid surface is important for the oil detachment from the solid surface. At the beginning of the simulations, ninety $C_{12}H_{26}$ molecules were put above the silica surfaces. Figure 1a illustrated the variation of energy LJ (SR) between silica surfaces and oil molecules during the adsorption process of $C_{12}H_{26}$. The energy of five systems decreased at the beginning of the simulations. After 10 ns, the LJ energy profile of the systems reached a plateau, indicating that the five systems all reached equilibrium. Figure 1b compared the interaction energy between $C_{12}H_{26}$ and silica surfaces as a function of time during the adsorption process. The interaction energy between $C_{12}H_{26}$ and Si-OH 0 was lowest. The oil detachment from Si-OH 0 might be the most difficult, which can be proved by the final configurations of $C_{12}H_{26}$ detachment simulations at 50 ns and the oil and water density profiles.

The morphology of the oil adsorbed is related to the types and topography of the solid surfaces. In previous works [13], the well-ordered and layered oil molecules formed a close-packed structure on hydrophilic surfaces. Zhu et al. [23] reported that the interaction energy between oil and silica surfaces increased with the depth of the grooves.

Side views of final configurations at different silica surfaces are shown in Figure 2. Five silica surfaces were used, with variable surface chemistry (Q^2, Q^3, Q^4 environments), and which were cleaved on different crystal planes and had different types and area densities of silanol groups. For all the systems, $C_{12}H_{26}$ molecules were adsorbed on the SiO_2 surfaces as layered structures. The first layer of $C_{12}H_{26}$ aligned parallel to the surface. There were four layers on the Si-OH 9.4/nm^2. Figure 3 shows the number density profiles of $C_{12}H_{26}$ in the direction (z) normal to the different silica surfaces at 60 ns. The density profile of Si-OH 9.4/nm^2 had four peaks. For Si-OH 6.9/nm^2, the Si-OH 4.7/nm^2 systems which were mixed crystalline, the density profiles had three peaks. The arrangement of $C_{12}H_{26}$ molecules can be divided into three layers (as shown in Figure 2b,c). The distribution of $C_{12}H_{26}$ molecules becomes more and more disordered on the second and third layers. For the Si-OH 2.4/nm^2 and Si-OH 0 systems, there were two broad peaks on the density profiles, and the adsorptions of oil molecules can be divided into two layers (Figure 2).

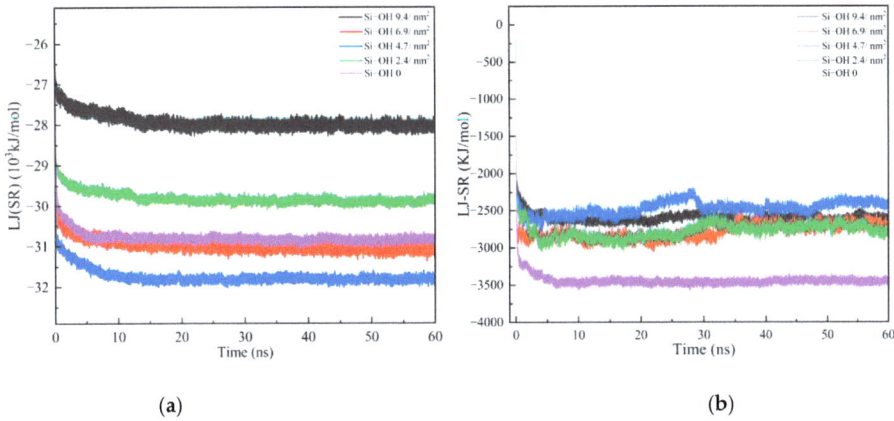

Figure 1. (**a**) The LJ (SR) energy profiles of $C_{12}H_{26}$ adsorption at different silica surfaces as a function of time. (**b**) The interaction energy between $C_{12}H_{26}$ and silica surfaces as a function of time during the adsorption process.

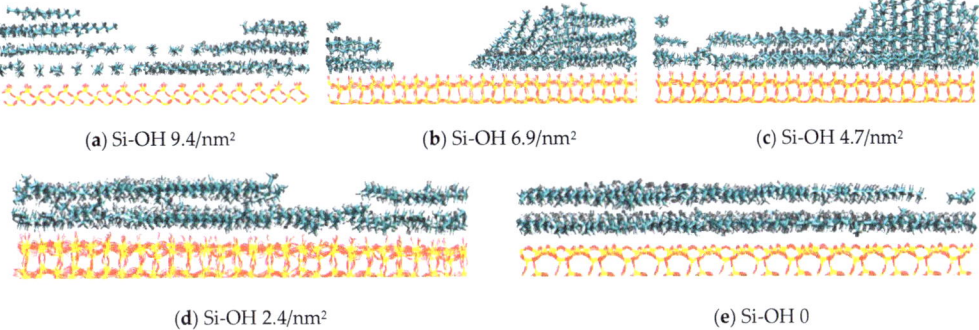

Figure 2. The side views of oil layers on different silica surfaces. Colors for atoms scheme are $C_{12}H_{26}$ (green), H (white), O (red), and Si (yellow).

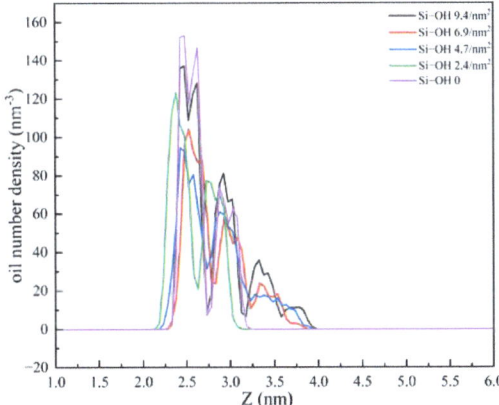

Figure 3. The number density profiles of $C_{12}H_{26}$ in the direction (z) normal to the different silica surfaces at 60 ns.

Top views of the final configurations of $C_{12}H_{26}$ after adsorption equilibrium are presented in Figure 4. Unlike the previous work [13], the first layer of oil molecules fully covered the silica surfaces. In this work, part of the silica surfaces was covered by $C_{12}H_{26}$ molecules. The rest of the oil molecules preferred to adsorb onto the first oil layer, rather than directly onto the silica surfaces. The distances between the first layers and surfaces were about 0.60 nm.

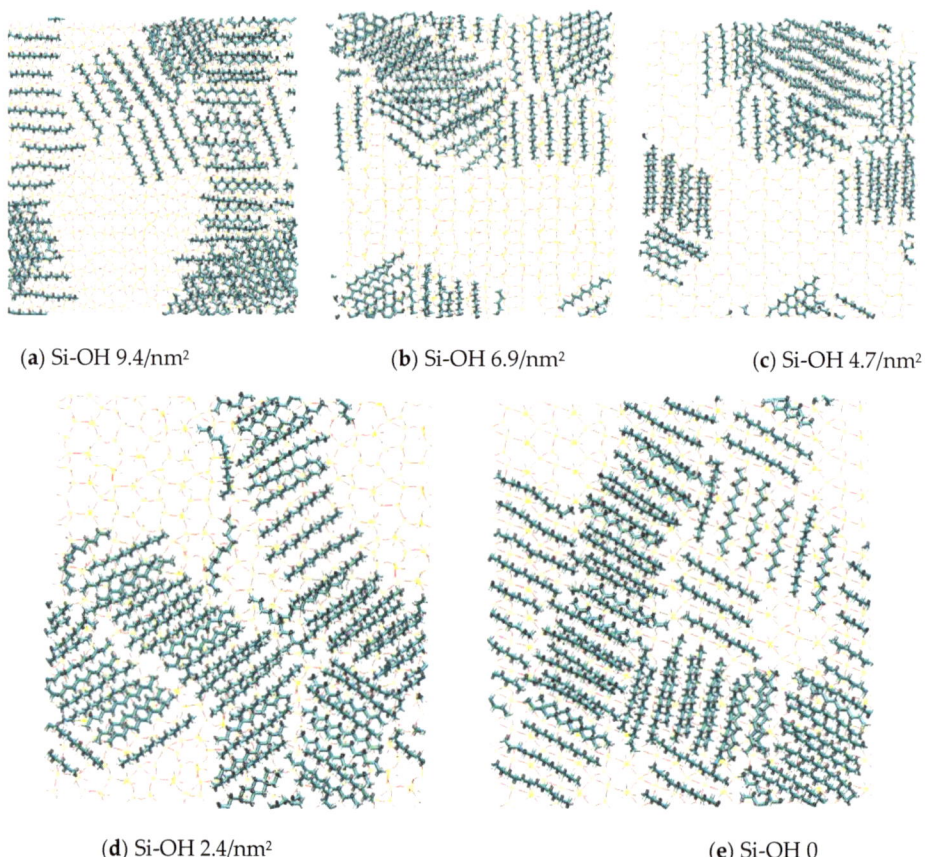

(a) Si-OH 9.4/nm²   (b) Si-OH 6.9/nm²   (c) Si-OH 4.7/nm²

(d) Si-OH 2.4/nm²   (e) Si-OH 0

Figure 4. Final adsorption configurations of $C_{12}H_{26}$ molecules on different silica surfaces (top view). Colors for atoms scheme are $C_{12}H_{26}$ (green), H (white), O (red), and Si (yellow).

2.2. Detachment of $C_{12}H_{26}$ on Different Silica Surfaces

After the $C_{12}H_{26}$ molecules adsorbed on the silica surfaces, SDS solutions were added above the oil layers. After 50 ns simulations, the final configurations of the five systems are shown in Figure 5. For Si-OH 9.4/nm², Si-OH 6.9/nm², Si-OH 4.7/nm² and Si-OH 2.4/nm², the oil molecules can be partly detached. The oil layer cannot be detached from the Si-OH 0 surface.

Kolev et al. [24] carried out experiments on the detachment of oil drops from glass substrates, and obtained results showing that the three-phase (solid–oil–water) contact line shrank spontaneously in anionic surfactant solutions, and eventually the oil drop detached from the substrate. The shrinking of the three-phase s contact line was due to the molecular penetration (diffusion) of the water molecules between the oil drop and the solid phase. The hydrogen bonds between the solid surface and water molecules are vital for the formation of water channels and the diffusion of water onto the solid surface. The

Si-OH 0 system is a Q^4 crystalline type which is prepared from the Q^3 surface by complete condensation of surface silanol groups at high temperature. No silanol groups were on the Si-OH 0 surface. Few H-bonds can form on the Si-OH 0 surface, so water molecules cannot penetrate through the oil layer and cannot diffuse at the Si-OH 0 surface. The oil arrangement of the first layer on the Si-OH 0 surface did not change after the addition of the surfactant solution. Oil molecules cannot be replaced by water molecules on the Si-OH 0 surface (as shown in Figure 5).

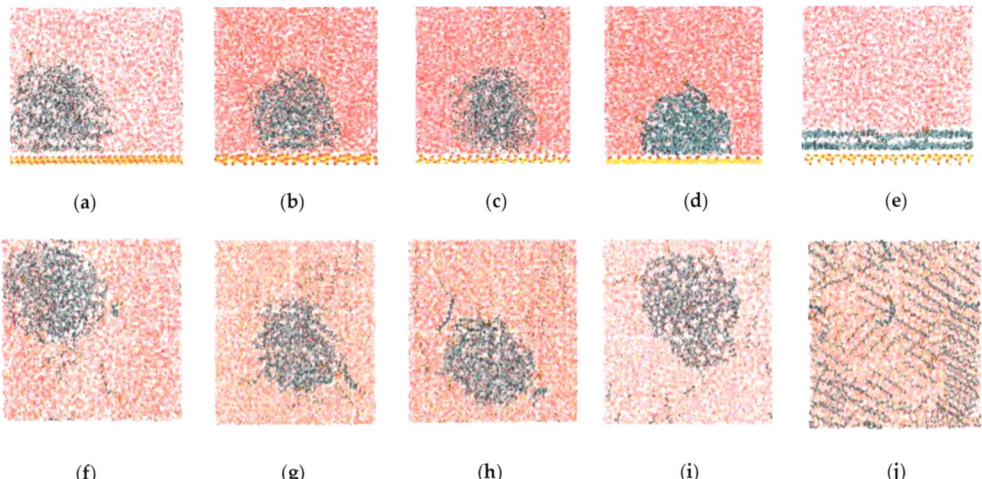

Figure 5. The final configurations of $C_{12}H_{26}$ detachment simulations at 50 ns (top view and side view). Colors for atoms scheme are $C_{12}H_{26}$ (green), H (white), O (red), and Si (yellow). Side views: (**a**) Si-OH 9.4/nm^2 (**b**) Si-OH 6.9/nm^2 (**c**) Si-OH 4.7/nm^2 (**d**) Si-OH 2.4/nm^2 (**e**) Si-OH 0. Top views: (**f**) Si-OH 9.4/nm^2 (**g**) Si-OH 6.9/nm^2 (**h**) Si-OH 4.7/nm^2 (**i**) Si-OH 2.4/nm^2 (**j**) Si-OH 0.

The number density profiles of the $C_{12}H_{26}$ molecules in the direction (z) perpendicular to the surface at different times are shown in Figure 6. For Si-OH 6.9/nm^2 and Si-OH 4.7/nm^2, the first peak of the oil number density curves decreased and disappeared at 5 ns. For Si-OH 9.4/nm^2 and Si-OH 2.4/nm^2, the first peaks of the oil number density curves decreased much slower than for the Si-OH 6.9/nm^2, Si-OH 4.7/nm^2. For Si-OH 0, the peak of oil number density at 3.15 nm did not decrease after 50 ns detachment simulations. These results agreed well with the final configurations of oil molecules at the end of the simulations. At the Si-OH 0, the arrangement of the first oil layer did not change.

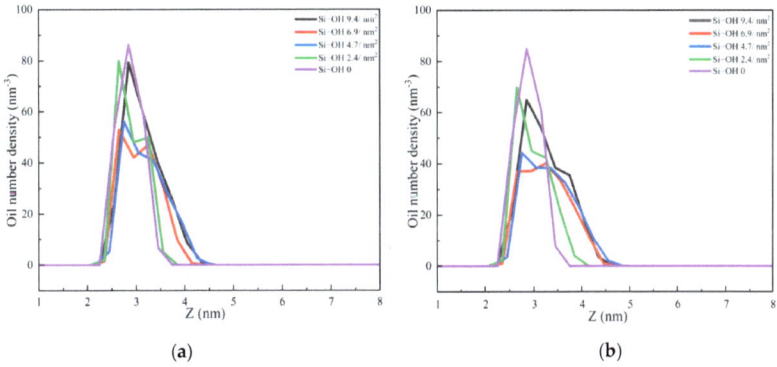

Figure 6. *Cont.*

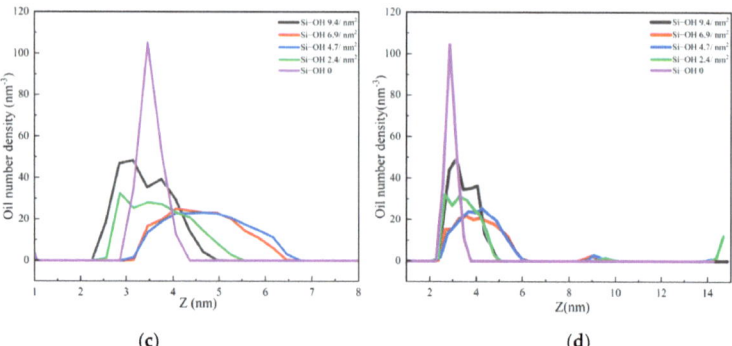

Figure 6. The number density profiles of the $C_{12}H_{26}$ molecules in the direction (z) perpendicular to the surface at different times. (**a**) 500 ps (**b**) 1 ns (**c**) 5 ns (**d**) 10 ns.

The number density profiles of water molecules in the direction (z) perpendicular to the surface at different times are shown in Figure 7. At the beginning of the simulations, the number densities of the water molecules were zero below the oil layer (z < 2.5 nm). The key to oil detachment on a hydrophilic surface is the formation of water channels and the diffusion of water on the three-phase contact line [13]. Once the water channels are formed, the oil will be detached rapidly, due to the weak interaction between the alkane molecules and silica surfaces. At 500 ps, new peaks emerged at silica surfaces for the Si-OH 6.9/nm², Si-OH 4.7/nm² systems. At 1ns, new peaks emerged for the Si-OH 9.4/nm² and Si-OH 2.4/nm². H-bonds are the main driving force for the formation of water channels and the diffusion of water molecules on the silica surfaces. Under the H-bonds between the water molecule and the silica surface, water penetrated through the oil layer and the water channel formed through the oil layers. The density of water molecules at the silica surface increased with the evolution of time.

Once the water-channel was formed, water molecules diffused on the solid surfaces which contained Q^2 and Q^3 surface environments. The solid–oil–water contact line was shrinking. The oil molecules were replaced by water molecules. The number densities of oil decreased as a function of time, as shown in Figure 6. The water channel became wider. More water molecules entered into the solid surface, and the new peaks of the water density profiles increased as a function of time (Figure 7c,d). Unlike the previous work [13], water number densities at the solid surfaces were lower than those in the bulk. The continuous water film could not form at the end of the simulations, and the oil did not completely detach. In contrast, a new peak of the water number density profile did not emerge at the Si-OH 0 surface. The water could not diffuse and could not replace the oil molecules on the Si-OH 0 surface.

Figure 8 shows the H-bond number profiles between the water molecules and silica surfaces as a function of time during the oil detachment simulations. Although Si-OH 9.4/nm² and Si-OH 6.9/nm² had more silanol groups pre nm², they all contained Q^2 surface environments, which had two silanol groups per superficial silicon atom (=Si(OH)₂). Figure 9 shows the H-bonds on the Si-OH 6.9/nm² surface at 50ns. Si-OH 6.9/nm² (Figure 10) contained mixed Q^2/Q^3(1:1) surface environments. As shown in Figure 9, Si(1) is a Q^2 type. Si(2), Si(3), Si(4), and Si(5) are Q^3 types. Silanol groups on Si(1) formed H-bonds with both water and silanol groups of Si(2), Si(5). However, the silanol groups of Si(3) and Si(4) only formed H-bonds with water. When the area density of the silanol groups increased from 4.7 to 9.4 per nm², the surface contained 50–100% Q^2 surface environments. The number of hydrogen bonds among the silanol groups increased when the surface contained more Q^2 surface (Figure 10), which hindered the diffusion of water. H-bonds formed among silanol groups at Si-OH 9.4/nm² were more than Si-OH 6.9/nm². The shrinking of the oil–water–solid contact line became slower as the silanol numbers and Q^2 contents increased.

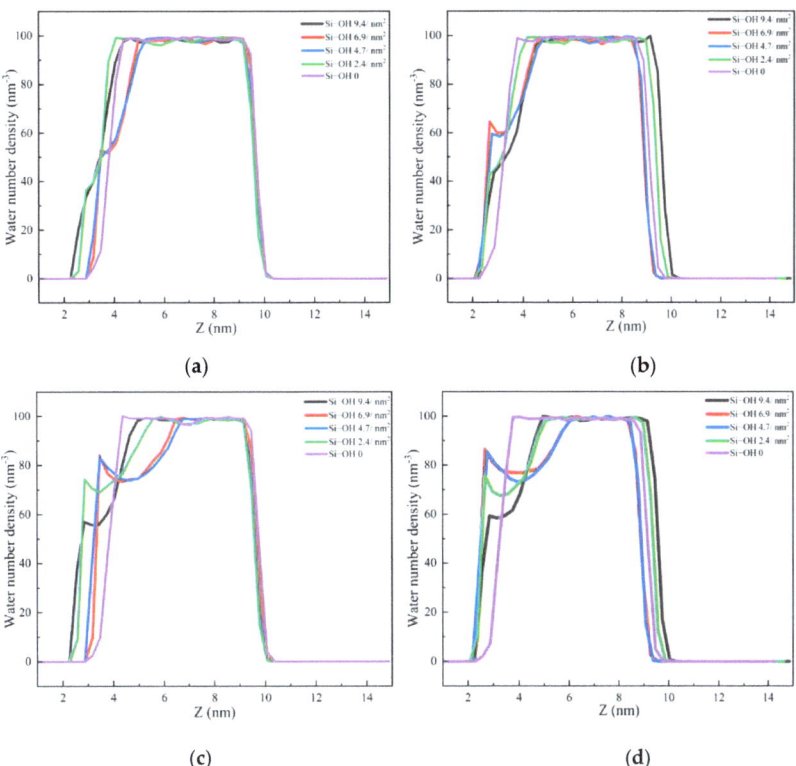

Figure 7. The number density profiles of the water molecules in the direction (z) perpendicular to the surfaces at different times. (**a**) 500 ps (**b**) 1 ns (**c**) 5 ns (**d**) 10 ns.

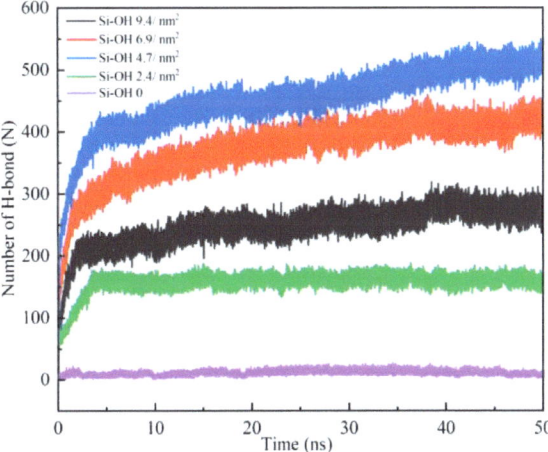

Figure 8. The number of hydrogen bonds formed between water molecules and SiO_2 surface as a function of time.

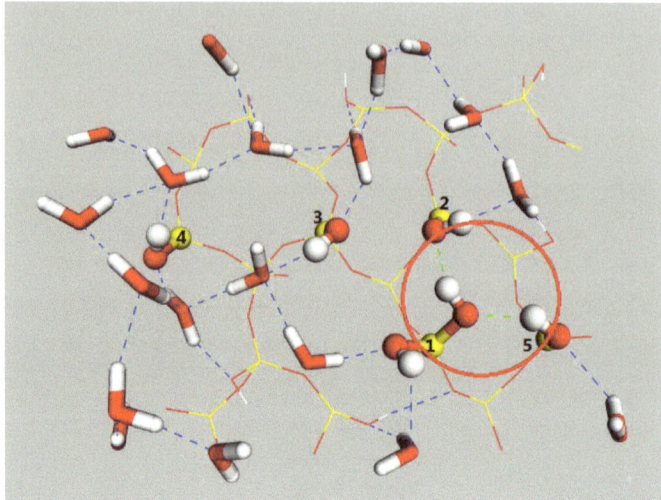

Figure 9. H-bonds formed on Si-OH 6.9/nm^2 surface at 50 ns. Colors for atoms scheme are H (white), O (red), and Si (yellow), H-bonds between water and silanol groups (blue), H-bonds among silanol groups (green). Si(1) is a Q^2 type. Si(2), Si(3), Si(4), and Si(5) are Q^3 types.

Si-OH 4.7/nm^2 are perfect Q^3 crystalline, which has one silanol group per superficial silicon atom ($\equiv$Si(OH)) type and the distribution of silanol groups are well-ordered (as shown in Figure 10). The H-bond number between Si-OH 4.7/nm^2 and water increased more quickly than that on Si-OH 9.4/nm^2 and Si-OH 6.9/nm^2. Oil molecules detached faster on Si-OH 4.7/nm^2. For Si-OH 2.4/nm^2, which contained Q^3/Q^4 mixed surface environments, the H-bond number among the silanol groups was zero. Si-OH 2.4/nm^2 had fewer silanol groups, the H-bond number between the water and silanol groups increased slowly, and the H-bond number was less than for Si-OH 9.4/nm^2, Si-OH 6.9/nm^2. Water molecules diffused faster on Si-OH 9.4/nm^2 and Si-OH 6.9/nm^2 than on Si-OH 2.4/nm^2. However, the Si-OH 0 surface had no silanol groups on the surface, and few H-bonds formed. Water diffusion was difficult on the Si-OH 0 surface, so the solid–oil–water contact line did not shrink. The water could not replace the oil molecules on the Si-OH 0 surface.

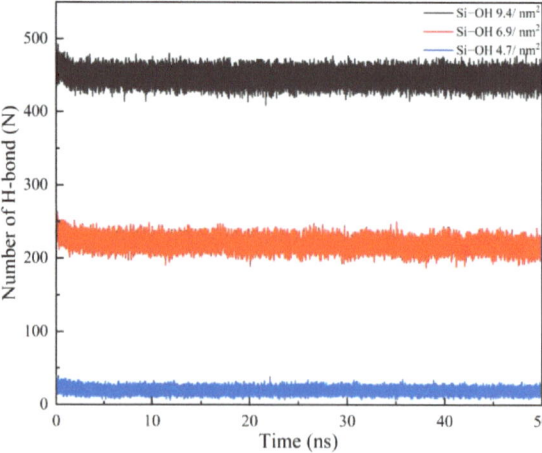

Figure 10. The number of hydrogen bonds among silanol groups as function of time.

3. Discussion

The shrinking of the oil–water–solid contact line was a key step in the oil detachment, due to water diffusion on the three-phase contact line. The formation of H-bonds was one of the driving forces for water diffusion on the solid surface. Silica exists in various crystalline polymorphs and amorphous phases. The type and area density of silanol groups depends on the crystal plane, particle size, synthesis protocol, thermal pretreatment, roughness and porosity.

In this work, silica surfaces were used with average area density of silanol groups of between 9.4 and 0 per nm^2 and contained variable Q^2, Q^3, Q^4 surface environments. Some hydrated cleavage planes of quartz, surfaces of large silica nanoparticles, and various forms of silica at high pH contain Q^2 surface environments (Figure 11a) and mixed Q^2/Q^3 surface environments (Figure 11b). The area density of the silanol groups is in the range of 9.4 to 4.7 per nm^2. The silanol groups of the Q^2 surface were (-Si(OH)$_2$), which was two silanol groups per superficial silicon atom. H-bonds could be formed among the silanol groups on the Q^2 surface. When the area density of the silanol groups increased from 4.7 to 9.4 per nm^2, the surface contained 50-100% Q^2 surface environments. The shrinking of oil–water–solid contact line became slower as the silanol numbers and Q^2 contents increased.

A perfect Q^3 surface contains one silanol group per superficial silicon atom (=Si(OH) and 4.7 Si-OH groups per nm^2 (Figure 11c). Few H-bonds could be formed among silanol groups on the Q^3 surface (as shown in Figure 10). H-bonds preferred to form between water and silanol groups (as shown in Figure 8), which promoted the diffusion of water on the Q^3 surface. The oil detachment was easier and faster on Si-OH 6.9/nm^2 and 4.7/nm^2 than on Si-OH 9.4/nm^2.

A pure Q^4 surface model (Figure 11e) was prepared from a Q^3 surface by complete condensation of the surface silanol groups at high temperature. The silanol density of the Q^4 surface was zero. Few H-bonds could be formed between the water and silanol groups on the Q^4 surface. Water cannot diffuse and oil cannot detach on the Q^4 surface.

Silica surfaces and nanoparticles annealed at 200–1000 °C comprise Q^2 and Q^4 environments, and the average area density of the surface silanol groups is in the range of 4.7 to 0 per nm^2. Si-OH 2.4/nm^2 were mixed Q^3/Q^4 environments (Figure 11d). H-bond numbers between the water and silanol groups of Si-OH 2.4/nm^2 were much less than for the Q^2/Q^3 surface, and so the three-phase contact line of Si-OH 2.4/nm^2 shrank slowly, and less oil could be removed.

(a) Si-OH 9.4/nm^2

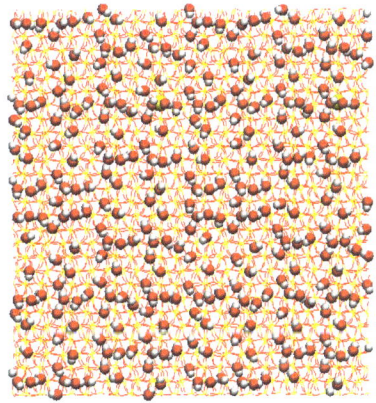
(b) Si-OH 6.9/nm^2

Figure 11. Cont.

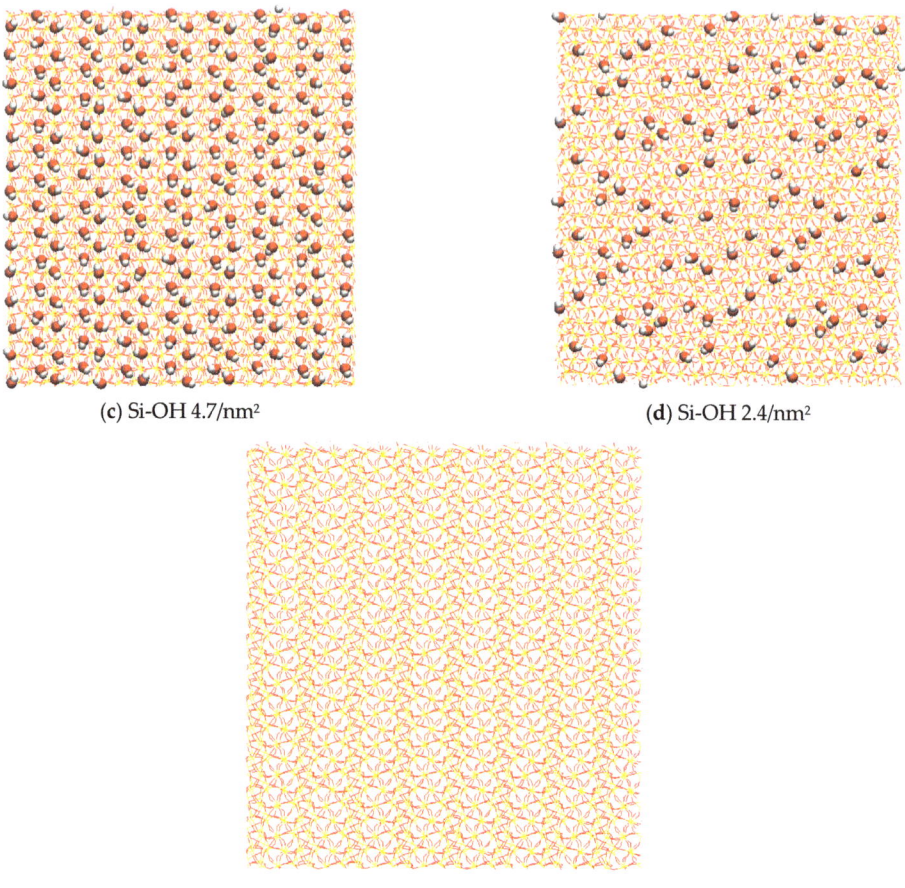

Figure 11. Top views of different silica surfaces. Colors for atoms scheme are H (white), O (red), and Si (yellow).

4. Materials and Methods

4.1. Model Systems

In this work, the $C_{12}H_{26}$ was used as an ideal model of the oil molecule. The $C_{12}H_{26}$ molecule and dodecyl sulfate ion were constructed using Avogadro-1.2.0n software. Silica surfaces were from Fateme's work [6] and neutrally charged. The Q^2 surface was derived from the (100) cleavage plane of α-quartz, and contained 9.4 Si-OH groups per nm^2. The Q^3 surface models were derived from the (101) cleavage plane of α-cristobalit, and contained 4.7 Si-OH groups per nm^2. The pure Q^4 surface model was prepared from the Q^3 surface by complete condensation of the surface silanol groups at high temperature. After energy minimization, some Si-O bonds stretched 10% and the density of the silanol groups was zero.

The SiO_2 surfaces are shown in Figure 11. Figure 11a–e indicated the hydroxylation densities of silicon surfaces of $9.4/nm^2$, $6.9/nm^2$, $4.7/nm^2$, $2.4/nm^2$ and 0, respectively. Si-OH $9.4/nm^2$ are Q^2 crystalline, and contained 9.4 Si-OH groups per nm^2. Si-OH $6.9/nm^2$ are Q^2 and Q^3 mixed crystalline, and contained 6.9 Si-OH groups per nm^2. Si-OH $4.7/nm^2$ are Q^3 crystalline, and contained 6.9 Si-OH groups per nm^2. Si-OH $2.4/nm^2$ was the Q^3 and Q^4 mixed crystalline type, and the Si-OH group density was 2.4 per nm^2. Si-OH 0 was the Q^4 crystalline type, and there were no Si-OH groups on the surface. The silica surface model box size was about $6.9 \times 6.9 \times 10.0$ nm^3. The energy minimization and equilibrium simulations were performed on five surfaces before the simulation process, to ensure the stability of the surface structure.

4.2. Computational Details

GROMMACS 2019.3 software package was employed to carry out all the MD simulations. The all-atom optimized performance for the liquid systems (OPLS-AA) force field [25–27] was adopted for all the potential. The simulation parameters for SDS and oil molecules used in this study were derived from the AMBER force field [28]. Water molecules were described by the simple point charge/extend (SPC/E) model [29]. The atomic types and atomic charges of SiO$_2$ in this paper are given in Table 1.

Table 1. Atom type symbols and atomic charges of silica used in this work.

Atom	Atom Type	Charge (e)
Si	SC4	1.1
Si-O-Si	OC23	−0.55
Si-O-H	OC24	−0.675
H	HOY	0.4

At the beginning of the simulations, ninety $C_{12}H_{26}$ molecules were put above each silica surface. After the adsorption of $C_{12}H_{26}$ molecules, the last frame of the trajectory was used in the detachment process. SDS solutions were added into the systems. Each of the systems was firstly minimized, using the steepest descent method. After the minimization, MD simulations (NVT) were carried out under canonical ensemble. Periodic boundary conditions were applied in the x, y, and z directions. For each system, the absorption and detachment process simulations were performed for 60 ns and 50 ns with a time step of 2 fs. The temperature was set to 300 K. The VMD 1.9.3 package was used for visualization More details of the parameters for the simulation systems are given in Table 2.

Table 2. Details of simulation systems.

	$C_{12}H_{26}$/N	SOL/N	SDS/N
Si-OH 9.4/nm^2	90	9964	5
Si-OH 6.9/nm^2	90	8941	5
Si-OH 4.7/nm^2	90	8890	5
Si-OH 2.4/nm^2	90	9650	5
Si-OH 0	90	9096	5

5. Conclusions

The area density and types of silanol groups played a role in the properties of the silica surfaces. In this paper, the adsorption and detachment processes of dodecane molecules were simulated on five silica surfaces with Q^2, Q^3, and Q^4 surface environments. The results demonstrated that more $C_{12}H_{26}$ were adsorbed on the Q^4 crystalline type. When the surfaces contained more Q^2 crystalline type which had ($\equiv$Si(OH)$_2$)-type silanol groups, less oil detached, due to the formations of H-bonds among the silanol groups. More oil detached on the Si-OH 4.7/nm^2, which were Q^3 crystalline type and had ($\equiv$Si(OH))-type silanol groups. There were no silanol groups on the Si-OH 0 surface, even though, in the presence of the surfactant solutions, the water cannot diffuse and the oil cannot detach from the Si-OH 0 surface. The area density and type of silanol groups both affect the formation of water channels and the diffusion of water on the silica surfaces, which are essential and fundamental steps for the liberation of oil from reservoir solids or host rocks.

Author Contributions: Conceptualization, B.J. and Q.L.; methodology, H.W.; software, H.H.; validation, B.J., H.H. and H.W.; formal analysis, Y.L.; investigation, B.Y.; resources, C.S.; data curation, B.J., H.H. and Y.L.; writing—original draft preparation, Q.L.; writing—review and editing, M.W. and C.S.; visualization, Q.L.; supervision, C.S.; project administration, M.W.; funding acquisition, C.S. All authors have read and agreed to the published version of the manuscript.

Funding: This is supported by Open Fund of State Key Laboratory of Water Resource Protection and Utilization in Coal Mining (Grant No. GJNY-18-73.13).

Institutional Review Board Statement: Not applicable.

Informed Consent Statement: Not applicable.

Data Availability Statement: The data that support the findings of this study are available from the corresponding author upon reasonable request.

Conflicts of Interest: The authors declare no conflict of interest.

Sample Availability: Not applicable.

References

1. Dudášová, D.; Simon, S.; Hemmingsen, P.V.; Sjöblom, J. Study of asphaltenes adsorption onto different minerals and clays: Part 1. Experimental adsorption with UV depletion detection. *Colloid Surf. A* **2008**, *317*, 1–9. [CrossRef]
2. He, L.; Lin, F.; Li, X.; Sui, H.; Xu, Z. Interfacial sciences in unconventional petroleum production: From fundamentals to applications. *Chem. Soc. Rev.* **2015**, *44*, 5446–5494. [CrossRef] [PubMed]
3. Dong, M.; Ma, S.; Liu, Q. Enhanced heavy oil recovery through interfacial instability: A study of chemical flooding for Brintnell heavy oil. *Fuel* **2009**, *88*, 1049–1056. [CrossRef]
4. Alvarado, V.; Manrique, E. Enhanced Oil Recovery: An Update Review. *Energies* **2010**, *3*, 1529–1575. [CrossRef]
5. Li, X.; Sun, W.; Wu, G.; He, L.; Li, H.; Sui, H. Ionic Liquid Enhanced Solvent Extraction for Bitumen Recovery from Oil Sands. *Energy Fuels* **2011**, *25*, 5224–5231. [CrossRef]
6. Emami, F.S.; Puddu, V.; Berry, R.J.; Varshney, V.; Patwardhan, S.V.; Perry, C.C.; Heinz, H. Force Field and a Surface Model Database for Silica to Simulate Interfacial Properties in Atomic Resolution. *Chem. Mater.* **2014**, *26*, 2647–2658, Correction in *Chem. Mater.* **2016**, *28*, 406–407. [CrossRef]
7. Johannessen, A.M.; Spildo, K. Enhanced Oil Recovery (EOR) by Combining Surfactant with Low Salinity Injection. *Energy Fuels* **2013**, *27*, 5738–5749. [CrossRef]
8. Adams, J.J. Asphaltene Adsorption, a Literature Review. *Energy Fuels* **2014**, *28*, 2831–2856. [CrossRef]
9. Liu, J.; Xu, Z.; Masliyah, J. Processability of Oil Sand Ores in Alberta. *Energy Fuels* **2005**, *19*, 2056–2063. [CrossRef]
10. Liu, J.; Xu, Z.; Masliyah, J. Studies on Bitumen−Silica Interaction in Aqueous Solutions by Atomic Force Microscopy. *Langmuir* **2003**, *19*, 3911–3920. [CrossRef]
11. Ahmadi, M.; Hou, Q.; Wang, Y.; Chen, Z. Interfacial and molecular interactions between fractions of heavy oil and surfactants in porous media: Comprehensive review. *Adv. Colloid Interface Sci.* **2020**, *283*, 102242. [CrossRef] [PubMed]
12. Liu, J.; Zhao, Y.; Ren, S. Molecular Dynamics Simulation of Self-Aggregation of Asphaltenes at an Oil/Water Interface: Formation and Destruction of the Asphaltene Protective Film. *Energy Fuels* **2015**, *29*, 1233–1242. [CrossRef]
13. Liu, Q.; Yuan, S.; Yan, H.; Zhao, X. Mechanism of Oil Detachment from a Silica Surface in Aqueous Surfactant Solutions: Molecular Dynamics Simulations. *J. Phys. Chem. B* **2012**, *116*, 2867–2875. [CrossRef]
14. Moncayo-Riascos, I.; Cortes, F.B.; Hoyos, B.A. Chemical Alteration of Wettability of Sandstones with Polysorbate 80. Experimental and Molecular Dynamics Study. *Energy Fuels* **2017**, *31*, 11918–11924. [CrossRef]
15. Moncayo-Riascos, I.; Hoyos, B.A. Fluorocarbon versus hydrocarbon organosilicon surfactants for wettability alteration: A molecular dynamics approach. *J. Ind. Eng. Chem.* **2020**, *88*, 224–232. [CrossRef]
16. Moncayo-Riascos, I.; de León, J.; Hoyos, B.A. Molecular Dynamics Methodology for the Evaluation of the Chemical Alteration of Wettability with Organosilanes. *Energy Fuels* **2016**, *30*, 3605–3614. [CrossRef]
17. Zhang, L.; Li, B.; Xia, Y.; Liu, S. Wettability modification of Wender lignite by adsorption of dodecyl poly ethoxylated surfactants with different degree of ethoxylation: A molecular dynamics simulation study. *J. Mol. Graph. Model.* **2017**, *76*, 106–117. [CrossRef] [PubMed]
18. Liu, Y.; He, Y.; Liu, Y.; Jiang, Y.; Zhang, Q.; Sun, Z.; Di, C. A Mechanistic Study of Wettability Alteration of Calcite as an Example of Carbonate Reservoirs Using Molecular Dynamics Simulation. *J. Energy Resour. Technol.* **2022**, *144*, 103006. [CrossRef]
19. Ahmadi, M.; Chen, Z. Molecular Dynamics Investigation of Wettability Alteration of Quartz Surface under Thermal Recovery Processes. *Molecules* **2023**, *28*, 1162. [CrossRef]
20. Mohammed, S.; Gadikota, G. Dynamic Wettability Alteration of Calcite, Silica and Illite Surfaces in Subsurface Environments: A Case Study of Asphaltene Self-Assembly at Solid Interfaces. *Appl. Surf. Sci.* **2020**, *505*, 144516. [CrossRef]
21. Li, X.; Xue, Q.; Zhu, L.; Jin, Y.; Wu, T.; Guo, Q.; Zheng, H.; Lu, S. How to select an optimal surfactant molecule to speed up the oil-detachment from solid surface: A computational simulation. *Chem. Eng. Sci.* **2016**, *147*, 47–53. [CrossRef]

22. Yang, Z.; Hu, R.; Chen, Y.-F. Molecular Origin of Wetting Characteristics on Mineral Surfaces. *Langmuir* **2023**, *39*, 2932–2942.
23. Zhu, X.; Chen, D.; Zhang, Y.; Wu, G. Insights into the Oil Adsorption and Cyclodextrin Extraction process on Rough Silica Surface by Molecular Dynamics Simulation. *J. Phys. Chem. C* **2018**, *122*, 2997–3005. [CrossRef]
24. Peter, A.K.; Krassimir, D.D.; Vesselin, L.K.; Theodor, D.G.; Mila, I.T.; Gunter, B. Detachment of Oil Drops from Solid Surfaces in Surfactant Solutions: Molecular Mechanisms at a Moving Contact Line. *Ind. Eng. Chem. Res.* **2005**, *44*, 1309–1321.
25. Koretsky, C.M.; Sverjensky, D.A.; Sahai, N. A model of surface site types on oxide and silicate minerals based on crystal chemistry; implications for site types and densities, multi-site adsorption, surface infrared spectroscopy, and dissolution kinetics. *Am. J. Sci.* **1998**, *298*, 349–438. [CrossRef]
26. Jorgensen, W.L.; Maxwell, D.S.; Tirado-Rives, J. Development and Testing of the OPLS All-Atom Force Field on Conformational Energetics and Properties of Organic Liquids. *J. Am. Chem. Soc.* **1996**, *118*, 11225–11236. [CrossRef]
27. Jorgensen, W.L.; Tirado-Rives, J. The OPLS [optimized potentials for liquid simulations] potential functions for proteins, energy minimizations for crystals of cyclic peptides and crambin. *J. Am. Chem. Soc.* **1988**, *110*, 1657–1666. [CrossRef] [PubMed]
28. Zhang, L.; Lu, X.; Liu, X.; Yang, K.; Zhou, H. Surface Wettability of Basal Surfaces of Clay Minerals: Insights from Molecular Dynamics Simulation. *Energy Fuels* **2016**, *30*, 149–160. [CrossRef]
29. Toukan, K.; Rahman, A. Molecular-dynamics study of atomic motions in water. *Phys. Rev. B* **1985**, *31*, 2643–2648. [CrossRef] [PubMed]

Disclaimer/Publisher's Note: The statements, opinions and data contained in all publications are solely those of the individual author(s) and contributor(s) and not of MDPI and/or the editor(s). MDPI and/or the editor(s) disclaim responsibility for any injury to people or property resulting from any ideas, methods, instructions or products referred to in the content.

Article

Harmonic Vibrational Frequency Simulation of Pharmaceutical Molecules via a Novel Multi-Molecular Fragment Interception Method

Linjie Wang [1], Pengtu Zhang [1], Yali Geng [1], Zaisheng Zhu [1] and Shiling Yuan [1,2,*]

[1] School of Chemical Engineering, Shandong Institute of Petroleum and Chemical Technology, Dongying 257061, China; linjiewang1989@hotmail.com (L.W.); gogogoooliver@163.com (P.Z.); yaligeng0313@163.com (Y.G.); longinces@126.com (Z.Z.)
[2] School of Chemistry and Chemical Engineering, Shandong University, Jinan 250199, China
* Correspondence: shilingyuan@sdu.edu.cn

Citation: Wang, L.; Zhang, P.; Geng, Y.; Zhu, Z.; Yuan, S. Harmonic Vibrational Frequency Simulation of Pharmaceutical Molecules via a Novel Multi-Molecular Fragment Interception Method. *Molecules* 2023, *28*, 4638. https://doi.org/10.3390/molecules28124638

Academic Editor: Francisco Torrens

Received: 16 May 2023
Revised: 5 June 2023
Accepted: 5 June 2023
Published: 8 June 2023

Copyright: © 2023 by the authors. Licensee MDPI, Basel, Switzerland. This article is an open access article distributed under the terms and conditions of the Creative Commons Attribution (CC BY) license (https:// creativecommons.org/licenses/by/ 4.0/).

Abstract: By means of a computational method based on Density Functional Theory (DFT), using commercially available software, a novel method for simulating equilibrium geometry harmonic vibrational frequencies is proposed. Finasteride, Lamivudine, and Repaglinide were selected as model molecules to study the adaptability of the new method. Three molecular models, namely the single-molecular, central-molecular, and multi-molecular fragment models, were constructed and calculated by Generalized Gradient Approximations (GGAs) with the PBE functional via the Material Studio 8.0 program. Theoretical vibrational frequencies were assigned and compared to the corresponding experimental data. The results indicated that the traditional single-molecular calculation and scaled spectra with scale factor exhibited the worst similarity for all three pharmaceutical molecules among the three models. Furthermore, the central-molecular model with a configuration closer to the empirical structure resulted in a reduction of mean absolute error (MAE) and root mean squared error (RMSE) in all three pharmaceutics, including the hydrogen-bonded functional groups. However, the improvement in computational accuracy for different drug molecules using the central-molecular model for vibrational frequency calculation was unstable. Whereas, the new multi-molecular fragment interception method showed the best agreement with experimental results, exhibiting MAE and RMSE values of 8.21 cm^{-1} and 18.35 cm^{-1} for Finasteride, 15.95 cm^{-1} and 26.46 cm^{-1} for Lamivudine, and 12.10 cm^{-1} and 25.82 cm^{-1} for Repaglinide. Additionally, this work provides comprehensive vibrational frequency calculations and assignments for Finasteride, Lamivudine, and Repaglinide, which have never been thoroughly investigated in previous research.

Keywords: DFT; molecular modeling; infrared spectroscopy; multi-molecular fragment interception

1. Introduction

Infrared (IR) spectroscopic analysis of biological samples commenced in the 1950s [1], and was speedily put to use in the realm of pharmaceutical research [2]. Infrared spectroscopy records the spectral pattern of samples and expresses the chemical composition as a function of wavenumbers between 400 and 4000 cm^{-1} [3,4]. With advancements in testing technology, Fourier transform infrared (FT–IR) spectroscopy emerged as a potent device for distinguishing and identifying an array of diverse samples, and, indeed, all forms of samples could be characterized thereby. It is a prompt and refined modality for sample characterization whereby the chemical configuration can be analyzed as simple molecules, revealing specific absorption bands in the FT–IR spectra [5,6].

The vibrational frequencies serve as a distinctive signature for chemical compounds, and are commonly employed in the identification and characterization of organic and inorganic specimens [7–9]. The infrared spectrum is derived from the vibrational motion of the molecules, and has significant implications in the qualitative and quantitative exploration of matter [10]. At present, infrared spectroscopy is a widely used and easily accessible characterization tool that plays a vital role in the understanding and utilization of materials [11]. Although advanced techniques, such as nuclear magnetic resonance and mass spectrometry, offer high-resolution molecular analysis, their prohibitive costs limit their accessibility. Therefore, obtaining material property information using simple and cost-effective methods is particularly meaningful for many research institutions and manufacturers. However, empirical guidance still dominates our interpretation of infrared spectra, presenting significant challenges in analyzing weak functional groups and fingerprint regions. Furthermore, infrared spectrometers have natural limitations, including peak overlapping caused by the proximity of strong vibration peaks, which impedes spectral analysis [12]. Therefore, it is urgent to develop a new perspective to address these problems effectively.

In recent years, the application of DFT in the field of pharmacy has experienced a rapid progression [13–15]. Quantum chemistry has played an important role in the field of drug research and development, and a large number of investigations have been conducted by means of computer-simulated infrared spectroscopy [16–18]. The force fields in these calculations provide potential energy surfaces, which map out the relative energy of a configuration of atoms in a molecule. One common type of force field is DFT-based, which includes the popular GGA [19]. The PBE exchange–correlation functional is a widely used form of GGA and has been shown to accurately predict and represent vibrational spectra [20–22]. In our work, we focus on the utilization of the GGA/PBE level to simulate vibrational spectra accurately.

However, up to the present, all of the research works have been optimized and frequency-calculated by constructing single-molecular structures with different functionals and basis sets [23,24]. The molecules were placed in a vacuum environment, and no heed was given to the spatial configuration and hydrogen bonding of the molecules. As a result, the outcomes were frequently imprecise, notably for molecules with intermolecular hydrogen bonds. We considered the reasons for this and found that the input single molecule models might differ significantly from the actual molecular state, which can be attributed to two factors. First, the first step of all calculations is model optimization, which is typically performed using the single-molecular model, neglecting the spatial hindrance caused by the crystal or amorphous state. This can lead to changes in molecular structure, which significantly affect vibrational frequencies and result in errors. Second, materials with functional groups in crystals or amorphous states are often connected through intermolecular hydrogen bonds (Figure S1), which have significant impacts on infrared spectral peaks. However, when performing frequency calculations with the single-molecular model as input, the molecular hydrogen-bond, or conjugation, information is lost, leading to increased computational errors. Therefore, when compared with experimental frequencies, theoretically-calculated harmonic frequencies were commonly discovered to be 10–15% excessively high, partly due to the exclusion of anharmonic effects and the incomplete incorporation of electron correlation [25]. To solve this problem, scale factors of different functionals were proposed [26–28].

The development of scale factor calculation has been divided into two stages. In the first stage, the factors were determined as an average of the experimental/theoretical ratios for individual modes utilizing numerous molecules. For instance, Hout, Levi, and Hehre (1982) [29] determined the scale factor of MP2/B3LYP (0.921, inverse of 1.086) using 36 molecules via this method. Secondly, the subsequent development of analytic second derivatives with finite difference techniques was proposed to increase the availability of theoretical frequencies. This method was validated by Simandiras, Handy and Amos [30], indicating higher accuracy. Other methods, such as employing the E(ZPE) equation, have also been proposed to calculate scale factors [25]. However, the methods and parameters in the equations provide empirical and statistical results from the modification of a large number of small molecular groups, which may not be applicable for molecules with distinct spatial configurations and intermolecular interactions. Furthermore, confusingly, different literature has shown different values of scale factor for the same DFT. For example, for HF/6-31G(d), Qi (2000) [31], Irikura (2005) [32] and Barone (2004) [33] demonstrated values of 0.9590, 0.8982 and 0.8929, respectively. Therefore, a new simulation method must be proposed to improve the calculation's accuracy and eliminate the perplexity of scale factor selection.

Harmonic approximation assumes that vibrations are harmonic and linear, which simplifies calculations but has severe limitations when it comes to describing real-world vibrational phenomena [34]. The harmonic model assumes that the potential energy function for a molecule is quadratic and that each vibrational mode oscillates with a single frequency. This assumption holds when the vibrations are sufficiently small that the restoring force in the bond potential can be approximated as linear, which makes it a reasonable starting point for predicting vibrational spectra. However, this approach breaks down when larger amplitude vibrations occur. Large-amplitude vibrations result in the stretching or bending of bonds beyond their equilibrium positions, which leads to the appearance of higher-order effects, including anharmonicity [35]. Anharmonicity describes the deviation of vibrational modes from harmonic behavior, occurring when interactions between different vibrational modes become significant, causing nonlinear coupling between the modes. This coupling can result in large deviations from harmonic behavior, altering vibrational frequencies and intensities [36]. In addition to anharmonicity, environmental factors can significantly affect the behavior of molecules in ways that the harmonic model does not account for. However, anharmonic analyses are computationally expensive and still not applicable to all types of molecules. Additionally, although some more sophisticated methods, such as B3LYP, M06-2X, and ω-B97X-D, can better describe anharmonic effects, the accuracy of the results is limited by various drawbacks of the single-molecular model. While the GGA/PBE method we use does not directly consider anharmonic effects in predicting vibrational frequencies, it is still a very common computational method in this field at present [20–22]. Therefore, in this paper, a new idea to construct a multi-molecular model containing molecular environment information and intermolecular conjugation information under harmonic approximation, so as to improve the accuracy of vibrational spectrum calculation and save computational resources, is proposed.

There are several reasons why including multiple molecules in our calculations can help reduce the limitations of the harmonic approximation and improve accuracy. Firstly, a multi-molecular model can take into account intermolecular interactions, which are crucial in many chemical processes, such as hydrogen bonding or protein–ligand binding. Neglecting such interactions can lead to significant errors in spectroscopic predictions. Secondly, in actual multi-molecular systems, the interactions between different molecules can change the frequency and intensity of vibration modes. For example, in a multi-molecular hydrogen-bond model, hydrogen bonding typically leads to a significant enhancement of anharmonic effects in the vibration modes and introduces higher-order vibration terms [37]. When we use a single-molecular model, we may neglect the influence of non-covalent interactions in multi-molecular systems and treat molecular vibrations as purely internal vibrations, which is a harmonic approximation and may lead to significant discrepancies with the experimental results. On the other hand, using a molecular model containing multiple molecules can better consider the non-covalent interactions and environmental effects between molecules, and, thus, more accurately describe the vibrational behavior and spectral characteristics of molecules. Additionally, the inclusion of multiple molecules in the model can lead to a more realistic representation of the sample material analyzed by spectroscopic techniques. We believe that including assets such as multiple molecules within the limitations of the harmonic approximation provides a useful and computationally efficient tool for analyzing vibrational spectra of light-to-medium-sized molecules.

In this paper, a novel vibrational simulation method was developed to facilitate the correct assignment of infrared spectra. The pharmaceutical molecules Finasteride (FIN), Lamivudine (LAM), and Repaglinide (REP) were employed as models for this purpose. The theoretical and experimental results of the vibrational harmonic spectra were compared to validate the new method. Three models were constructed in the Material Studio program, and vibrational simulations of single molecules, multi-molecular fragments, and central molecules were conducted and compared with the scaled spectrum.

2. Results and Discussion

2.1. Traditional Single-Molecular Vibration Analysis

Frequency simulations conducted thus far have only involved single molecules and scaling with scale factors. However, calculated spectra with scale factors still exhibit significant deviation from experimental results, and the scale effect varies across different molecules and functional groups, even with the same factor. To verify this problem, we extracted FIN, LAM, and REP single molecules as model molecules, optimized and calculated their harmonic vibration frequencies and infrared spectra, and compared these with the experimental spectra. The infrared spectra for the experimental spectra, the simulated single molecule, and the simulated single molecule after scaling for FIN, LAM, and REP are displayed in Figure 1. A portion of observed and calculated vibrational frequencies, along with their respective dominant normal modes, for FIN, LAM, and REP are listed in Table 1. The scaled wavenumbers are presented after being scaled by a scaling factor of 0.99 [38]. All assignments are provided in the Supplementary Materials (Tables S1–S3).

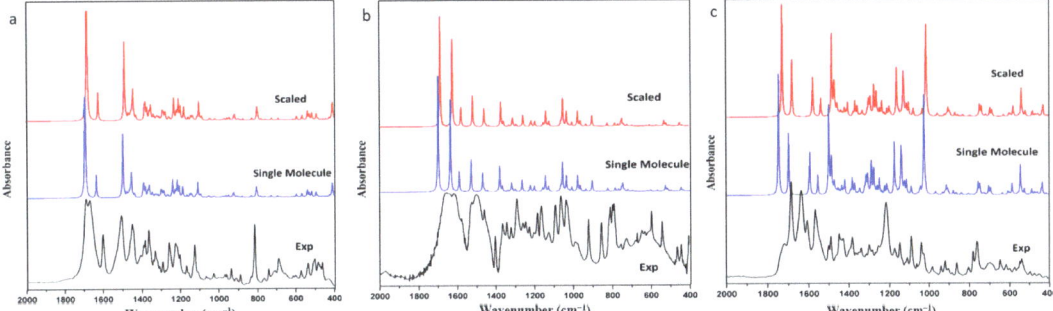

Figure 1. The infrared spectra of experimental, simulated single molecule, and simulated single molecule after scaling for (**a**) FIN, (**b**) LAM, and (**c**) REP.

Table 1. A portion of observed and calculated vibrational frequencies with their respective dominant normal modes for FIN, LAM, and REP [a].

Name	Assignment	Exp	Single Molecular	Single Molecular after Scaled by Scale Factor
FIN	$\nu N_2 H_{35}$	3429	3517.12	3481.95
	A:$\nu N_1 H$	3349	3514.26	3479.12
	$\nu C_{13} H_{14}$	2914	2924.11	2894.87
	$\nu C_3 = O_1$; $\nu C_1 = C_2$; $\beta N_1 H_{36}$	1688	1695.16	1678.21
	$\nu C_{19} = O_2$; $\beta N_2 H_{35}$	1668	1693.63	1676.69
	$\beta C_{8,16} H_{9,19}$	1277	1268.44	1255.76
	$\rho C_{14-15} H_{15-18}$; $\gamma N_2 C_{19}$	766	766.83	759.16
	MAE (for all data) [b]		12.99	16.99
	RMSE (for all data) [c]		29.80	27.14
LAM	$\nu_{as} N_3 H_{3-4}$	3383	3656.99	3620.42
	$\nu_s N_3 H_{3-4}$	3328	3518.43	3483.25
	$\nu C_1 O_1$	1651	1701.03	1684.02

Table 1. Cont.

Name	Assignment	Exp	Single Molecular	Single Molecular after Scaled by Scale Factor
LAM	$\nu C_3 N_2$; $\beta N_3 H_3$	1498	1531.51	1516.19
	$\gamma C_2 H_1$; $\rho C_8 H_{9-10}$	1030	1041.08	1033.67
	$\rho N_3 H_{3-4}$; $\beta C_5 H_5$; $\gamma C_{4,6} H_{2,6}$; $\gamma C_8 H_{9-10}$; ring prckering vibration;	538	533.74	528.40
	MAE (for all data) [b]		52.17	50.56
	RMSE (for all data) [c]		109.50	101.75
REP	$\nu O_4 H_{36}$	3428	3640.80	3566.16
	$\nu N_2 H_{35}$	3307	3551.09	3478.29
	$\nu C_{12,16} H$	2804	2863.56	2804.86
	$\nu C_{25} O_3$; $\beta O_4 H_{36}$	1686	1747.65	1711.82
	$\nu C_{17} O_1$; $\beta N_2 H_{35}$; $\gamma C_{18} H_8$	1635	1699.94	1665.09
	$\nu C_{7,18} H$; $N_2 H$	1300	1299.66	1273.02
	$\gamma N_2 H_{35}$; $\omega C_{18} H_{8-9}$; $\gamma C_{7-11} H$	474	475.74	465.99
	MAE (for all data) [b]		26.48	29.39
	RMSE (for all data) [c]		57.51	44.34

[a] Frequencies are in cm^{-1}; [b] MAE in cm^{-1}; [c] RMSE in cm^{-1}; ν: stretching; β: in-plane bending; γ: out-of-plane bending; ρ: in-plane rocking.

As depicted in Figure 1, the calculated and scaled infrared spectra of single molecules for FIN, LAM, and REP were similar to the experimental results. The simulated and experimental spectra exhibited similar peak shapes at corresponding wavenumbers, indicating high calculation accuracy. MAE and RMSE were used in the linear regression analysis to assess the deviations between the theoretical and experimental structural parameters. The MAE and RMSE between the calculated, scaled, and experimental infrared spectra for each molecule were compared, yielding values of 12.99 cm^{-1} and 29.80 cm^{-1}, 52.17 cm^{-1} and 109.50 cm^{-1}, 26.48 cm^{-1} and 57.51 cm^{-1} for the single-molecular simulation; and 16.99 cm^{-1} and 27.14 cm^{-1}, 50.56 cm^{-1} and 101.75 cm^{-1}, 29.39 cm^{-1} and 44.34 cm^{-1} for the scaled spectra of FIN, LAM, and REP, respectively. Therefore, after scaling, the infrared spectrum of LAM exhibited greater similarity to the experimental data, while the opposite was observed for FIN and REP. Thus, for macro-molecules, particularly those with complex configurations, the scale factor may not be entirely applicable. Nonetheless, at hydrogen-bonding sites, the scaled vibrations outperformed the single-molecular simulations. In the case of FIN (Table 1), the N-H and C=O stretching were calculated as 3517.12 cm^{-1}, 3514.26 cm^{-1}, 1695.16 cm^{-1}, 1693.63 cm^{-1} for the single-molecular simulation and 3481.95 cm^{-1}, 3479.12 cm^{-1}, 1678.21 cm^{-1}, 1676.69 cm^{-1} for the scaled spectra, respectively. In these bands, the scaled vibrations exhibited better fit to the experimental spectra, which was consistent with the behaviors observed in LAM and REP (Table 1). This result correlates with a slight reduction in the value of the RMSE after scaling. Consequently, the scale factor was primarily employed to adjust the vibrations of hydrogen-bonded functional groups, and had minimal impact on the overall molecule, particularly in the skeleton vibration and fingerprint region.

This could be due to the fact that the optimization was searching for the smallest potential point by changing the atomic situation around the input structure. However, most single-molecular models are derived from crystal lattices, which simplifies the structural optimization process, but results in a loss of information regarding intermolecular interactions and spatial arrangements. Consequently, this leads to inaccuracies in calculation and vibration assignment. Although the use of a scale factor improves the similarity between the calculated and experimental results to some extent, it is an empirical and statistical value that is insensitive to intermolecular interactions and spatial configurations. This shortcoming may lead to unsatisfactory outcomes. Therefore, it can be concluded that the traditional single-molecular simulation method, even with the inclusion of a scale factor, is not a reliable approach for accurately calculating molecular frequencies.

2.2. Central-Molecular Simulation Analysis

In order to investigate the impact of spatial configuration on vibrational simulation, we constructed the minimum repeating space structure unit to preserve the configuration as much as possible while minimizing computational resources. To accurately depict the true configuration and intermolecular interactions, the crystal cells of FIN, LAM, and REP were imported in the MS program and supercells of $2 \times 1 \times 1$, $3 \times 3 \times 1$, $2 \times 1 \times 3$ were constructed, respectively. The structures of supercells are shown in Figure S1. We extracted the minimum repeating unit from the supercell of molecules and optimized it using the GGA/PBE functional. After optimization, the central single molecule was then extracted and calculated the frequencies at the same level of theory. The minimum multi-molecular repeating units for FIN, LAM, and REP are shown in Figure 2. The scaled, central-molecular calculated and experimental spectra for the three drugs are also shown in Figure 3. A portion of available experimental and theoretical vibrational frequencies with their respective dominant normal modes for FIN, REP, and LAM are listed in Table 2. All the assignments are included in the Supplemental Materials (Tables S1–S3).

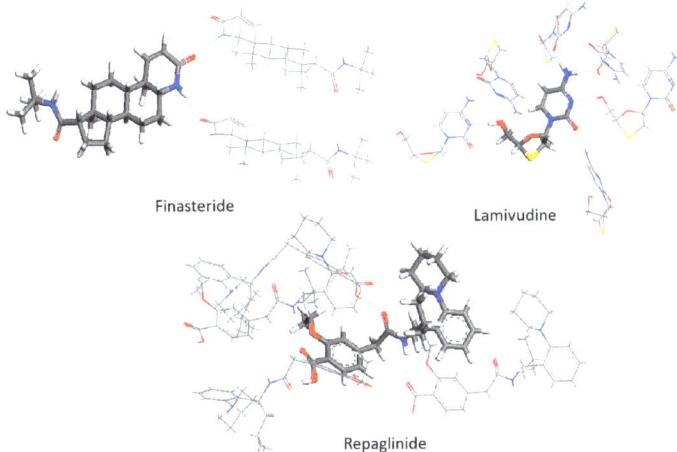

Figure 2. The minimum multi-molecular repeating units of FIN, LAM, and REP (Hydrogen bonds are illustrated by blue dashed lines). N atoms are colored blue, O atoms are colored red, H atoms are colored white, C atoms are shown in gray, and S atoms are shown in yellow.

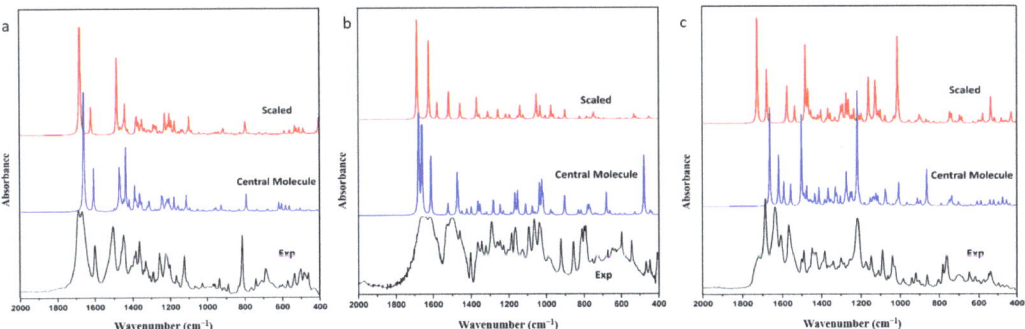

Figure 3. The infrared spectra of experimental, simulated central-molecular and simulated single-molecular models after scaling for (**a**) FIN, (**b**) LAM, and (**c**) REP.

Table 2. Observed and calculated vibrational frequencies and their dominant normal modes for FIN, LAM, and REP [a].

Name	Assignment	Exp.	Central Molecular	Single Molecular after Scaled by Scale Factor
FIN	νN_2H_{35}	3429	3522.08	3481.95
	$A:\nu N_1H$	3349	3437.05	3479.12
	$\nu C_{13}H_{14}$	2914	2920.99	2894.87
	$\nu C_3=O_1; \nu C_1=C_2; \beta N_1H_{36};$	1688	1664.17	1678.21
	$\nu C_{19}=O_2; \beta N_2H_{35};$	1668	1662.83	1676.69
	$\beta C_{8,13,16}H_{9,14,19}$	1225	1224.87	1218.55
	$\beta N_1H_{36}; \beta C_{6,10}H_{6-7,10-11}; \delta asC_{17,18}H_{20-25}$	890	882.29	872.55
	MAE (for all data) [b]		9.34	16.99
	RMSE (for all data) [c]		21.81	27.14
LAM	$\nu_{as}N_3H_{3-4}$	3383	3484.85	3620.42
	$\nu_s N_3H_{3-4}$	3327.9	3331.83	3483.25
	νC_1O_1	1615	1671.64	1684.02
	$\beta N_3H_3; \nu C_3N_2$	1498	1467.85	1516.19
	$\beta C_{5,7}H_{5,8}; \omega C_6H_{6-7}; \gamma C_2H_1$	1184	1182.63	1151.56
	$\omega C_6H_{6-7}; \nu C_6S_1$	752	757.16	750.60
	MAE (for all data) [b]		20.36	50.56
	RMSE (for all data) [c]		30.40	101.75
REP	νO_4H_{36}	3428	3162.17	3566.16
	νN_2H_{35}	3307	3424.00	3478.29
	$\nu C_{25}O_3; \beta O_4H_{36}$	1686	1666.48	1711.82
	$\nu C_{17}O_1; \beta N_2H_{35}; \gamma C_{18}H_8$	1635	1621.51	1665.09
	$\nu C_{7,18}H; \nu N_2H$	1300	1304.87	1273.02
	$\gamma N_2H_{35}; \gamma C_{2,3,4,5}H_{6,7,5,4}; \gamma C_{12-16}H; \nu C_1N_2$	619	610.53	597.38
	MAE (for all data) [b]		23.12	29.39
	RMSE (for all data) [c]		52.21	44.34

[a] Frequencies are in cm^{-1}; [b] MAE in cm^{-1}; [c] RMSE in cm^{-1}; ν: stretching; β: in-plane bending; γ: out-of-plane bending; δ: formation; ω: out-plane rocking.

From Figure 3, it is apparent that the spectra of the central molecule were superiorly compatible than those of the single molecule and the scaled spectrum. This observation strongly implies that spatial configuration played a pivotal role in the frequency calculations of all three molecules. Furthermore, the MAEs and RMSEs between the experimental infrared spectra and the simulated and scaled spectra of the central molecule for FIN, LAM, and REP were compared. The results showed that the MAE and RMSE for the central-molecular model were 9.34 cm^{-1} and 21.81 cm^{-1}, 20.36 cm^{-1} and 30.40 cm^{-1}, 23.12 cm^{-1} and 52.21 cm^{-1} for FIN, LAM, and REP, respectively, whereas the corresponding values for the scaled spectra were 16.99 cm^{-1} and 27.14 cm^{-1}, 50.56 cm^{-1} and 101.75 cm^{-1}, 29.39 cm^{-1} and 44.34 cm^{-1}, as listed in Table 2. The significant reduction observed in the RMSE of LAM was primarily due to the precise calculation of the infrared vibration at the hydrogen-bonding site ($\nu_{as}N_3H_{3-4}$) by the central-molecular model. Therefore, the central-molecular model for vibrational simulation was found to be more accurate than the single-molecular and scaled results for all three pharmaceutics. Moreover, for the hydrogen-bonded functional groups, the central-molecular model was also found to yield better fits to the experimental spectra.

After optimization, the geometrical parameters (bond length, bond angle, and dihedral angle) of the single molecule may differ from those of the central molecule due to the lack of intermolecular interactions in the former case. For instance, in the case of the REP molecule, the ethoxy and carboxyl groups were twisted by 4.06 and 6.22 degrees, respectively, after optimization by the single-molecular and multi-molecular repeating units (as observed in Figure 4). Therefore, multi-molecular units can retain the configuration when optimized with functional theory, leading to more accurate calculations of frequencies, which are closer to experimental values.

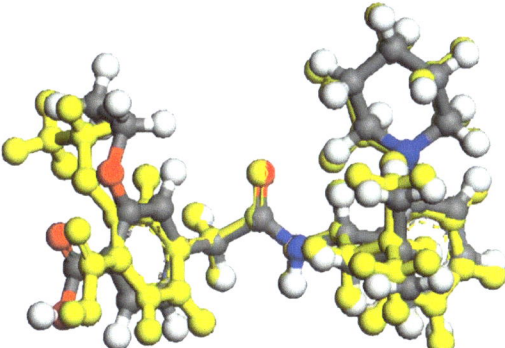

Figure 4. The structural comparison of REP molecule after optimization with single-molecular and multi-molecular repeating units. The multi-molecular optimized structure is colored yellow.

One may be inclined to assume that the single molecule extracted from the crystal cell has the same configuration as the real molecule and performs optimally. However, it was observed that the calculated spectra of the single molecule without optimization differed greatly from the experimental results, and even displayed imaginary frequencies. This could be attributed to the fact that frequency simulations in Gaussian are based on the first derivative of the potential function being zero. As such, structures without optimization cannot be processed using the Schrodinger equation.

Therefore, when compared to the single-molecular model and scaled spectra, the central-molecular model that was optimized using multi-molecular repeating units exhibited a better fit to the experimental spectrum. However, the central-molecular model only provided configuration information and lacked information on intermolecular interactions. As a result, the calculated frequencies at the hydrogen-bonding donor and receptor sites still deviated significantly from the experimental results. For instance, for LAM (Table 2), the experimental value of N_3-H_3 stretching was 3383 cm^{-1}, which was calculated as 3620.42 cm^{-1} for scaled spectra and 3484.85 cm^{-1} for the central-molecular calculation, respectively. For REP (Table 2), the experimental value of O_4H_{36} peak was at 3428 cm^{-1}, which corresponded to 3162.17 cm^{-1} and 3566.16 cm^{-1} for the central-molecular and scaled spectra, respectively. In general, the accuracy of the central-molecular model was similar to that of the scale factor method. To overcome this limitation, a new multi-molecular fragment interception method was proposed.

2.3. Multi-Molecular Fragment Interception Simulation Analysis

The discussion of the central-molecular model analyzed the influence of configuration on vibrational simulation. However, the calculated frequencies of hydrogen-bonded functional groups exhibited significant deviation from experimental results. To solve this problem, a multi-molecular fragment interception method was proposed. This model was derived from the optimized multi-molecular repeating unit by eliminating long-distance molecular fragments with low conjugation and charge effects around the central molecule. To accurately depict the true configuration and intermolecular interactions, the crystal cells of FIN, LAM, and REP were imported into the MS program and supercells of 2 × 1 × 1, 3 × 3 × 1, 2 × 1 × 3 were constructed, respectively (Figure S1).

To retain the space configuration and hydrogen-bond information as far as possible, we optimized the repeating unit using the GGA/PBE functional, rather than a single molecule. Additionally, to reserve the computing resources, the long-distance molecular fragments with low conjugation and charge effect around the central molecule were intercepted to construct the multi-molecular fragment (see Figure 5). The frequency of the molecules was then calculated using the same functional, and the infrared spectrum cooperation are shown in Figure 6. Upon comparison of the calculated spectra with the experimental spectra, numerous matching bands in this region were identified and are presented in Table 3, for FIN, LAM, and REP.

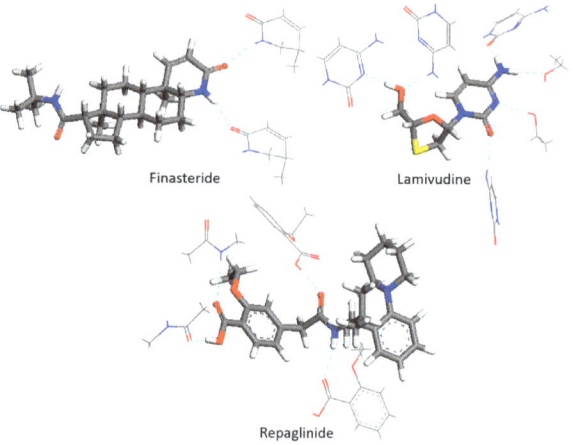

Figure 5. The multi-molecular fragment interception models of the three drugs.

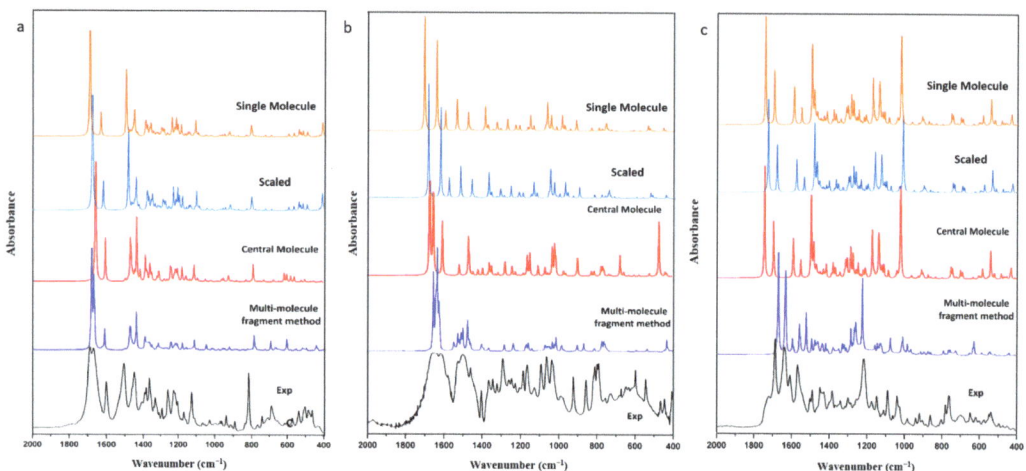

Figure 6. The infrared spectrum of experimental, simulated multi-molecular fragment, simulated central molecule, simulated single molecule and scaled single molecule for (**a**) FIN, (**b**) LAM, and (**c**) REP.

The calculated spectra of the multi-molecular fragment exhibited greater similarity to the experimental spectra in terms of frequencies and peak shapes for all three drug molecules (see Figure 6). This phenomenon suggests that the multi-molecular fragment model, which includes the original space configuration and intermolecular interactions, is superior to traditional scaled single-molecular calculations for frequency simulation. Both MAE and RMSE were calculated to evaluate the performance of the central-molecular, single-molecular, and multi-molecular fragment models. The MAE values for the central-molecular, single-molecular, and multi-molecular fragment models were 9.34 cm^{-1}, 20.36 cm^{-1}, and 23.12 cm^{-1} for FIN, and 8.21 cm^{-1}, 15.95 cm^{-1}, and 12.10 cm^{-1} for the multi-molecular fragment model. In addition, the RMSE values were also calculated and compared with the MAE values. The RMSE values for FIN were 29.80 cm^{-1}, 21.81 cm^{-1}, and 18.35 cm^{-1} for the single-molecular, central-molecular, and multi-molecular fragment models, respectively. For LAM, the RMSE values were 109.50 cm^{-1}, 30.40 cm^{-1}, and 26.46 cm^{-1}, and for REP, the RMSE values were 57.51 cm^{-1}, 52.21 cm^{-1}, and 25.82 cm^{-1}. Therefore, the utilization of the multi-molecular fragment model resulted in a considerable reduction in the MAE and RMSE values. These results indicate that the multi-molecular fragment interception model outperformed both the single-molecular and central-molecular models in terms of ability to accurately predict vibrational frequencies. Therefore, the multi-molecular fragment interception model for vibrational simulation was more accurate than the other models described above for all three pharmaceutics. Additionally, the frequencies calculation with the multi-molecular fragment model for hydrogen-bonded functional groups also demonstrated the best performance relative to empirical spectra. For instance, for FIN in Table 3, the C_3-O_1 and C_{19}=O_2 stretching vibrations were calculated to be 1664.17 cm^{-1} and 1662.83 cm^{-1} for the central-molecular model, and 1681.01 cm^{-1} and 1665.21 cm^{-1} for the multi-molecular fragment model, respectively. The experimental frequencies for these two vibrations were 1688 cm^{-1} and 1668 cm^{-1}, indicating that the multi-molecular fragment interception model was a better fit to the experimental spectrum. Similar conclusions could be drawn for the REP and LAM molecules.

Table 3. Vibrational frequencies and dominant normal modes of multi-molecular fragment model and central molecular model for FIN, LAM, and REP [a].

Name	Assignment	Exp.	Multi-Molecular Fragment Model	Central Molecular
FIN	$\nu N_2 H_{35}$	3429	3520.79	3522.08
	A:$\nu N_1 H$	3349	3399.51	3437.05
	$\nu C_3=O_1$; $\nu C_1=C_2$; $\beta N_1 H_{36}$;	1688	1681.01	1664.17
	$\nu C_{19}=O_2$; $\beta N_2 H_{35}$;	1668	1665.21	1662.83
	$\beta C_{8,13,16} H_{9,14,19}$	1225	1225.5	1224.87
	$\gamma N_1 H_{36}$; $\rho C_{15} H_{17-18}$; $\delta_{as} C_{17} H_{20-22}$	600	589.69	580.71
	MAE (for all data) [b]		8.21	9.34
	RMSE (for all data) [c]		18.35	21.81
LAM	$\nu_{as} N_3 H_{3-4}$	3383	3393.37	3484.85
	$\nu_s N_3 H_{3-4}$	3327.9	3232.27	3331.83
	$\nu C_1 O_1$	1651	1653.55	1671.64
	$\beta N_3 H_3$; $\nu C_3 N_2$	1498	1498.19	1467.85
	$\gamma C_2 H_1$; $\rho C_8 H_{9-10}$	1030	1033.46	1031.56
	$\gamma N_3 H_3$; $\gamma C_5 H_5$	752	757.55	757.16
	MAE (for all data) [b]		15.95	20.36
	RMSE (for all data) [c]		26.46	30.40
REP	$\nu O_4 H_{36}$	3428	3415.14	3162.17
	$\nu N_2 H_{35}$	3307	3395.98	3424.00
	$\nu C_{25} O_3$; $\beta O_4 H_{36}$	1686	1689.41	1666.48
	$\nu C_{17} O_1$; $\beta N_2 H_{35}$; $\gamma C_{18} H_8$	1635	1632.85	1621.51
	$\beta C_{7-12,14} H$; $\beta N_2 H_{35}$	1112	1125.59	1117.81
	$\gamma N_2 H_{35}$; $\gamma O_4 H_{36}$; $\gamma C_{26} H$; ring prckering vibration	763	758.78	743.32
	MAE (for all data) [b]		12.10	23.12
	RMSE (for all data) [c]		25.82	52.21

[a] Frequencies are in cm^{-1}; [b] MAE in cm^{-1}; [c] RMSE in cm^{-1}; ν: stretching; β: in-plane bending; γ: out-of-plane bending; δ: formation; ρ: in-plane rocking.

The reason for the improved accuracy of the multi-molecular fragment model may be attributed to the fact that the model was obtained by intercepting a fragment from the optimized minimum multi-molecular repeating units of hydrogen bonds, which retained both the hydrogen-bonded information and the original configuration. Compared to the central-molecular model, the multi-molecular fragment model removed the long-distance molecular fragments with low conjugation and charge effect around the central molecule, and, thereby still retaining the hydrogen-bonded information as much as possible. Moreover, during the optimization and interception processes, the central molecule of the multi-molecular fragment model was able to maintain its configuration nature and intermolecular interactions, unlike a single molecule in vacuum. Consequently, using a single-molecular model simplified the atomic positions, velocities, and potential energy and yielded a considerable discrepancy between the calculated results and the actual situations. Additionally, a single-molecular model is incapable of describing intermolecular interactions and environmental factors, such as steric hindrance and lattice constraints. In contrast, a multi-molecular model addresses intermolecular interactions between neighboring molecules, encompasses hydrogen bonding, electrostatic, and van der Waals forces, and calculates infrared resonance for complex systems. In addition, the multi-molecular model also considers the interactions between the molecule and its surrounding environment. Compared to the single-molecular model, the multi-molecular model provides better characterization of the molecular spatial structure, and functional group structure, and offers a realistic portrayal of macroscopic substance states, enhancing computational accuracy. The multi-molecular model also encompasses intramolecular resonance effects, dense packing effects, and nonlinear effects, enabling a more comprehensive depiction of molecular vibration and physical–chemical processes, thereby improving the prediction of vibrational frequencies.

As a result, the new method of multi-molecular fragment interception effectively overcomes the limitations of traditional single-molecular vibrational simulation and outperforms other methods with the smallest simulation deviation. It essentially preserves the configuration information and intermolecular interactions, and significantly improves the calculation accuracy, offering a more critical and time-efficient approach for frequency simulation.

2.4. Absolute Error Analysis

To investigate the sources of absolute errors (AEs) among different models, we classified the vibrational frequencies of three types of drug molecules into functional group vibrations and non-functional group vibrations. We then studied in detail the changes in AEs caused by different models, as shown in Figure 7.

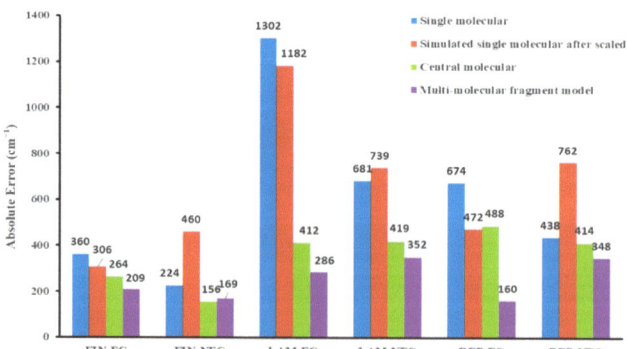

Figure 7. Absolute errors in functional group (FG) and non-functional group (NFG) vibrations of FIN, LAM, and REP.

The results showed that for functional group vibrations, the scale factor had some accuracy improvement effect on the correction of the single-molecular model results, but the effect was generally poor and fluctuated greatly. In the non-functional group set, the scale factor correction increased the calculation error, instead of lowering it. Therefore, the mechanism of adding scale factors sacrificed the calculation accuracy of skeletal vibrations to promote a higher fit between functional group vibrations and experimental values, although this increase in accuracy was not significant (see in Figure 7).

In addition, for functional group vibrations, although the center-molecular model comes from the optimized structure of multi-molecular models, it does not contain intermolecular hydrogen-bonding, or conjugated, information, but the center molecule model has a significantly reduced error compared to the single molecule model. This may be due to two reasons. First, the center molecule model inherits the advantages of the multi-molecule model, and the averaged molecular structure composed of multiple molecules is more similar to the actual molecular structure. Therefore, by averaging multiple molecules and taking the center molecule, the compression or deformation of the actual molecular structure caused by strong interaction forces such as hydrogen bonding can be eliminated. Secondly, considering the intermolecular interactions together with the individual atomic structure in the optimization process can improve the accuracy of calculating the frequency of hydrogen-bonding functional groups in the center molecule model. However, the center molecules of REP did not show small relative AE values, and the center molecule models of FIN and LAM molecules showed great differences in accuracy improvement. This indicates that there is still considerable instability in using center molecule models for vibrational frequency calculation. For the non-functional group region, the analysis of the AEs between the center molecule and the multi-molecule fragment model shows that the AEs of the two models are very similar, and the decrease in AEs compared to the single molecule model is also similar. This indicates that the improvement in calculation accuracy of the non-functional group region is mainly due to the fact that the molecular structure is closer to the true state after optimizing the overall molecular structure of the multi-molecule repeating unit, and the effect of hydrogen bonding information on the vibrational calculation optimization of the non-functional group region is not significant. This conclusion is also easy to understand because there is no hydrogen-bond generation in the non-functional group region, so the addition of multi-molecule hydrogen bonds has a very small effect on vibrational calculation.

Additionally, we can also illustrate that for FIN, LAM, and REP, the reduction in MAE values of the multi-molecular fragment model and single-molecular scale model were 8.07 cm^{-1}, 8.96 cm^{-1} and 57.04 cm^{-1}, 18.94 cm^{-1} and 28.36 cm^{-1}, 13.36 cm^{-1} in the functional group and non-functional group regions, respectively. For LAM and REP, the improvement of computational accuracy in the functional group regions had a more significant effect on reducing the overall MAE, whereas for FIN, the contribution of both functional group and non-functional group vibrations to the overall error was similar. Interestingly, for all three drug molecules, the magnitude of MAE reduction in the multi-molecular fragment model in the functional group region (LAM > REP > FIN) was consistent with the variational trend of intermolecular hydrogen bonds or interactions. This indicates that, for multi-molecular models, the more intermolecular hydrogen bonds present, the greater the reduction in errors observed in functional group vibrations. This may be due to the retention of hydrogen bonding and conjugation information in the multi-molecular model, leading to higher computational accuracy at the hydrogen-bonding connection sites during vibrational frequency calculation.

In conclusion, the improvement in accuracy of the center-molecular and multi-molecular fragment models in the non-functional group region was mainly due to the molecular structure being closer to the true state, and the molecular conjugation information having no significant effect on the accuracy improvement of this part. In the functional group region, both molecular structure and intermolecular conjugation information had significant impacts on the calculation accuracy of vibrational frequencies. The more hydrogen bonding among molecules in the original structure, the greater the contribution of intermolecular conjugational systems to the reduction of MAE. However, the center-molecular model did not have good accuracy performance in all drug molecule models, while the multi-molecular fragment model showed the lowest error values among all models for all drug molecules.

2.5. Computational Time Comparison

To provide a quantitative comparison of the computational performance among different models, we conducted structural optimization and vibrational frequency calculations for all models using the same computer and computational resources. The time required for each calculation is summarized in Table 4.

Table 4. Comparison of calculation times among different models.

Drug	Process	Single Molecule Model (h)	Central Molecule Model (h)	Multi-Molecule Fragment Model (h)
FIN	Configuration optimization	0.27	4.18	4.18
	Frequency calculation	2.51	2.44	4.10
LAM	Configuration optimization	0.10	15.40	15.40
	Frequency calculation	0.44	0.48	3.26
REP	Configuration optimization	0.26	26.21	26.21
	Frequency calculation	2.45	2.42	7.76

Traditional single-molecular models have the smallest number of atoms, and requires the least amount of time for configuration optimization and vibrational frequency calculation, with a total calculation time not exceeding 3 h. In our paper, the central-molecular and multi-molecular fragment models both used minimal molecular repeat units with hydrogen bonds as initial structures for configuration optimization, causing a significant increase in the time required for this step. Compared to the single-molecular model, FIN, LAM, and REP required an additional 3.91 h, 15.30 h, and 27.95 h, respectively. In the vibrational frequency calculation stage, the central-molecular model took a similar amount of time as the single-molecular model because only one molecular unit was involved. Whereas, due to the inclusion of a large number of hydrogen bonds and conjugated structural fragments in the multi-molecular fragment model, its frequency calculation took approximately 2 h longer than that of the single-molecular model.

However, when selecting drug molecules as calculation models to demonstrate the applicability of new methods more convincingly, we specially selected drug molecules with a high number of atoms, complex structures, and representative hydrogen-bond numbers. For small drug molecules, commonly used for vibrational analysis verification, such as aspirin, the calculation time using the multi-molecular fragment model was only 0.29 h. Moreover, we were pleased to find that although the overall calculation time was longer due to various reasons, including the lack of professional computing servers, our calculations could still be completed within roughly one day. Achieving high computational accuracy is critical in ensuring precise spectral analysis, as even small infrared spectral peak shifts can cause significant analytical errors. Thus, considering the significant reduction in MAEs and the instability observed in the prediction results of the central-molecular model (Figure 7), we regarded the additional time cost as being justified.

The key findings from our study also have several important implications for molecular simulation and drug design. Firstly, the improved accuracy of the multi-molecular fragment interception model suggests that accurately describing intermolecular interactions is critical for accurate vibrational frequency calculations. This finding highlights the importance of advanced modeling approaches that accurately capture the complex interplay between molecules in biological systems. Secondly, this method can be applied to a wide range of molecular systems and provide insights into their structural and functional properties. By accurately calculating vibrational frequencies, the new approach can aid in the interpretation of IR spectra and elucidation of molecular structure and function. This capability is particularly relevant in drug design, where understanding the vibrational properties of drug molecules is critical for predicting their behaviors in vivo. Additionally, this method provides highly accurate guidance for the determination of supramolecular structures, such as co-crystals or co-amorphous systems, enabling precise structural characterization. Thirdly, it has potential applications in optimizing drug formulations and identifying potential drug candidates with desired properties. By accurately describing the intermolecular interactions that contribute to the stability and efficacy of drug formulations, the new approach can aid in the development of more effective and efficient drug delivery systems. Additionally, by accurately describing the vibrational modes associated with specific functional groups, this method can facilitate the identification of potential drug candidates with specific properties or functional groups. We believe that this approach has significant implications for molecular simulation and drug design, and has the potential to contribute to the development of more effective and efficient drug design and development strategies.

3. Materials and Methods

3.1. Materials

Finasteride, Repaglinide, and Lamivudine were obtained from Hunan Qianjin Xiangjiang Pharm. Inc. (Zhuzhou, China). Ethanol was obtained from Sinopharm Chemical Reagent Co., Ltd. (Shanghai, China). The chemicals were used as received from the companies without further purifications.

3.2. Preparation of Drug Crystalline Forms

The Finasteride form I [39] and Repaglinide form I [40] were obtained by recrystallization from absolute ethanol by means of the slow evaporation method and ethanol/water (2:1) by the solvent–antisolvent method, respectively. The received samples were dried under vacuum for 48 h at 313 K.

The Lamivudine form II [41] was produced by heating a suspension of Lamivudine in industrial methylated spirit to reflux to obtain a clear solution. The solution was filtered while hot and half the amount of the solvent from the filtrate was distilled. Then, heating was stopped and the concentrated solution was seeded with authentic form II crystals. The seeded solution was then cooled from 353 K to 298 K for one hour. After further cooling of the the suspension to 288 K and stirring it for an hour, it was filtered and washed with IMS and then dried to give the Form II crystals.

The precipitates were stored in a desiccator until use in the experiment.

3.3. FT–IR

FT–IR patterns were recorded using a NICOLET 380 FT-IR spectrometer (Townsend, MA, USA) at wavelengths of 4000–400 cm^{-1}. Samples were prepared in KBr pellets by grinding Ca. 1 mg of the drug with KBr and the resolution was 2 cm^{-1}.

3.4. Computer Details

The spatial configuration and atom numbering schemes of Finasteride, Repaglinide and Lamivudine molecules are given in Figure 8. Initial crystalline cells of Finasteride (form I, CCDC Code: WOLXOK02), Repaglinide (form I, CCDC Code: JOHKUM) and Lamivudine (form II, CCDC Code: RUKHAG) were obtained from the Cambridge Crystallographic Database (CCDC, Cambridge, UK) with Conquest 1.8 software (CSD, version 5.27, November 2006 plus 31 updates, Conquest version 1.8).

For each pharmaceutical molecule, we created three distinct models: the single-molecular model, the central-molecular model, and the multi-molecular fragment model, with the single-molecular model serving as a control group. The single-molecular model was a single molecular structure obtained by optimizing the molecular structure extracted from the crystal cell. The central-molecular and multi-molecular fragment models were optimized with multi-molecular hydrogen-bonded repeating units. The minimum multi-molecular hydrogen-bonding repeating unit refers to the random selection of a central molecule and the extraction of all the molecules that were hydrogen-bonded to it to form a multi-molecular hydrogen-bonding unit, with the aim of preserving the molecular spatial structure and hydrogen-bonding information as much as possible. We obtained the central molecule by taking the central molecule after optimizing the multi-molecular repeating unit with the GGA/PBE functional [19], a widely used and extensively tested method with proven accuracy in various applications [42–44]. The vibrational frequencies were then calculated to investigate the influence of proximity to the original material structure. Additionally, for the multi-molecular fragment model, we further processed the optimized multi-molecular repeating unit by removing structures farther away from the central molecule and only preserving the portion connected to it through hydrogen bonds or conjugated structures, thus, saving computational resources. The construction process of the three models is depicted in Figure 9.

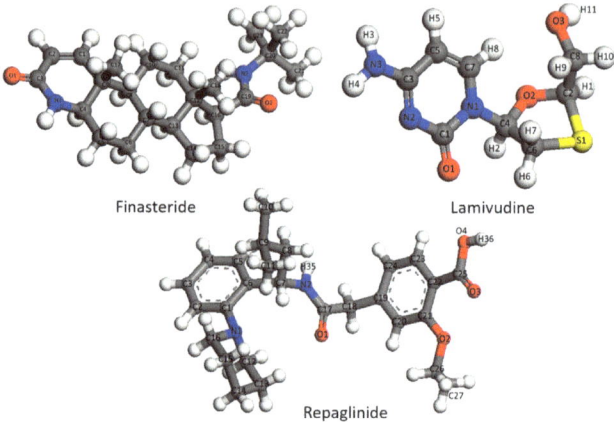

Figure 8. The spatial configuration and atom numbering schemes of Finasteride, Lamivudine, and Repaglinide.

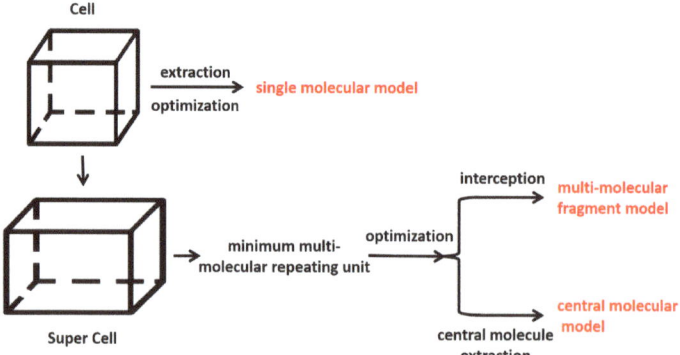

Figure 9. The construction process of the three models.

The structural optimization of the single-molecular and minimum repeating units was performed at the GGA/PBE functional level under vacuum conditions. PBE is the recommended default exchange–correlation functional, especially for studies of molecules interacting with metal surfaces, but also widely-used and reliable for bulk calculations [19]. Convergence criteria were specified independently for maximum energy change, maximum force and maximum displacement, these being 1×10^{-5} Hartree, 0.002 Hartree Å^{-1}

and 0.005 Å, respectively. The vibrational frequencies of the single-molecular and multi-molecular fragment were simulated at the same level of theory as the geometrical optimizations, and the MAEs between the calculated and experimented frequencies were compared to validate the accuracy of the new method. All calculations were conducted by the DMol3 module in Material Studio 2017 program package [45], utilizing a Dell Precision 7670 Workstation.

4. Conclusions

In summary, our study presents a novel methodology for simulating equilibrium geometry harmonic vibrational frequencies using multi-molecular fragment interception models. The results showed that the traditional single-molecular simulation and scaling method deviated significantly from the experimental results, while the central-molecular model was better but still not accurate enough, especially for hydrogen-bonded functional groups. Moreover, using the central-molecular model for vibrational frequency calculation resulted in unstable computational accuracy for different drug molecules. The multi-molecular fragment model, on the other hand, demonstrated the highest calculation accuracy with minimum simulation deviation, by preserving the configuration information and intermolecular interactions. The study also provided complete vibrational frequencies calculations and assignments for the three pharmaceutics, which could provide more accurate interpretations of infrared spectra and have never been thoroughly investigated in previous research. Overall, the proposed method has the potential to subvert the current frequency simulation method and improve the accuracy of vibrational spectrum interpretation.

Despite the potential applications of the method, there are limitations and challenges that must be overcome to further enhance its accuracy and applicability. One limitation is the computational cost of the multi-molecular fragment interception model. The model requires a large number of calculations to account for intermolecular interactions, which can increase computational time and limit its applicability to larger systems. To overcome this limitation, one possible improvement would be to incorporate machine learning approaches to optimize the multi-molecular fragment models and improve computational efficiency. For example, artificial neural networks could be trained to estimate intermolecular potentials, reducing the number of calculations required to accurately describe intermolecular interactions. Another limitation is the scalability of the methodology to larger molecular systems. The multi-molecular fragment interception model is more accurate than traditional single-molecular simulations and scaling methods but has limitations in accurately capturing intermolecular interactions in large systems. To overcome this limitation, one possible future direction would be to explore the use of hybrid functionals, which combine the strengths of different exchange–correlation functionals, to improve the accuracy of vibrational frequency calculations for complex molecular systems. By incorporating these hybrid functionals, we may be able to accurately describe intermolecular interactions in larger systems, expanding the applicability of the method to a wider range of molecular systems.

Supplementary Materials: The following supporting information can be downloaded at: https://www.mdpi.com/article/10.3390/molecules28124638/s1, Figure S1: The supercell structures of $2 \times 1 \times 1$ FIN supercell (a), $3 \times 3 \times 1$ LAM supercell (b), and (c) $2 \times 1 \times 3$ REP supercell (Hydrogen bonds are illustrated by blue dashed lines); Table S1: Observed and calculated vibrational frequencies of finasteride with different simulation models; Table S2: Observed and calculated vibrational frequencies of lamivudine with different simulation models; Table S3: Observed and calculated vibrational frequencies of repaglinide with different simulation models.

Author Contributions: Conceptualization, S.Y.; Formal analysis, L.W. and P.Z.; Funding acquisition, P.Z. and Z.Z.; Investigation, L.W., P.Z., Y.G. and Z.Z.; Methodology, L.W. and P.Z.; Project administration, S.Y.; Software, S.Y.; Supervision, L.W., P.Z., Y.G. and Z.Z.; Validation, L.W., P.Z., Y.G. and Z.Z.; Writing—original draft, L.W. and S.Y. All authors have read and agreed to the published version of the manuscript.

Funding: This research was funded by the Scientific Development Foundation of Dongying, grant number DJ2021016 and DJB2022017.

Institutional Review Board Statement: Not applicable.

Informed Consent Statement: Not applicable.

Data Availability Statement: Not applicable.

Conflicts of Interest: The authors declare no conflict of interest.

Sample Availability: Not applicable.

References

1. Barer, R.; Cole, A.R.H.; Thompson, H.W. Infra-red spectroscopy with the reflecting microscope in physics, chemistry and biology. *Nature* **1949**, *163*, 198–201. [CrossRef] [PubMed]
2. Thomas, L.C.; Greenstreet, J.E.S. The identification of micro-organisms by infrared spectrophotometry. *Spectrochim. Acta* **1954**, *6*, 302–319. [CrossRef]
3. de Peinder, P.; Neil, J.; Everall, J.M.C.; Griffiths, P.R. Vibrational spectroscopy of polymers: Principles and practice. *Anal. Bioanal. Chem.* **2008**, *392*, 567–568. [CrossRef]
4. Salzer, R.; Peter, R.; Griffiths, J.; de Haseth, A. Fourier transform infrared spectrometry. *Anal. Bioanal. Chem.* **2008**, *391*, 2379–2380. [CrossRef]
5. Wenning, M.; Scherer, S. Identification of microorganisms by FTIR spectroscopy: Perspectives and limitations of the method. *Appl. Microbiol. Biotechnol.* **2013**, *97*, 7111–7120. [CrossRef]
6. Legner, N.; Meinen, C.; Rauber, R. Root differentiation of agricultural plant cultivars and proveniences using FTIR spectroscopy. *Front. Plant Sci.* **2018**, *9*, 748. [CrossRef]
7. Grimme, S.; Bannwarth, C.; Shushkov, P. A robust and accurate tight-binding quantum chemical method for structures, vibrational frequencies, and noncovalent interactions of large molecular systems parametrized for all spd-block elements (Z = 1–86). *J. Chem. Theory Comput.* **2017**, *13*, 1989–2009. [CrossRef]
8. Baiz, C.R.; Błasiak, B.; Bredenbeck, J.; Cho, M.; Choi, J.H.; Corcelli, S.A.; Zanni, M.T. Vibrational spectroscopic map, vibrational spectroscopy, and intermolecular interaction. *Chem. Rev.* **2020**, *120*, 7152–7218. [CrossRef]
9. Lu, G.; Haes, A.J.; Forbes, T.Z. Detection and identification of solids, surfaces, and solutions of uranium using vibrational spectroscopy. *Coord. Chem. Rev.* **2018**, *374*, 314–344. [CrossRef]
10. Falk, M.; Ford, T.A. Infrared spectrum and structure of liquid water. *Can. J. Chem.* **1966**, *44*, 1699–1707. [CrossRef]
11. Wiercigroch, E.; Szafraniec, E.; Czamara, K.; Marta, Z.P.; Katarzyna, M.; Kamila, K.; Agnieszka, K.; Malgorzata, B.; Kamilla, M. Raman and infrared spectroscopy of carbohydrates: A review. *Spectrochim. Acta Part A Mol. Biomol. Spectrosc.* **2017**, *185*, 317–335. [CrossRef] [PubMed]
12. Gastegger, M.; Behler, J.; Marquetand, P. Machine learning molecular dynamics for the simulation of infrared spectra. *Chem. Sci.* **2017**, *8*, 6924–6935. [CrossRef] [PubMed]
13. van de Streek, J.; Neumann, M.A. Validation of molecular crystal structures from powder diffraction data with dispersion-corrected density functional theory (DFT-D). *Acta Crystallogr. Sect. B Struct. Sci. Cryst. Eng. Mater.* **2014**, *70*, 1020–1032. [CrossRef] [PubMed]
14. Fathi Azarbayjani, A.; Aliasgharlou, N.; Khoshbakht, S.; Ghanbarpour, P.; Rahimpour, E.; Barzegar-Jalali, M.; Jouyban, A. Experimental Solubility and Density Functional Theory Studies of Deferasirox in Binary Solvent Mixtures: Performance of Polarizable Continuum Model and Jouyban–Acree Model. *J. Chem. Eng. Data* **2019**, *64*, 2273–2279. [CrossRef]
15. Halim, K.N.; Rizk, S.A.; El-Hashash, M.A.; Ramadan, S.K. Straightforward synthesis, antiproliferative screening, and density functional theory study of some pyrazolylpyrimidine derivatives. *J. Heterocycl. Chem.* **2021**, *58*, 636–645. [CrossRef]
16. Govindasamy, P.; Gunasekaran, S.; Ramkumaar, G.R. Natural bond orbital analysis, electronic structure and vibrational spectral analysis of N-(4-hydroxyl phenyl) acetamide: A density functional theory. *Spectrochim. Acta Part A Mol. Biomol. Spectrosc.* **2014**, *130*, 621–633. [CrossRef]
17. Karakaya, M.; Kurekci, M.; Eskiyurt, B.; Sert, Y.; Cırak, C. Experimental and computational study on molecular structure and vibrational analysis of an antihyperglycemic biomolecule: Gliclazide. *Spectrochim. Acta Part A Mol. Biomol. Spectrosc.* **2015**, *135*, 137–146. [CrossRef]
18. Polat, T.; Yildirim, G. Investigation of solvent polarity effect on molecular structure and vibrational spectrum of xanthine with the aid of quantum chemical computations. *Spectrochim. Acta Part A Mol. Biomol. Spectrosc.* **2014**, *123*, 98–109. [CrossRef]
19. Perdew, J.P.; Burke, K.; Ernzerhof, M. Generalized gradient approximation made simple. *Phys. Rev. Lett.* **1996**, *77*, 3865. [CrossRef]
20. Liu, W.; Zhao, Y.; Zhang, Y.; Yan, J.; Zhu, Z.; Hu, J. Study on theoretical analysis of C4F7N infrared spectra and detection method of mixing ratio of the gas mixture. *J. Mol. Spectrosc.* **2021**, *381*, 111521. [CrossRef]
21. Rahmani, R.; Boukabcha, N.; Chouaih, A.; Hamzaoui, F.; Goumri-Said, S. On the molecular structure, vibrational spectra, HOMO-LUMO, molecular electrostatic potential, UV–Vis, first order hyperpolarizability, and thermodynamic investigations of 3-(4-chlorophenyl)-1-(1yridine-3-yl) prop-2-en-1-one by quantum chemistry calculations. *J. Mol. Struct.* **2018**, *1155*, 484–495.

22. Araújo, R.L.; Vasconcelos, M.S.; Barboza, C.A.; Neto, J.X.L.; Albuquerque, E.L.; Fulco, U.L. DFT calculations of the structural, electronic, optical and vibrational properties of anhydrous orthorhombic L-threonine crystals. *Comput. Theor. Chem.* **2019**, *1170*, 112621. [CrossRef]
23. Sert, Y.; Puttaraju, K.B.; Keskinoğlu, S.; Shivashankar, K.; Ucun, F. FT-IR and Raman vibrational analysis, B3LYP and M06-2X simulations of 4-bromomethyl-6-tert-butyl-2H-chromen-2-one. *J. Mol. Struct.* **2015**, *1079*, 194–202. [CrossRef]
24. Wang, L.J.; Zeng, W.B.; Gao, S. Vibrational spectroscopic investigation and molecular structure of a 5α-reductase inhibitor: Finasteride. *Química Nova* **2020**, *43*, 586–592. [CrossRef]
25. Pople, J.A.; Scott, A.P.; Wong, M.W.; Radom, L. Scaling factors for obtaining fundamental vibrational frequencies and zero-point energies from HF/6–31G* and MP2/6–31G* harmonic frequencies. *Isr. J. Chem.* **1993**, *33*, 345–350. [CrossRef]
26. Laury, M.L.; Carlson, M.J.; Wilson, A.K. Vibrational frequency scale factors for density functional theory and the polarization consistent basis sets. *J. Comput. Chem.* **2012**, *33*, 2380–2387. [CrossRef]
27. Merrick, J.P.; Moran, D.; Radom, L. An evaluation of harmonic vibrational frequency scale factors. *J. Phys. Chem. A* **2007**, *111*, 11683–11700. [CrossRef]
28. Alecu, I.M.; Zheng, J.; Zhao, Y.; Truhlar, D.G. Computational thermochemistry: Scale factor databases and scale factors for vibrational frequencies obtained from electronic model chemistries. *J. Chem. Theory Comput.* **2010**, *6*, 2872–2887. [CrossRef]
29. Hout, J.R.F.; Levi, B.A.; Hehre, W.J. Effect of electron correlation on theoretical vibrational frequencies. *J. Comput. Chem.* **1982**, *3*, 234–250. [CrossRef]
30. Simandiras, E.D.; Handy, N.C.; Amos, R.D. On the high accuracy of mp2-optimised geometmes and harmonic frequencies with large basis sets. *Chem. Phys. Lett.* **1987**, *133*, 324–330. [CrossRef]
31. Oi, T. Calculations of reduced partition function ratios of monomeric and dimeric boric acids and borates by the ab initio molecular orbital theory. *J. Nucl. Sci. Technol.* **2000**, *37*, 166–172. [CrossRef]
32. Irikura, K.K.; Johnson, R.D.; Kacker, R.N. Uncertainties in scaling factors for ab initio vibrational frequencies. *J. Phys. Chem. A* **2005**, *109*, 8430–8437. [CrossRef] [PubMed]
33. Barone, V. Vibrational zero-point energies and thermodynamic functions beyond the harmonic approximation. *J. Chem. Phys.* **2004**, *120*, 3059–3065. [CrossRef] [PubMed]
34. Van-Oanh, N.T.; Falvo, C.; Calvo, F.; Lauvergnat, D.; Basire, M.; Gaigeotde, M.; Parneix, P. Improving anharmonic infrared spectra using semiclassically prepared molecular dynamics simulations. *Phys. Chem. Chem. Phys.* **2012**, *14*, 2381–2390. [CrossRef] [PubMed]
35. Tanimoto, M.; Kuchitsu, K.; Morino, Y. Molecular structure and the effect of large-amplitude vibration of carbon suboxide as studied by gas electron diffraction. *Bull. Chem. Soc. Jpn.* **1970**, *43*, 2776–2785. [CrossRef]
36. Carnimeo, I.; Puzzarini, C.; Tasinato, N.; Stoppa, P.; Charmet, A.P.; Biczysko, M.; Cappelli, C.; Barone, V. Anharmonic theoretical simulations of infrared spectra of halogenated organic compounds. *J. Chem. Phys.* **2013**, *139*, 074310. [CrossRef]
37. Watson, J.K.G.; Aliev, M.R. Higher-Order Effects in the Vibration—Rotation Spectra of Semirigid Molecules. *Mol. Spectrosc. Mod. Res. V3* **2012**, *3*, 1.
38. Palafox, M.A. DFT computations on vibrational spectra: Scaling procedures to improve the wavenumbers. *Phys. Sci. Rev.* **2018**, *3*. [CrossRef]
39. Othman, A.; Evans, J.S.; Evans, I.R.; Harris, R.K.; Hodgkinson, P. Structural study of polymorphs and solvates of finasteride. *J. Pharm. Sci.* **2007**, *96*, 1380–1397. [CrossRef]
40. Rani, D.; Goyal, P.; Chadha, R. Conformational flexibility and packing plausibility of repaglinide polymorphs. *J. Mol. Struct.* **2018**, *1157*, 263–275. [CrossRef]
41. Harris, R.; Yeung, R.; Brian Lamont, R.; Lancaster, R.; Lynn, S.; Staniforth, S. 'Polymorphism' in a novel anti-viral agent: Lamivudine. *J. Chem. Soc. Perkin Trans. 2* **1997**, *12*, 2653–2660. [CrossRef]
42. Zou, B.; Dreger, K.; Mück-Lichtenfeld, C.; Grimme, S.; Schäfer, H.J.; Fuchs, H.; Chi, L. Simple and complex lattices of N-alkyl fatty acid amides on a highly oriented pyrolytic graphite surface. *Langmuir* **2005**, *21*, 1364–1370. [CrossRef] [PubMed]
43. Piacenza, M.; Grimme, S. Van der Waals interactions in aromatic systems: Structure and energetics of dimers and trimers of pyridine. *ChemPhysChem* **2005**, *6*, 1554–1558. [CrossRef] [PubMed]
44. Parac, M.; Etinski, M.; Peric, M.; Grimme, S. A theoretical investigation of the geometries and binding energies of molecular tweezer and clip host− guest systems. *J. Chem. Theory Comput.* **2005**, *1*, 1110–1118. [CrossRef]
45. Delley, B. From molecules to solids with the DMol 3 approach. *J. Chem. Phys.* **2000**, *113*, 7756–7764. [CrossRef]

Disclaimer/Publisher's Note: The statements, opinions and data contained in all publications are solely those of the individual author(s) and contributor(s) and not of MDPI and/or the editor(s). MDPI and/or the editor(s) disclaim responsibility for any injury to people or property resulting from any ideas, methods, instructions or products referred to in the content.

Article

Prediction of Ethanol-Mediated Growth Morphology of Ammonium Dinitramide/Pyrazine-1,4-Dioxide Cocrystal at Different Temperatures

Yuanping Zhang [1], Boyu Ma [1], Xinlei Jia [2,*] and Conghua Hou [3]

[1] School of Coal Engineering, Shanxi Datong University, Datong 037003, China; 18234095862@163.com (Y.Z.); 17835702113@163.com (B.M.)
[2] Department of Chemical Engineering and Safety, Binzhou University, Binzhou 256603, China
[3] School of Environment and Safety Engineering, North University of China, Taiyuan 030051, China; houconghua@163.com
* Correspondence: 18434362466@163.com

Citation: Zhang, Y.; Ma, B.; Jia, X.; Hou, C. Prediction of Ethanol-Mediated Growth Morphology of Ammonium Dinitramide/Pyrazine-1,4-Dioxide Cocrystal at Different Temperatures. *Molecules* **2023**, *28*, 4534. https://doi.org/10.3390/molecules28114534

Academic Editor: Borislav Angelov

Received: 11 April 2023
Revised: 25 May 2023
Accepted: 30 May 2023
Published: 3 June 2023

Copyright: © 2023 by the authors. Licensee MDPI, Basel, Switzerland. This article is an open access article distributed under the terms and conditions of the Creative Commons Attribution (CC BY) license (https://creativecommons.org/licenses/by/4.0/).

Abstract: The crystal morphology of high energetic materials plays a crucial role in aspects of their safety performance such as impact sensitivity. In order to reveal the crystal morphology of ammonium dinitramide/pyrazine-1,4-dioxide (ADN/PDO) cocrystal at different temperatures, the modified attachment energy model (MAE) was used at 298, 303, 308, and 313 K to predict the morphology of the ADN/PDO cocrystal under vacuum and ethanol. The results showed that under vacuum conditions, five growth planes of the ADN/PDO cocrystal were given, which were (1 0 0), (0 1 1), (1 1 0), (1 1 −1), and (2 0 −2). Among them, the ratios of the (1 0 0) and (0 1 1) planes were 40.744% and 26.208%, respectively. In the (0 1 1) crystal plane, the value of S was 1.513. The (0 1 1) crystal plane was more conducive to the adsorption of ethanol molecules. The order of binding energy between the ADN/PDO cocrystal and ethanol solvent was (0 1 1) > (1 1 −1) > (2 0 −2) > (1 1 0) > (1 0 0). The radial distribution function analysis revealed that there were hydrogen bonds between the ethanol and the ADN cations, van der Waals interactions with the ADN anions. As the temperature increased, the aspect ratio of the ADN/PDO cocrystal was reduced, making the crystal more spherical, which helped to further reduce the sensitivity of this explosive.

Keywords: ADN/PDO cocrystal; attachment energy model; molecular dynamics; growth morphology

1. Introduction

High-energy compounds are a class of compounds containing explosive groups that can independently undergo chemical reactions and output energy. They are widely used in exploration, aviation, and other fields. Ammonium dinitramide (ADN) [1] is an ionic energetic compound, which is in the α crystal form at normal temperature and pressure, with a crystal density of 1.812 g/cm^3 [2], an enthalpy of formation of −1.22 mJ/kg [3], and an oxygen balance of 25.8% [4]. Compared with ammonium perchlorate (AP), ADN has a higher enthalpy of formation [5], which can provide higher specific impulse when used in propellants. Since ADN is free of chlorine, its combustion products are clean, and its characteristic signal, such as primary smoke, etc., is low when it burns. However, the application of ADN is limited by the high hygroscopicity at room temperature [6,7].

Cocrystallization of energetic materials can effectively improve the physical and chemical properties [8,9]. Previous studies [10,11] have shown that cocrystallization is an effective way to improve the hygroscopicity of ADN. The moisture absorption rate experiment with an AD/18C6 (18-crown-6) cocrystal [12] showed that the hygroscopicity of the ADN/18C6 cocrystal was significantly lower than that of the ADN. The modified attachment energy was adopted by Xie [13] to explore the influence of ethanol on the morphology of an ADN/18C6 cocrystal at different temperatures. The results revealed that the morphology of the ADN/18C6 cocrystal was close to spherical at 293 K, and the cocrystal of the ADN/18C6 had lower hygroscopicity than the ADN at this temperature. Pyrazine-1,4-dioxide (PDO) was selected by Michael K [14] to construct an ADN/PDO cocrystal with a molar ratio of 2:1. The hygroscopicity test showed that the hygroscopicity of the ADN decreased significantly after the cocrystallization of the ADN and PDO. The thermal stability of the ADN/PDO cocrystal was higher than the ADN. Meanwhile, the energetic properties were improved over the ADN.

As an important indicator of crystals, their morphology has been widely considered [15–17]. The morphology of an energetic material can significantly affect its sensitivity [18], which is affected by saturation, solvent, temperature, etc. [19,20]. In this work, an "explosive/solvent" double-layer model was constructed, and four temperatures were adopted to perform the molecular dynamics, which was carried out at 298, 303, 308, and 313 K. The morphology of the ADN/PDO cocrystal was predicted with the help of a modified attachment energy model. The binding energy, radial distribution function, and diffusion coefficient were used to reveal the growth morphology of the ADN/PDO cocrystal in an ethanol environment.

2. Results and Discussion

2.1. Force Field Verification

As an ionic compound, ADN has strong electrostatic interactions between anions and cations. In order to accurately calculate the interaction energy between the ADN/PDO and ethanol, the DMol^3 program was selected to calculate the Mulliken charge of the ADN/PDO unitcell and ethanol, which was adopted as the atomic charge for subsequent calculations. The Perdew–Burke–Ernzerhof (PBE) correlation was used to calculate the exchange-correlation energy. The all-electron method was implemented to treat the core electrons. The double numerical plus polarization numerical basis set (DNP) was adapted. The Mulliken charge is shown in Figure 1.

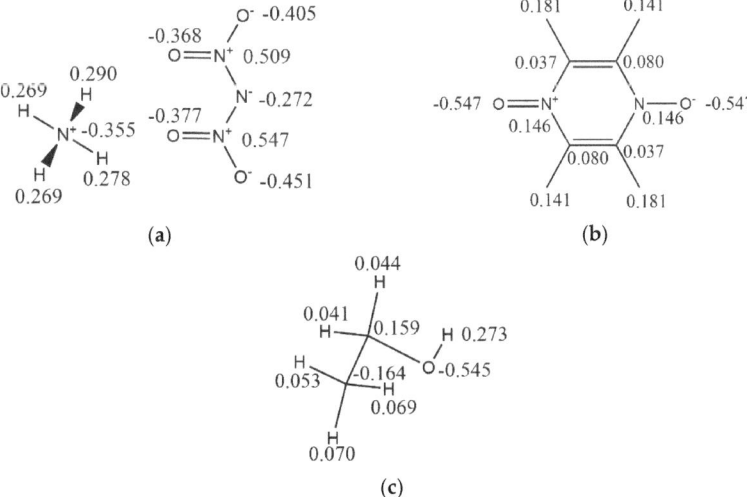

Figure 1. Mulliken charges of the ADN, PDO, and ethanol. (**a**) ADN (**b**) PDO (**c**) EtOH.

In classical molecular dynamics, the force field parameters directly determine the accuracy of calculation results. In order to accurately describe the interaction in the ADN/PDO cocrystal, COMPASSIII, PCFF, CVFF, Universal, and Dreiding force fields and the Mulliken charge were adopted to describe the properties of the atom in the whole simulation model. The optimized unit cell parameters of the ADN/PDO cocrystal are given in Table 1. The relative error (RE) of the unit cell parameters in Table 1 revealed that the RE optimized by CVFF was smaller than the results under other force fields, and it was closer to the experimental value. Meanwhile, the calculation of Yusop [21] revealed that the CVFF force field was suitable for the molecular dynamics simulation of ethanol. Therefore, the CVFF force field was adopted to perform the molecular dynamics simulation.

Table 1. ADN/PDO unit cell parameters and relative errors optimized by five force fields.

Parameters	Exp	COMPASSIII	PCFF	CVFF	Universal	Dreiding
a/Å	11.592	10.946	11.533	11.626	11.819	11.347
b/Å	8.188	8.143	9.167	8.504	8.472	7.960
c/Å	7.227	6.832	6.946	7.217	7.687	7.696
α/°	90.000	90.000	90.000	90.000	90.000	90.000
β/°	101.236	99.546	93.744	101.050	100.974	100.628
γ/°	90.000	90.000	90.000	90.000	90.000	90.000
RE_a	0.00%	−5.57%	−0.51%	0.29%	1.96%	−2.12%
RE_b	0.00%	−0.54%	11.96%	3.86%	3.47%	−2.78%
RE_c	0.00%	−5.47%	−3.89%	−0.14%	6.36%	6.49%
RE_α	0.00%	0.00%	0.00%	0.00%	0.00%	0.00%
RE_β	0.00%	−1.67%	−7.40%	−0.18%	−0.26%	−0.60%
RE_γ	0.00%	0.00%	0.00%	0.00%	0.00%	0.00%

Note: RE is the relative error.

2.2. Morphology of the ADN/PDO in Vacuum

The morphology of the ADN/PDO in vacuum was calculated based on the AE model, as shown in Figure 2. Under vacuum conditions, the important growth planes and area ratios of the ADN/PDO cocrystal are shown in Table 2. It can be seen from Figure 2 that the vacuum morphology of the ADN/PDO cocrystal was approximately hexagonal prism, and the aspect ratio of the crystal was 2.569. Its crystal morphology was mainly composed of five growth crystal planes (1 0 0), (0 1 1), (1 1 0), (1 1 −1), and (2 0 −2). Among them, the (1 0 0) surface had the largest exposed area, accounting for 40.744%, and had the greatest morphological importance. According to Formula (3), it can be seen that the absolute value of the attachment energy was proportional to the relative growth rate of the corresponding crystal plane. The order of the relative growth rate of each crystal plane of the ADN/PDO was (2 0 −2) > (1 1 −1) > (1 1 0) > (0 1 1) > (1 0 0).

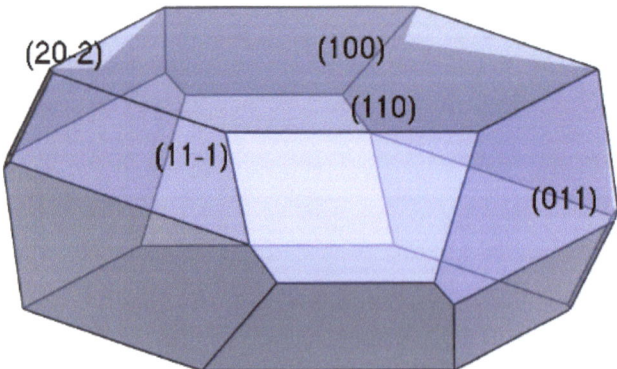

Figure 2. The morphology of the ADN/PDO cocrystal under vacuum (attachment energy model).

Table 2. The crystal habits parameters of the ADN/PDO cocrystal under vacuum (attachment energy model).

(h k l)	d_{hkl}	Surface Area/Å^2	$E_{att\ (total)}$/(kcal·mol^{-1})	Total Facet Area/%	Aspect Ratio
(1 0 0)	11.411	61.370	−61.570	40.744	
(0 1 1)	5.442	128.669	−117.417	26.208	
(1 1 0)	6.818	102.703	−109.348	18.177	2.569
(1 1 −1)	5.220	134.151	−119.789	14.450	
(2 0 −2)	3.306	105.900	−148.904	0.422	

In the modified attachment energy (MAE) model, S was used to characterize the surface roughness of the (h k l) crystal plane [22]. The larger the value of S, the rougher the surface of the corresponding crystal plane, and the more conducive to the adsorption of solvents [18,23]. It can be seen that the values of S corresponding to (1 0 0), (0 1 1), (1 1 0), (1 1 −1), and (2 0 −2) crystal planes were 1.223, 1.513, 1.359, 1.389, and 1.836, respectively. The packing patterns of five crystal planes of the ADN/PDO and the corresponding Connolly surfaces are shown in Figure 3. Based on Figure 3, more chemical groups were exposed in the (0 1 1) and (2 0 −2) crystal planes compared with the other three crystal planes. It can be seen from Figure 3b,e that the (0 1 1) crystal plane was mainly exposed to the PDO and the cations of the ADN, and the (2 0 −2) crystal plane was exposed to both the cations and anions of the ADN and PDO. This results in the (2 0 −2) crystal plane were more bumpy and rougher than the (0 1 1) crystal plane. The (2 0 −2) plane was the roughest and had more adsorption sites. It was conducive to absorbing the molecule of solvent. This had a certain hindering effect on the growth of the crystal plane.

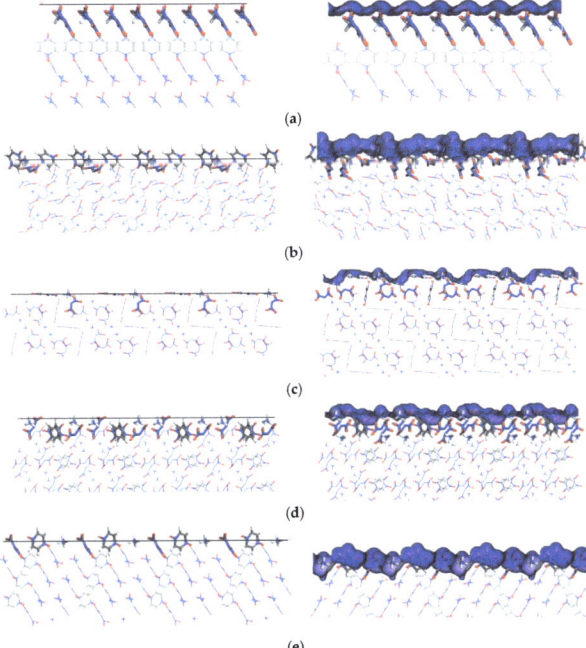

Figure 3. The molecular stacking structure (**left**) and Connolly surface (**right**) of the important growth plane of the ADN/PDO. (**a**) (1 0 0) growth plane (**b**) (0 1 1) growth plane (**c**) (1 1 0) growth plane (**d**) (1 1 −1) growth plane (**e**) (2 0 −2) growth plane.

2.3. Morphology of the AND/PDO in Solvent

2.3.1. Binding Energy

When a crystal grows in a solution, the solute molecules first diffuse to the crystal surface through diffusion, then overcome the desorption energy barrier of the solvent, and finally adsorb to the crystal surface to complete the crystal growth [23,24]. Therefore, the crystal morphology can be significantly affected by the adsorption of solvents on the crystal surface. The greater the binding energy between solvent and crystal plane, the stronger the binding effect, the greater the desorption energy barrier that the solute needs to overcome, and the stronger the inhibitory effect of the solvent on the growth of crystal plane. The formula for calculating the binding energy is

$$E_{bind} = -E_{int} = -(E_{tot} - E_{cry} - E_{sol}) \tag{1}$$

where E_{bind}, E_{int} are the binding energy and interaction energy between the crystal plane and the solvent, respectively. E_{tot} is the total energy of the mixed system of crystal plane and solvent, E_{cry} is the energy of the crystal plane in the model, E_{sol} is the energy of the solvent in the model. The calculation results are shown in Table 3.

Table 3. Energy details of the layered model for each crystal face of ADN/PDO-EtOH at different temperatures.

T/K	(h k l)	E_{tot}/ (kcal·mol^{-1})	E_{sol}/ (kcal·mol^{-1})	E_{cry}/ (kcal·mol^{-1})	E_{bind}/ (kcal·mol^{-1})
298	(1 0 0)	−31,481.17	−977.69	−30,057.88	445.60
	(0 1 1)	−30,628.73	−1821.46	−27,671.08	1136.19
	(1 1 0)	−34,837.67	−1936.03	−32,224.39	677.25
	(1 1 −1)	−36,757.43	−1570.34	−34,100.57	1086.52
	(2 0 −2)	−13,735.02	−1740.63	−11,101.82	892.57
303	(1 0 0)	−31,274.01	−817.80	−29,976.35	479.86
	(0 1 1)	−30,552.83	−1790.10	−27,651.20	1111.53
	(1 1 0)	−34,632.60	−1770.36	−32,133.13	729.11
	(1 1 −1)	−36,837.35	−1703.37	−34,060.68	1073.30
	(2 0 −2)	−13,690.91	−1658.16	−11,116.84	915.91
308	(1 0 0)	−31,210.31	−829.50	−29,875.30	505.50
	(0 1 1)	−30,341.76	−1626.49	−27,541.56	1173.72
	(1 1 0)	−34,523.77	−1706.96	−32,087.83	728.98
	(1 1 −1)	−36,633.24	−1583.95	−33,947.88	1101.40
	(2 0 −2)	−13,497.78	−1561.47	−11,070.35	865.96
313	(1 0 0)	−31,010.92	−714.65	−29,865.60	430.68
	(0 1 1)	−30,239.24	−1560.73	−27,580.70	1097.81
	(1 1 0)	−34,250.38	−1550.09	−31,982.37	717.92
	(1 1 −1)	−36,502.03	−1486.06	−33,840.10	1175.87
	(2 0 −2)	−13,428.79	−1492.03	−11,066.84	869.92

According to Table 3, it can be seen that the binding energy between the ADN/PDO cocrystal and the ethanol solvent were positive on all important growth crystal planes, indicating that the interaction between the solute and the solvent was dominated by attraction. The variations of the binding energy between the ADN/PDO and the ethanol on each important crystal face of the ADN/PDO cocrystal are shown in Figure 4, which were obtained under different temperatures. The order of binding energy between five crystal planes of the ADN/PDO cocrystal and ethanol was (0 1 1) > (1 1 −1) > (2 0 −2) > (1 1 0) > (1 0 0). It can be revealed that the binding effect between the ethanol and the ADN/PDO was the strongest on the (0 1 1) crystal plane. Meanwhile, ethanol had the strongest inhibitory effect on the growth of this crystal plane. It can be seen from Figure 4 that the binding energy between the ADN/PDO crystal plane and ethanol first increased and then decreased as the temperature increased. Its value was always positive, indicating that the process of ethanol adsorption onto the ADN/PDO crystal plane is an exothermic process.

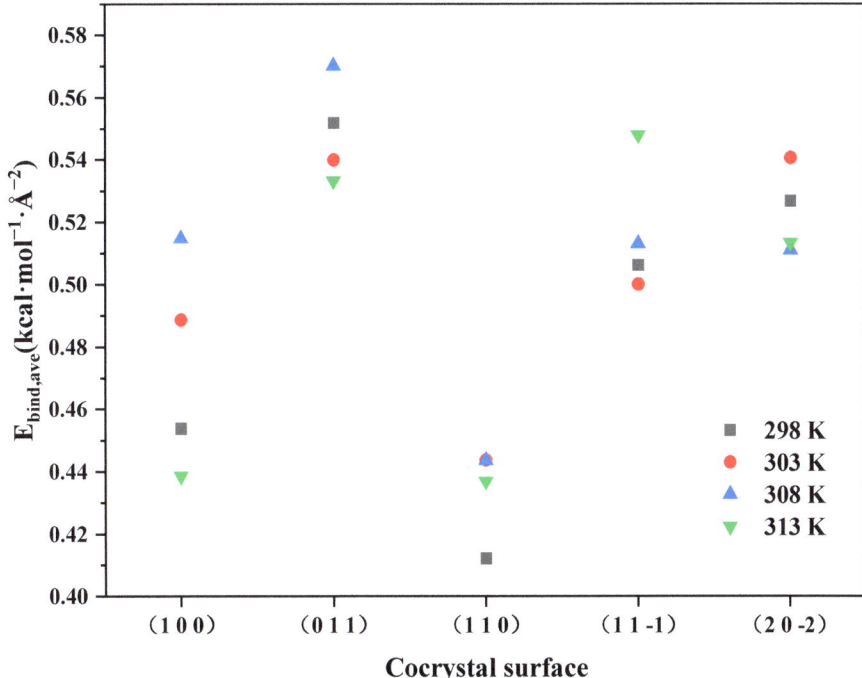

Figure 4. Binding energy between the ADN/PDO and ethanol at different temperatures.

2.3.2. Radial Distribution Function

The radial distribution function (RDF) is often adopted to analyze non-bonding interactions between atomic pairs. It represents the probability of finding atom B at a distance r from atom A. In general, non-bonding interactions between atomic pairs include hydrogen bonding (less than 3.1 Å), van der Waals forces (3.1–5.0 Å), and electrostatic forces (greater than 5.0 Å).

Taking the (0 1 1) crystal plane as an example, the RDF results are shown in Figure 5. The black, red, and blue curves corresponded to the RDF between the oxygen atom in ethanol (EtOH-O) and the nitrogen atom connected to two nitro groups in the ADN anion (ADN-N), the RDF between the EtOH-O and the oxygen atom in the PDO (PDO-O), and the RDF between the EtOH-O and the nitrogen atom in the ADN cations (ADN-N1), respectively. The coordinates of the local maximum value of each curve are given in the Figure 5. Based on Figure 5a, the RDF of the EtOH-O and ADN-N1 had the first peak at r = 2.85 Å at 298 K, indicating that there was a hydrogen bond between the ethanol and ADN cations. The peak of the black curve appeared at the position of r = 4.91 Å, indicating that there was a van der Waals interaction between the ethanol and the ADN anions. There were two peaks in the radial distribution function between the EtOH-O and PDO-O, which appeared at r = 2.67 and 4.67 Å respectively. This revealed the presence of hydrogen bonds and van der Waals interactions between the ADN and the PDO.

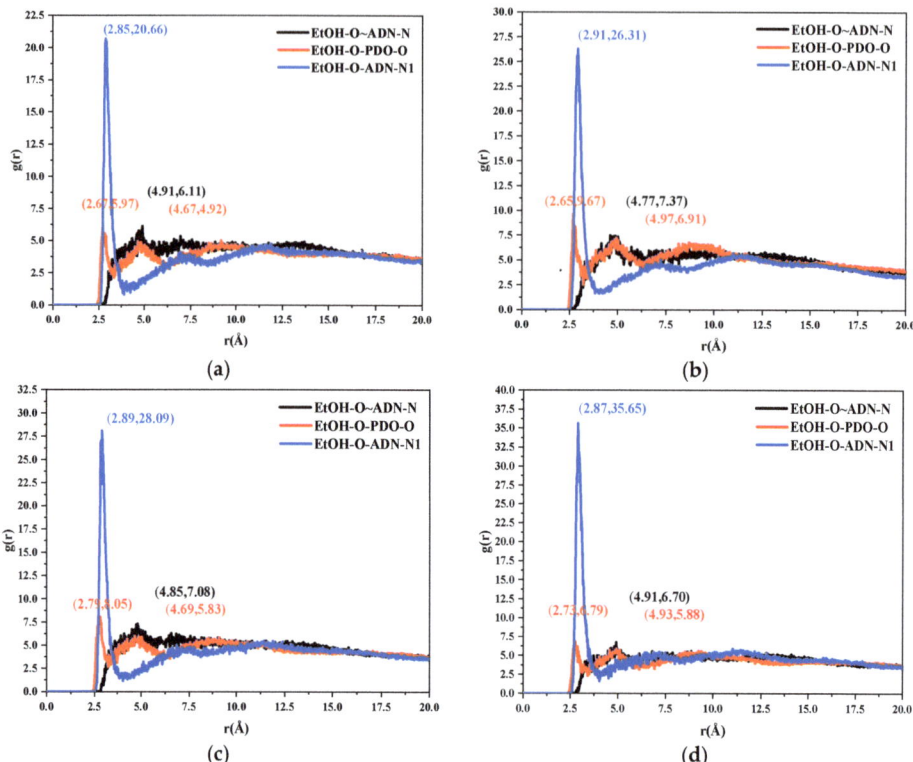

Figure 5. Radial distribution functions between the oxygen atoms in the ethanol and the oxygen atoms in the PDO, the nitrogen atoms in the ADN in the (0 1 1) plane at different temperatures. (**a**) 298 K (**b**) 303 K (**c**) 308 K (**d**) 313 K.

Similarly, when the temperatures were 303, 308, and 313 K, there were hydrogen bonds and van der Waals interactions between the ethanol and the PDO. There were hydrogen bonds and van der Waals interactions between the ethanol and the cation and anion of the ADN, respectively. As the temperature increased, the number of local maximum values of the three RDFs remained unchanged, while the abscissa of the local maximum point moved to the right as a whole. This revealed that the increase in temperature led to the change of the hydrogen bond and van der Waals interaction between the ethanol and the ADN, PDO, which first increased and then decreased. It was consistent with the binding energy results in Figure 4.

2.3.3. Diffusion Coefficients

Generally, the diffusion of a solvent is greatly affected by temperature. The diffusion of a solvent on the crystal surface may cause it to adsorb on the crystal surface. Absorption of solute molecules to the crystal surface is blocked by this process. The growth of the corresponding crystal planes is also inhibited. To reveal the effect of temperature on the diffusion of ethanol molecules on the AND/PDO crystal plane, the calculated trajectories were processed, and the corresponding mean square displacement curves were calculated, which is given in Figure 6.

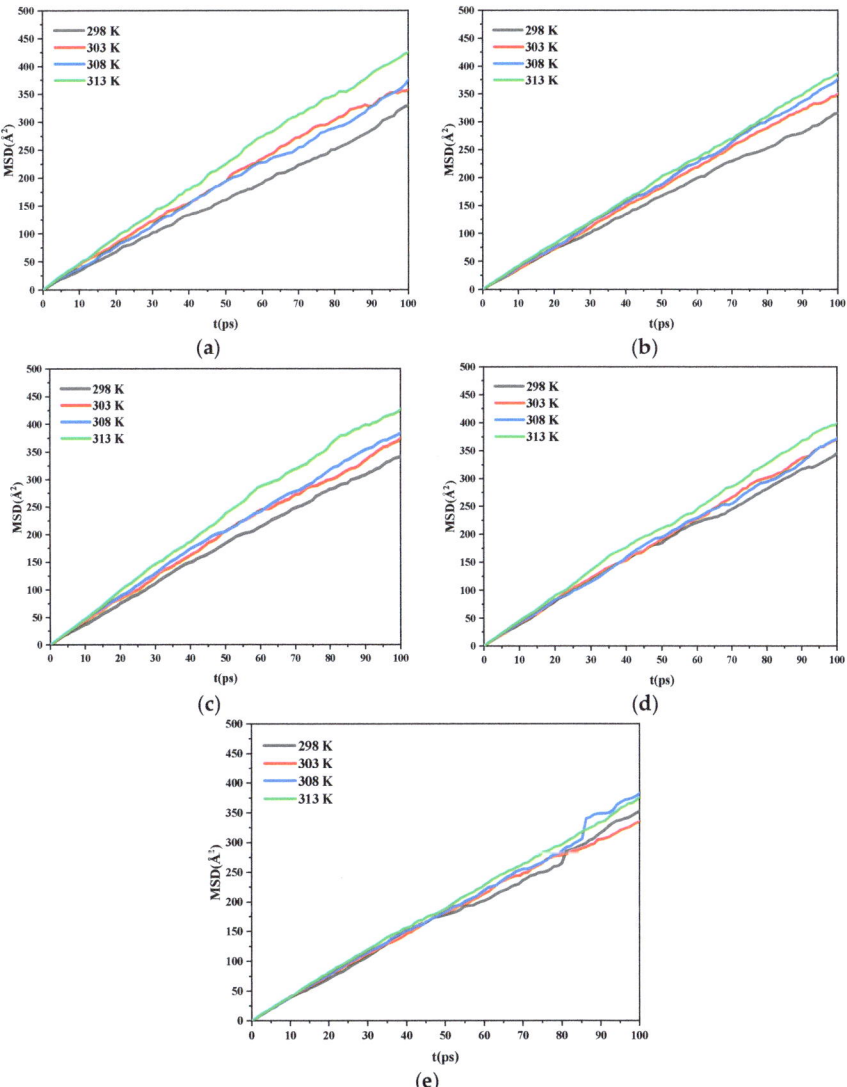

Figure 6. Mean square displacement curves of the solvent in the ADN/PDO-EtOH layered model at different temperatures. (**a**) (1 0 0) (**b**) (0 1 1) (**c**) (1 1 0) (**d**) (1 1 −1) (**e**) (2 0 −2).

According to Einstein's law of diffusion, the formula for calculating the diffusion coefficient is

$$\lim_{l \to \infty} \left\langle \left| \vec{r}(t) - \vec{r}(0) \right|^2 \right\rangle = \lim_{l \to \infty} \text{MSD} = 6Dt \qquad (2)$$

where D is the diffusion coefficient. Therefore, the diffusion coefficient is one-sixth of the slope of the mean square displacement curve.

The diffusion coefficients of the ethanol on each crystal surface at different temperatures are given in Table 4. Taking the (0 1 1) surface as an example, when the temperature was 298 K, the diffusion coefficient of the ethanol in the system was 0.52×10^{-8} m$^2 \cdot$s^{-1}. When the temperature rose to 313 K, the diffusion coefficient of the system was 0.64×10^{-8} m$^2 \cdot$s^{-1}. As the temperature increased, the diffusion coefficient of the ethanol under the same crystal plane increased gradually. At the same temperature, the largest diffusion coefficient of the ethanol in the layered model was obtained under the (1 1 0) crystal plane. However, the final morphology of the crystals in the solvent was determined by the diffusion of the solute and solvent and the competitive adsorption of the solute and solvent on the crystal surface.

Table 4. Diffusion coefficient of the solvent on each crystal face at different temperatures.

(h k l)	$D/(\times 10^{-8}\ m^2 \cdot s^{-1})$			
	298 K	303 K	308 K	313 K
(1 0 0)	0.53	0.61	0.60	0.71
(0 1 1)	0.52	0.59	0.62	0.64
(1 1 0)	0.57	0.61	0.63	0.73
(1 1 −1)	0.56	0.61	0.60	0.66
(2 0 −2)	0.57	0.56	0.62	0.61

2.3.4. Morphology Analysis

According to the modified AE model, the corrected attachment energy and aspect ratio between the crystal plane of the ADN/PDO cocrystal and ethanol are shown in Table 5, which was calculated under different temperatures. The calculated morphology of the ADN/PDO cocrystal in ethanol at different temperatures and the corresponding experimental crystal morphology of the ADN/PDO cocrystal [25] are shown in Figure 7. Based on Table 5, it can be seen that the (0 1 1) crystal surface of the ADN/PDO cocrystal had a greater interaction with the ethanol than the (1 0 0) plane at the four temperatures (298, 303, 308, and 313 K). This indicated that the ethanol had a strong inhibitory effect on the growth of the ADN/PDO (0 1 1) crystal surface.

Table 5. Modified attachment energies and related parameters of the ADN/PDO cocrystals in ethanol at different temperatures.

T/K	(h k l)	$E_{int}/$ (kcal·mol^{-1})	$E_s/$ (kcal·mol^{-1})	$E_{att}'/$ (kcal·mol^{-1})	Total Facet Area/%	Aspect Ratio
298	(1 0 0)	−445.60	−27.85	−27.51	15.37	
	(0 1 1)	−1136.19	−71.01	−10.00	84.63	
	(1 1 0)	−677.25	−42.33	−51.81	-	3.324
	(1 1 −1)	−1086.52	−67.91	−25.46	-	
	(2 0 −2)	−892.57	−55.79	−46.46	-	
303	(1 0 0)	−479.86	−29.99	−24.90	19.85	
	(0 1 1)	−1111.54	−69.47	−12.33	80.15	
	(1 1 0)	−729.11	−45.57	−47.40	-	2.653
	(1 1 −1)	−1073.30	−67.08	−26.61	-	
	(2 0 −2)	−915.91	−57.24	−43.78	-	
308	(1 0 0)	−505.50	−31.59	−22.94	12.33	
	(0 1 1)	−1173.72	−73.36	−6.45	87.67	
	(1 1 0)	−728.98	−45.56	−47.41	-	4.090
	(1 1 −1)	−1101.40	−68.84	−24.17	-	
	(2 0 −2)	−865.96	−54.12	−49.52	-	
313	(1 0 0)	−430.68	−26.92	−28.65	10.70	
	(0 1 1)	−1097.81	−68.61	−13.63	61.50	
	(1 1 0)	−717.92	−44.87	−48.35	-	2.353
	(1 1 −1)	−1175.87	−73.49	−17.71	27.80	
	(2 0 −2)	−869.92	−54.37	−49.06	-	

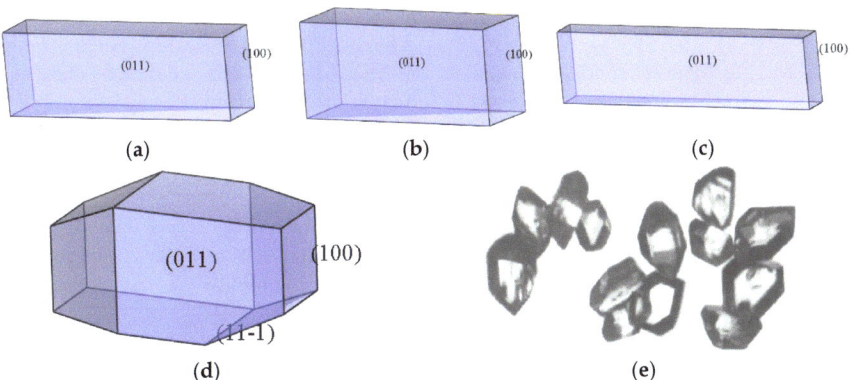

Figure 7. Morphology of the ADN/PDO cocrystal in ethanol and the ADN/PDO cocrystal morphology at different temperatures. (**a**) 298 K (**b**) 303 K (**c**) 308 K (**d**) 313 K (**e**) experiment.

According to the results of S on each crystal plane of the ADN/PDO cocrystal, the (2 0 −2) crystal plane was the roughest, which was the most favorable for the adsorption of solvents. However, the binding energy between the solvent and the (2 0 −2) crystal plane was weak, and the diffusion coefficient of ethanol on the (2 0 −2) plane was small. Under the comprehensive influence of various factors, the relative growth rate of the (2 0 −2) plane was large, which made the (2 0 −2) crystal surface disappear eventually during the growth process. Similarly, the roughness of the (0 1 1) surface and the binding energy between explosives and ethanol were large. This gave the (0 1 1) surface the largest area ratio in the ethanol solvent.

When the temperature was lower than 308 K, the morphology of the ADN/PDO cocrystal in the ethanol solvent was a quadrangular prism. The three crystal planes (1 1 0), (1 1 −1), and (2 0 −2) eventually disappeared during the growth process owing to the high relative growth rate. When the temperature increased from 298 to 308K, the (0 1 1) surface area ratio of the ADN/PDO cocrystal was greater than 80%, and the aspect ratios of the ADN/PDO cocrystal were 3.324, 2.653, and 4.090, respectively. When the temperature was 313 K, the morphology of the ADN/PDO cocrystal was irregular prism, and the aspect ratio was 2.353. The ADN/PDO cocrystal morphology calculated by the modified AE model was the closest to the experimental one at 313 K. The presented modeling and simulations can also be applied for the prediction of the drug crystal topologies in a pharmaceutical application employing drug delivery carriers [26–28].

3. Modeling and Simulation

3.1. Modified Attachment Energy Model

The Attachment Energy Model (AE model) was developed by Hartman and Bennema [25,29] on the basis of the Periodic Bond Chain (PBC) theory. In the AE model, the relative growth rate of each crystal plane of the crystal is proportional to the absolute value of the crystal plane attachment energy.

$$R_{hkl} \propto |E_{att}| \tag{3}$$

where R_{hkl} is the relative growth rate, and E_{att} is the attachment energy of the crystal plane, kcal·mol^{-1}. The attachment energy (E_{att}) is defined as the energy released by a wafer with a thickness of d_{hkl} attached to the (h k l) crystal plane. The formula is

$$E_{att} = E_{latt} - E_{slice} \tag{4}$$

where E_{att}, E_{latt}, E_{slice} are the attachment energy, lattice energy of crystal, and the energy of growth slice, respectively.

By calculating the attachment energy of the crystal face, the habit of the crystal can be predicted with the help of the AE model. However, the external environment of crystal growth is not considered by the AE model. It is difficult to accurately reveal the actual growth process of crystals in solution [30,31]. Therefore, the AE model should be corrected, and the modified formula for calculating the attachment energy is

$$E'_{att} = E_{att} - S \cdot E_s \tag{5}$$

where S is used to describe the surface characteristics and is defined as $S = \frac{A_{acc}}{A_{hkl}}$.

Among them, A_{acc} is the accessible area of solvent on the crystal plane unit (h k l), A_{hkl} is the cross-sectional area of the crystal plane unit (h k l). E_s represents the effect of the solvent on the crystal plane growth, which is defined as

$$E_s = E_{int} \cdot \frac{A_{hkl}}{A_{box}} \tag{6}$$

where A_{box} and E_{int} are the cross-sectional area of the simulation box and the interaction energy between the crystal layer and the solvent, respectively.

E_{int} is defined as

$$E_{int} = E_{tot} - (E_{cry} + E_{sol}) \tag{7}$$

where E_{tot}, E_{cry}, E_{sol} are the total energy of the mixed system of crystal plane and solvent, the energy of crystal plane in the calculation model, and the energy of the solvent in this model, respectively. Meanwhile, under solvent conditions, the relative growth rate of the crystal is proportional to the absolute value of the corrected attachment energy.

$$R'_{hkl} \propto \left| E'_{hkl} \right| \tag{8}$$

3.2. Computational Methods

The initial unit cell for the ADN/PDO cocrystal was obtained from the Cambridge Crystallographic Data Centre (CCDC) [14]. The initial unit cell parameters of the ADN/PDO cocrystal were a = 11.592 Å, b = 8.188 Å, c = 7.227 Å, α = γ = 90°, β = 101.236°, and the space group was P21/c, belonging to the monoclinic crystal system. The molecular structure of the ADN and PDO and the unit cell structure of the ADN/PDO cocrystal are shown in Figure 8. In Figure 8, the gray, white, red, and blue balls correspond to carbon, hydrogen, oxygen, and nitrogen atoms, respectively.

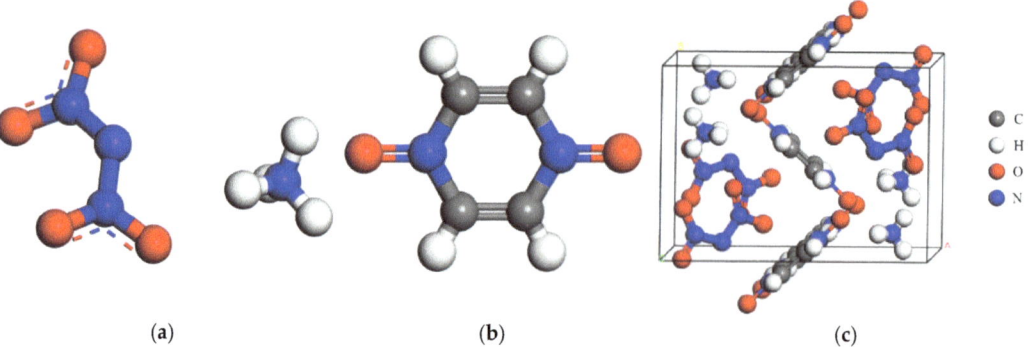

Figure 8. ADN molecular structure, PDO molecular structure, and unit cell structure of the ADN/PDO cocrystal. (**a**) ADN molecular structure (**b**) PDO molecular structure (**c**) Unit cell structure.

The flowchart of the calculation model is given in Figure 9. At first, the stable structure of the ADN/PDO unit cell was obtained after geometry optimization. The Smart optimization algorithm was used, which is the built-in algorithm in Materials Studio 2020.

The Ewald summation method was selected as the electrostatic force summation method, and the precision was set to 0.001 kcal·mol^{-1}. The atom-based method was selected as the van der Waals force summation method, and the cut-off radius was set to 12.5 Å. The morphology of the AND/PDO in vacuum was predicted with the help of the AE model. The important growth planes under vacuum conditions were determined. Then, the important crystal planes of the ADN/PDO were cut to expand it into a 4 × 4 × 4 supercell structure. A solvent layer containing 400 ethanol molecules was constructed using the Amorphous Cell module. The Build Layers function was used to construct an "explosive/solvent" double-layer model with a 50 Å vacuum layer above the solvent layer.

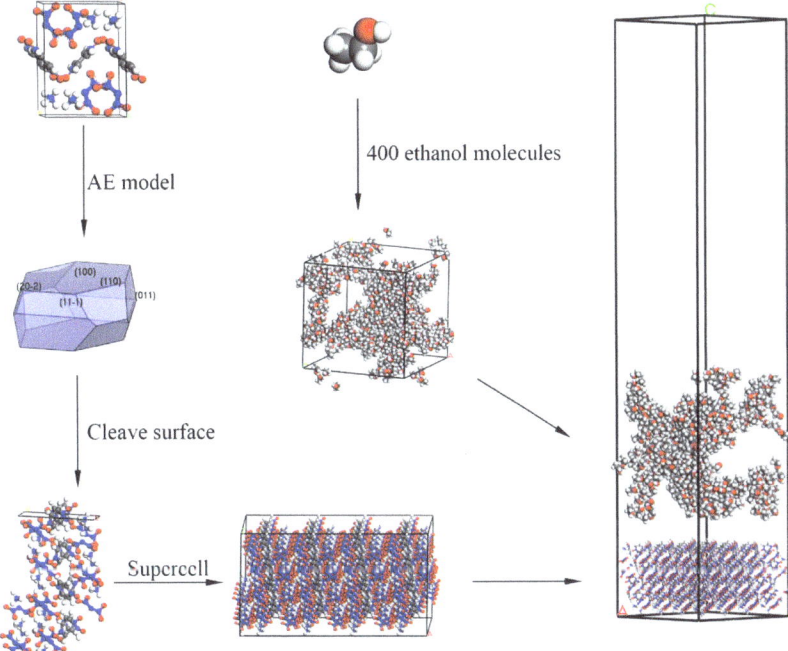

Figure 9. Schematic diagram of the flow chart of calculation model construction.

The model was optimized using the Forcite module of Materials Studio 2020, followed by molecular dynamics calculations at specified temperatures (298, 303, 308, and 313 K). When molecular dynamics calculations were performed, the ensemble and the simulation time were set as NVT and 500 ps, respectively. A total of 500,000 calculation steps were performed, and the timestep was set as 1 fs. The Andersen method was adopted to control the temperature through the whole simulation process. The initial velocity of the particle was randomly assigned according to a Gaussian distribution. Finally, the homemade script was used to calculate the interaction energy between the ADN/PDO explosive and ethanol, the radial distribution function, the mean square displacement, and the mass density distribution based on the stable part of the calculated trajectory.

4. Conclusions

In summary, the modified attachment energy (MAE) model was adopted to investigate the morphology of the ADN/PDO cocrystal in vacuum and ethanol at different temperatures (298, 303, 308, and 313 K). The CVFF forcefield and Mulliken charges were used to perform the molecular dynamics simulation. The main conclusions of this work are as follows:

(1) The growth morphology of the ADN/PDO is hexagonal prism in vacuum, and the five main crystal surfaces are (1 0 0), (0 1 1), (1 1 0), (1 1 −1), and (2 0 −2). Among them, the (1 0 0) surface has the largest exposed area, accounting for 40.744%.

(2) The binding energy between the ADN/PDO cocrystal and ethanol solvent is positive on all important growth planes. The order of binding energy is (0 1 1) > (1 1 −1) > (2 0 −2) > (1 1 0) > (1 0 0). The binding effect between the ethanol and ADN/PDO is strongest on the (0 1 1) crystal plane.

(3) The radial distribution function analysis of the (0 1 1) crystal plane showed that there are hydrogen bonds between the ethanol and ADN cations, van der Waals interactions with the ADN anions, and hydrogen bonds and van der Waals interactions with the PDO at the same time. In the (0 1 1) crystal plane, the value of S is 1.513, which indicates that this surface has a large roughness. This is more conducive to the adsorption of ethanol molecules.

(4) As the temperature increases, the diffusion coefficient of the ethanol under the same crystal plane increases gradually. Meanwhile, the morphology analysis indicated that increasing the temperature is beneficial to reducing the aspect ratio of the crystal. This is conducive to the reduction of explosive sensitivity.

Author Contributions: Software, B.M.; Validation, C.H.; Data curation, X.J.; Writing—original draft, Y.Z. All authors have read and agreed to the published version of the manuscript.

Funding: This research was funded by [the Scientific and Technological Innovation Programs of Higher Education Institutions in Shanxi Province] grant number [2022L450], [the Initiation Fund for Doctoral Research of Shanxi Datong University] grant number [2020-B-03], [Shandong Youth Natural Science Foundation Project] grant number [ZR2020QE111], and [National Postdoctoral Science Foundation] grant number [2020M681073].

Institutional Review Board Statement: Not applicable.

Informed Consent Statement: Not applicable.

Data Availability Statement: Not applicable.

Conflicts of Interest: The authors declare no conflict of interest.

Sample Availability: Not applicable.

References

1. Östmark, H.; Bemm, U.; Langlet, A.; Sandén, R.; Wingborg, N. The Properties of Ammonium Dinitramide (ADN): Part 1, Basic Properties and Spectroscopic Data. *J. Energ. Mater* **2000**, *18*, 123–138. [CrossRef]
2. Venkatachalam, S.; Santhosh, G.; Ninan Ninan, K. An Overview on the Synthetic Routes and Properties of Ammonium Dinitramide (ADN) and Other Dinitramide Salts. *Propell. Explos. Pyrot.* **2004**, *29*, 178–187. [CrossRef]
3. Kon'kova, T.S.; Matyushin, Y.N.; Miroshnichenko, E.A.; Vorob'ev, A.B. Thermochemical properties of dinitramidic acid salts. *Russ. Chem. Bulletin.* **2009**, *58*, 2020–2027. [CrossRef]
4. Chen, F.; Xuan, C.; Lu, Q.; Xiao, L.; Yang, J.; Hu, Y.; Zhang, G.-P.; Wang, Y.; Zhao, F.; Hao, G.; et al. A Review on the High Energy Oxidizer Ammonium Dinitramide: Its Synthesis, Thermal Decomposition, Hygroscopicity, and Application in Energetic Materials. *Def. Technol.* **2023**, *19*, 163–195. [CrossRef]
5. Kumar, P. An Overview on Properties, Thermal Decomposition, and Combustion Behavior of ADN and ADN Based Solid Propellants. *Def. Technol.* **2018**, *14*, 661–673. [CrossRef]
6. Cui, J.; Han, J.; Wang, J.; Huang, R. Study on the Crystal Structure and Hygroscopicity of Ammonium Dinitramide. *J. Chem. Eng. Data* **2010**, *55*, 3229–3234. [CrossRef]
7. Chen, X.; He, L.; Li, X.; Zhou, Z.; Ren, Z. Molecular Simulation Studies on the Growth Process and Properties of Ammonium Dinitramide Crystal. *J. Phys. Chem. C* **2019**, *123*, 10940–10948. [CrossRef]
8. Zhang, J.; Shreeve, J.M. Time for pairing: Cocrystals as advanced energetic materials. *CrystEngComm* **2016**, *18*, 6124–6133. [CrossRef]
9. Muravyev, N.V.; Fershtat, L.L.; Dalinger, I.L.; Suponitsky, K.Y.; Ananyev, I.V.; Melnikov, I.N. Rational Screening of Cocrystals using Thermal Analysis: Benchmarking on Energetic Materials. *Cryst. Growth Des.* **2022**, *22*, 7349–7362. [CrossRef]
10. Hu, D.; Wang, Y.; Xiao, C.; Hu, Y.; Zhou, Z.; Ren, Z. Studies on Ammonium Dinitramide and 3, 4-Diaminofurazan Cocrystal for Tuning the Hygroscopicity. *Chinese J. Chem. Eng.* **2023**, *in press*. [CrossRef]

11. Ren, Z.; Chen, X.; Yu, G.; Wang, Y.; Chen, B.; Zhou, Z. Molecular Simulation Studies on the Design of Energetic Ammonium Dinitramide Co-Crystals for Tuning Hygroscopicity. *CrystEngComm* **2020**, *22*, 5237–5244. [CrossRef]
12. Qiao, S.; Li, H.; Yang, Z. Decreasing the Hygroscopicity of Ammonium Dinitramide (ADN) through Cocrystallization. *Energetic Mater. Front.* **2022**, *3*, 84–89. [CrossRef]
13. Xie, H.; Gou, R.; Zhang, S. Theoretical Study on the Effect of Solvent Behavior on Ammonium Dinitramide (ADN)/1,4,7,10,13,16-Hexaoxacyclooctadecane (18-Crown-6) Cocrystal Growth Morphology at Different Temperatures. *Cryst. Res. Technol.* **2021**, *56*, 2000203. [CrossRef]
14. Bellas, M.K.; Matzger, A.J. Achieving Balanced Energetics through Cocrystallization. *Angew. Chem. Int. Ed.* **2019**, *58*, 17185–17188. [CrossRef] [PubMed]
15. Liu, Y.; Niu, S.; Lai, W.; Yu, T.; Ma, Y.; Gao, H.; Zhao, F.; Ge, Z. Crystal Morphology Prediction of Energetic Materials Grown from Solution: Insights into the Accurate Calculation of Attachment Energies. *CrystEngComm* **2019**, *21*, 4910–4917. [CrossRef]
16. Esfahani, R.E.; Zahedi, P.; Zarghami, R. 5-Fluorouracil-Loaded Poly(Vinyl Alcohol)/Chitosan Blend Nanofibers: Morphology, Drug Release and Cell Culture Studies. *Iran Polym. J.* **2021**, *30*, 167–177. [CrossRef]
17. Sun, L.; Wang, Y.; Yang, F.; Zhang, X.; Hu, W. Cocrystal Engineering: A Collaborative Strategy toward Functional Materials. *Adv. Mater.* **2019**, *31*, 1902328. [CrossRef]
18. Liu, Y.; Yu, T.; Lai, W.; Ma, Y.; Kang, Y.; Ge, Z. Adsorption Behavior of Acetone Solvent at the HMX Crystal Faces: A Molecular Dynamics Study. *J. Mol. Graph. Model.* **2017**, *74*, 38–43. [CrossRef]
19. Lovette, M.A.; Browning, A.R.; Griffin, D.W.; Sizemore, J.P.; Snyder, R.C.; Doherty, M.F. Crystal Shape Engineering. *Ind. Eng. Chem. Res.* **2008**, *47*, 9812–9833. [CrossRef]
20. Dandekar, P.; Kuvadia, Z.B.; Doherty, M.F. Engineering Crystal Morphology. *Annu. Rev. Mater. Res.* **2013**, *43*, 359–386. [CrossRef]
21. Yusop, S.N.; Anuar, N.; Salwani, M.A.N.; Abu Bakar, N.H. Molecular Dynamic Investigation on the Dissolution Behaviour of Carbamazepine Form III in Ethanol Solution. *Key Eng. Mater.* **2019**, *797*, 149–157. [CrossRef]
22. Liu, Y. Crystal Morphology Prediction Method of Energetic Materials: Attachment Energy Model and Its Development. *Chin. J. Explos. Propellant* **2021**, *44*, 578–588. [CrossRef]
23. Liu, Y.; Lai, W.; Yu, T.; Ma, Y.; Kang, Y.; Ge, Z. Understanding the Growth Morphology of Explosive Crystals in Solution: Insights from Solvent Behavior at the Crystal Surface. *RSC Adv.* **2017**, *7*, 1305–1312. [CrossRef]
24. Liu, Y.; Lai, W.; Ma, Y.; Yu, T.; Kang, Y.; Ge, Z. Face-Dependent Solvent Adsorption: A Comparative Study on the Interfaces of HMX Crystal with Three Solvents. *J. Phys. Chem. B* **2017**, *121*, 7140–7146. [CrossRef] [PubMed]
25. Hartman, P. The Attachment Energy as a Habit Controlling Factor: III. Application to Corundum. *J. Cryst. Growth.* **1980**, *49*, 166–170. [CrossRef]
26. Bonaccorso, A.; Gigliobianco, M.R.; Pellitteri, R.; Santonocito, D.; Carbone, C.; Di Martino, P.; Puglisi, G.; Musumeci, T. Optimization of curcumin nanocrystals as promising strategy for nose-to-brain delivery application. *Pharmaceutics* **2020**, *12*, 476. [CrossRef]
27. Rakotoarisoa, M.; Angelov, B.; Espinoza, S.; Khakurel, K.; Bizien, T.; Drechsler, M.; Angelova, A. Composition-switchable liquid crystalline nanostructures as green formulations of curcumin and fish oil. *ACS Sustain. Chem. Eng.* **2021**, *9*, 14821–14835. [CrossRef]
28. Rakotoarisoa, M.; Angelov, B.; Espinoza, S.; Khakurel, K.; Bizien, T.; Angelova, A. Cubic Liquid Crystalline Nanostructures Involving Catalase and Curcumin: BioSAXS Study and Catalase Peroxidatic Function after Cubosomal Nanoparticle Treatment of Differentiated SH-SY5Y Cells. *Molecules* **2019**, *24*, 3058. [CrossRef]
29. Hartman, P.; Bennema, P. The Attachment Energy as a Habit Controlling Factor: I. Theoretical Considerations. *J. Cryst. Growth.* **1980**, *49*, 145–156. [CrossRef]
30. Liu, Y.; Bi, F.; Lai, W.; Wei, T.; Ma, Y.; Ge, Z. Crystal Morphology Prediction od Dihydroxylammonium 5,5'-Bistetrazole-1,1'-diolate Under Different Growth Conditions. *Chin. J. Energetic Mater.* **2018**, *26*, 210–217. [CrossRef]
31. Zhou, T.; Chen, F.; Li, J.; Cao, D.; Wang, J. Growth Morphology of TKX-50 in Formic Acid/Water Mixed Solvent by Molecular Dynamics Simulation. *Chin. J. Energetic Mater.* **2020**, *28*, 865–873. [CrossRef]

Disclaimer/Publisher's Note: The statements, opinions and data contained in all publications are solely those of the individual author(s) and contributor(s) and not of MDPI and/or the editor(s). MDPI and/or the editor(s) disclaim responsibility for any injury to people or property resulting from any ideas, methods, instructions or products referred to in the content.

Article

A MD Simulation Prediction for Regulation of *N*-Terminal Modification on Binding of CD47 to CD172a in a Force-Dependent Manner

Yang Zhao, Liping Fang, Pei Guo, Ying Fang * and Jianhua Wu *

Institute of Biomechanics, School of Biology and Biological Engineering, South China University of Technology, Guangzhou 510006, China
* Correspondence: yfang@scut.edu.cn (Y.F.); wujianhua@scut.edu.cn (J.W.)

Abstract: Cancer cells can evade immune surveillance through binding of its transmembrane receptor CD47 to CD172a on myeloid cells. CD47 is recognized as a promising immune checkpoint for cancer immunotherapy inhibiting macrophage phagocytosis. *N*-terminal post-translated modification (PTM) via glutaminyl cyclase is a landmark event in CD47 function maturation, but the molecular mechanism underlying the mechano-chemical regulation of the modification on CD47/CD172a remains unclear. Here, we performed so-called "ramp-clamp" steered molecular dynamics (SMD) simulations, and found that the *N*-terminal PTM enhanced interaction of CD172a with CD47 by inducing a dynamics-driven contraction of the binding pocket of the bound CD172a, an additional constraint on CYS15 on CD47 significantly improved the tensile strength of the complex with or without PTM, and a catch bond phenomenon would occur in complex dissociation under tensile force of 25 pN in a PTM-independent manner too. The residues GLN52 and SER66 on CD172a reinforced the H-bonding with their partners on CD47 in responding to PTM, while ARG69 on CD172 with its partner on CD47 might be crucial in the structural stability of the complex. This work might serve as molecular basis for the PTM-induced function improvement of CD47, should be helpful for deeply understanding CD47-relevant immune response and cancer development, and provides a novel insight in developing of new strategies of immunotherapy targeting this molecule interaction.

Keywords: CD47; CD172a; post-translated modification; molecular dynamics simulation; structure–function relation; mechano-chemical regulation

1. Introduction

The transmembrane protein CD47, also known as an integrin-associated protein (IAP) and served as a self-recognition marker [1], is ubiquitously expressed in human cells, various hematologic and solid tumors [2], and involved in cell apoptosis, proliferation, adhesion, migration and tumor development, as well as immune responses [3–7]. Binding to CD172a (the signal regulatory protein α, SIRPα) on macrophage, CD47 on tumor cells can send the inhibitory signal "do not eat me" to macrophages so as to maintain the immune tolerance of own cells, prevent tumor cells from being phagocytized, and inhibit tumor antigen presentation [8,9]. The overexpression of CD47 has become a key strategy of the tumor cell in evading macrophage-mediated phagocytosis [10,11]. Thus, CD47 has emerged as a promising checkpoint molecule for anti-tumor therapy [12,13].

CD47 is a 50 kDa membrane receptor consisting of an extracellular *N*-terminal IgV-like domain, five transmembrane helices, and a short C-terminal intracellular tail [14,15]. The extracellular IgV-like domain contains five N-glycosylation sites, while there exist four kinds of spliceosomes in the intracellular regions of 4–36 amino acids (Figure 1) [15–17]. The distribution and expression of these spliceosomes are different in vivo; spliceosome 2 of sixteen amino acids is the most widely expressed and distributed on the surfaces of almost all hematopoietic cells, endothelial cells, and epithelial cells [16]. Crystallographic studies show very little ligand-induced change in CD47 conformation [17]. The disulfide bond between CYS15 in the extracellular domain and CYS245 in the transmembrane domain of CD47 is susceptible to the redox potential change, and leads CD47 to take an orientation in favor of ligation binding and the subsequent signal transduction [14,17,18]. CD172a is a crucial ligand of CD47, belongs to the immunoglobular superfamily, and has an extracellular region of three Ig-like domains, in which only the first Ig-like domain participates in binding with CD47 (Figure 1), and an intracellular domain of two immunoreceptor tyrosine-based inhibitory motifs (ITIMs) [19]. CD47 promotes the tyrosine phosphorylation of CD172a ITIMs, causing recruitment of Src homology region 2 domain-containing phosphatase-1 and 2 (SHP-1 and 2) and inhibiting phagocytosis [20,21].

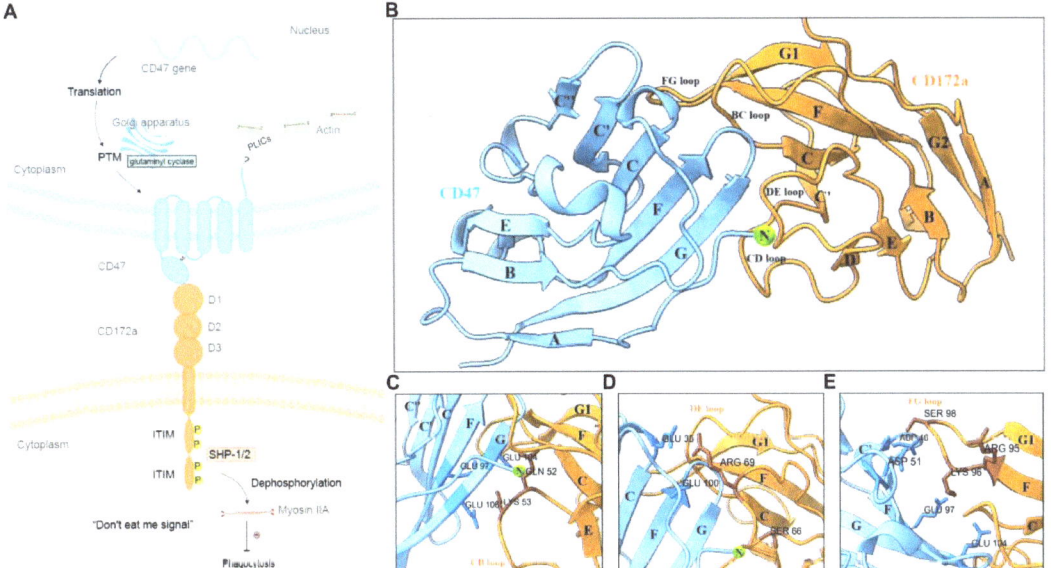

Figure 1. The signaling pathway, crystal structure and binding site of the CD47/CD172a complex. (**A**) The schematic diagram for CD47-CD172a axis. Interaction of CD47 and CD172a induces the phosphorylation of two tyrosine residues within the immunoreceptor tyrosine-based inhibition motif (ITIM) of SIRPα's cytoplasmic domain. This event subsequently initiates the recruitment and activation of SHP1 and SHP2, leading to a signaling cascade that culminates in the dephosphorylation of myosin IIA and resulting in the inhibition of phagocytosis. Upregulation of CD47 expression is one of the mechanisms underlying the increased activity of the CD47-CD172a axis, while the post-translational modification of CD47 is involved in this aforementioned process, too. (**B**) The crystal structure of the complex of CD47 with CD172a. The crystal structure of the complex (PDB ID: 2JJT) is shown using a new cartoon representation. The CD47 extracellular IgV-like domain (in cyan) (the 1st–115th residue) is responsible for binding with CD172a. It contains three short β-strands (A, B and E), two helices, and five long β-strands (C, C′, C″, F and G) [17]. The first extracellular Ig subdomain (the 2nd–115th) (in orange) of CD172a binds to CD47, is consisting of eight β-strands (A, B, C, C″, D, E, F, G1, G2 and F) and their linker loops. The Cα-atom of pyroglutamic acid (green) at position 1 in the residue sequence of the wild-type (WT) CD47 is represented in van der Waals mode. The crystal structure of CD47 (in cyan) bound to CD172a (in orange) is shown in a new cartoon representation. (**C–E**) The binding site of CD47/CD172a complex. The labeled residues in CD47 or in the CD loop (**C**), DE loop (**D**) and FG loop (**E**) of CD172a were involved in binding to CD47.

Passing through Golgi apparatus, CD47 will be catalyzed by glutaminyl cyclase, which converts a glutamine at the *N*-terminus to a pyroglutamic acid (PCA) and is dominant for CD47 signaling [22,23]. This post-translational modification (PTM) significantly improves CD47's affinity with CD172a, possibly by increasing protein resistance to enzymatic degradation [24]. However, the dynamics mechanism for regulation of the N-terminal modification on CD47 affinity with CD172a is unclear, even though the solved crystal structure of CD47/CD172a complex has shown two hydrogen bonds between SER66 on CD172a and the pyroglutamic acid, the N-terminal residue of CD47 [17]. Additionally, several lines of evidence indicate biochemical regulation in the interaction of CD47 with CD172a. Cytokine stimulation in HLH patients reduces CD47 expression [25], interleukin 4 (IL-4), IL-7, and IL-13 influence CD47 expression on Sezary cells [26], and the expression level of CD47 in cancer cells is regulated by miR-133a, MYC oncogene, and the ERK signaling pathway [27–29]. A transformation from the "do not eat me" signal to "eat me" one is believed to relevant to a conformation change of CD47 for the experimental aging red blood cells [30].

Numerous studies have shown mechano-sensing properties of immunoreceptors [31]. Periodic mechanical force stimulation of human chondrocytes can elicit a CD47-dependent change in the membrane potential, showing an actor of CD47 as a force sensor [32]. CD47-antibody-induced apoptosis occurs in tumor cells on hard substrates rather than soft gels [33], suggesting a requirement of substrate stiffness for CD47-modulated cancer development. The Protein Linking Integrin-Associated Complexes (PLICs) can act as a linker of the CD47 cytoplasmic tail and the cell skeleton [34], transmitting transmembrane mechanochemical signaling. Additionally, interaction of CD47 with either thrombospondin and/or CD172a can promote the integrin-dependent recruitment of T cells towards endothelial cells [35,36]. These studies mentioned above do suggest a possible deep involvement of mechano-microenvironment in CD47-relevant cellular physio-pathological processes, but there is less knowledge of the mechanism of mechanical regulation on interaction of CD47 with CD172a.

Herein, we investigate the effects of PTM and tensile force on the interaction of CD47 with CD172a through molecular dynamics (MD) simulations with two molecular systems using CD172a bound to CD47 with or without PTM. Our results state that the PTM-induced local flexibility change in CD172a might be responsible for the upregulation of CD47 affinity to CD172a, and, the N-terminal PTM and tensile force may improve CD47 affinity to CD172a by stabilizing the binding pocket of CD172a. The present study provides a novel insight into the dynamics mechanism and molecular basis for CD47/CD172a interaction under tensile force with or without PTM. This should assist with our understanding of CD47-involved cellular immune response and tumor development and should be useful for developing novel therapeutic strategies targeting CD47 in cancers.

2. Results

2.1. N-Terminal Post-Translational Modification Improved CD47 Affinity to CD172a

To assess the role of post- translational modification (PTM) on binding of CD47 with CD172a at the atom level, free MD simulation of 40 ns was performed thrice for each of two different equilibrated CD172a/CD47 complexes, which took its wild-type (WT) or GLN-type (GT), respectively (Figure 1) (Materials and Methods). The WT complex had a PTM on N-terminus of CD47 and lacked the GT one.

We found from the time courses of Cα-RMSD (Figure 2A) that the Cα-RMSD pattern of WT complex has less difference from that of GT complex, while Cα-RMSD fluctuated slightly, increased slowly with time, and then reached a plateau of about 1.75 Å, while it varied within a range of 1 Å throughout the simulation time no matter whether CD47 was WT or GT, showing a high thermal stability of CD172a/CD47 complex in a PTM-independent manner. N_{HB}, the mean number of hydrogen bonds across the interfacial surface, was calculated to be 9.8 ± 0.27 for the WT complex and 8.7 ± 0.16 for the GT complex (Figure 2B), exhibiting a PTM-enhanced H-bonding between CD172a and CD47. As a result, a lower dissociation probability f_D (= $(6.77 \pm 7.48) \times 10^{-7}$) was predicted for the WT complex in comparison with f_D (= $(1.30 \pm 0.63) \times 10^{-5}$) of the GT complex (Figure 2C). This PTM-induced upregulation of CD47 affinity to CD172a was reflected by the PTM-induced increase in the mean buried SASA and decrease in both the mean binding energy and the mean Rgyr (Figure 2D–F). The present data showed that the mean binding energies (E_B) were -553.7 ± 35.13 and −520.4 ± 16.09 kcal/mol, the mean Buried-SASA values were 9.9 ± 0.01 and 9.1 ± 0.16 nm^2, and the mean rotation radius (Rgyr) were 19.9 ± 0.03 and 20.2 ± 0.07 Å for the WT and GT complexes, respectively (Figure 2D–F). These data had a PTM-upregulated affinity of CD172a to CD47. This result matched well with a previous report that the N-terminal modification of CD47 could enhance the binding ability of CD47 to CD172a [23].

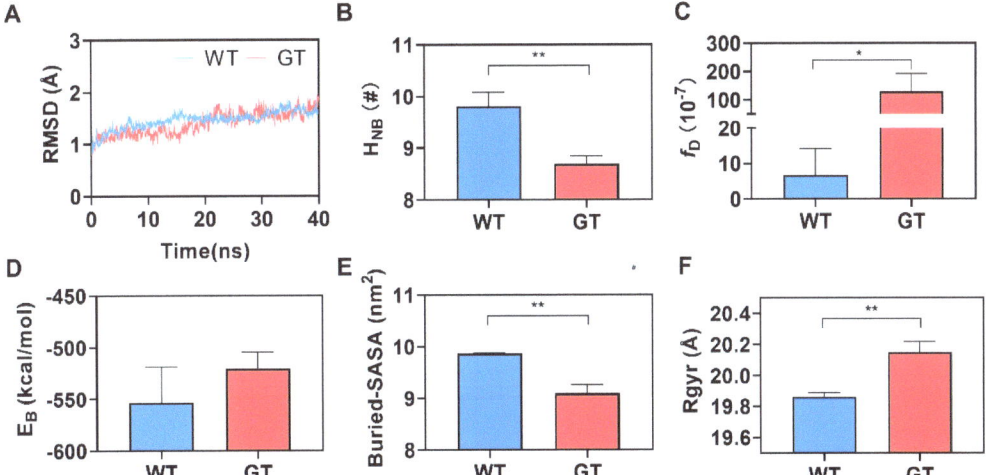

Figure 2. Effect of N-terminal modification on interaction of CD47 with CD172a. (**A**) The mean time-courses of Cα-RMSD, (**B**) the mean interfacial H-bond number (N_{HB}), (**C**) the dissociation probabilities (f_D), (**D**) the mean binding energy (E_B), (**E**) the mean buried SASA, and (**F**) the mean rotation radius (Rgyr) of either WT or GT CD47/CD172a complex over 40 ns production simulation. The p-values of the unpaired two-tailed Student's t test were shown to indicate the statistical difference significance (* $p < 0.05$; ** $p < 0.01$), or lack thereof. Data were shown with mean ± S.D of three runs.

2.2. PTM Enhanced Interfacial H-Bonding and Induced Contraction of Binding Pocket of Bound CD172a

To investigate the structural basis for the PTM-enhanced CD47 affinity with CD172a, we further examined the effect of CD47 N-terminal PTM on the complex conformation through free MD simulations. Three geometrical characteristics, consisting of the distance (H) between the centroids of CD and DE loops in CD172 (Figure 3A), the angle (α) between the C strand of CD172a and the straight line extending from the Cα atom of GLN52 to the Cα atom of LYS53 in CD172a (Figure 3B), and the angle (θ) between the G strand in CD47 and the C strand in CD172a (Figure 3C), were used to quantify conformational evolution under thermal excitation.

We found that high conformational conservation lay in the bound CD47 with or without PTM, while a significant PTM-induced conformation change occurred in the binding pocket of the bound CD172a. This PTM-induced allostery of CD172a was demonstrated by the PTM-induced decreases in H and α (Figure 3D,E). The values of H and α were 16.1 ± 0.4 Å and 69.8° ± 2.7° for the WT complex and 17.4 ± 0.3 Å and 90.0° ± 2.7° for the GT complex, respectively, while a reduction of θ from 95.4° ± 1.3° to 89.5° ± 0.4° occurred (Figure 3F). This allostery stated a PTM-induced contraction of the binding pocket of the bound CD172a, brought the CD loop closer to the BC and DE loops for the bound CD172a (Figures 1 and 3A,B), and led to closer contact between CD47 and CD172a by causing a roll of CD172a on CD47 (Figure 3C,F), saying that contraction of the binding pocket of CD172a was involved in PTM-induced enhancement of CD47 affinity to CD172a.

Additionally, the mean RMSF pattern of CD172a showed that flexibilities of the CD loop (the 50th–60th residues) in the WT complex were smaller than that of the GT complex (Figure 3G), suggesting a more stable CD172a binding pocket or a stronger CD47-related constraint on CD172a in the WT complex instead of the GT one.

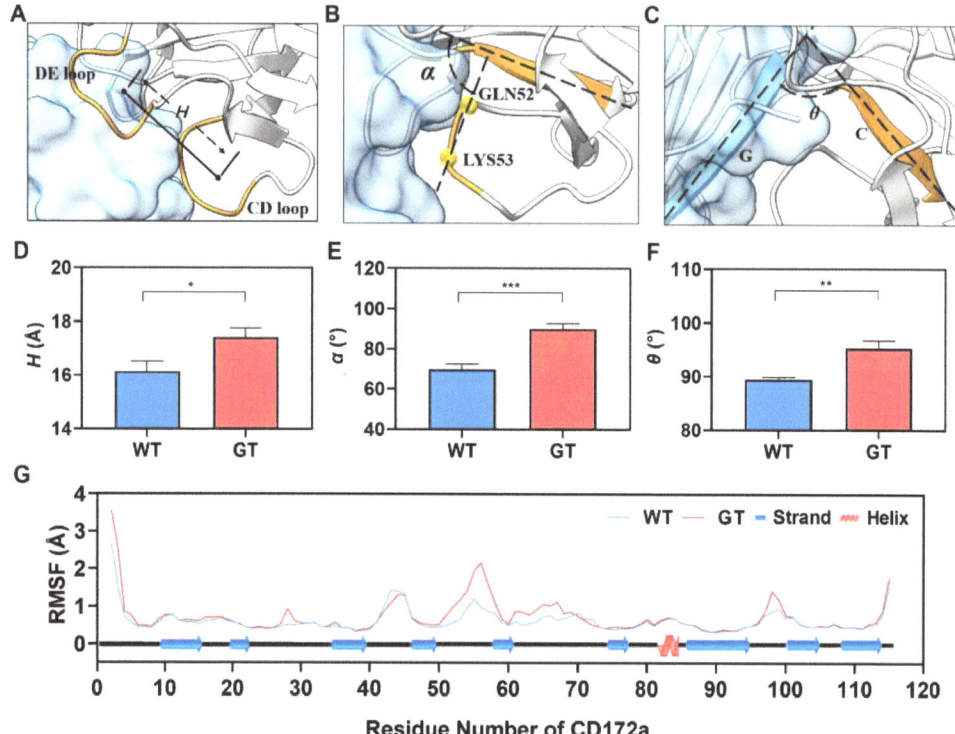

Figure 3. PTM-induced Allostery of the bound CD172a. The schematic diagram for (**A**) the distance H between the centroids of CD loop and DE loop in CD172a, (**B**) the angle α between the C strand in CD172a and the straight line extending from the Cα-atom of GLN52 to the Cα atom of LYS53 in CD172a, and (**C**) the angle θ between the G strand in CD47 and the C strand in CD172a. The plots of the distance H (**D**), the angle α (**E**), and θ (**F**) over thrice 40 ns simulations. (**G**) The mean RMSF of the bound CD172a over three 40 ns simulation. WT or GT denoted the bound CD47 with or without PTM, respectively. The p-values of the unpaired two-tailed Student's t test were shown to indicate the statistical difference significance (* $p < 0.05$; ** $p < 0.01$; *** $p < 0.001$), or lack thereof. Data were shown with mean ± S.D of three runs.

Meanwhile, we detected twelve hydrogen bonds in the binding site from thrice 40ns free MD simulations (Table 1, Figure 1C–E). Of all these bonds, seven were enhanced significantly by the N-terminal PTM of CD47, and the involved residues were ASP57, GLU97, GLU100, GLU104, GLU106, GLU100, and PCA1 on CD47, as well as their respective partners, such as GLN52, LYS53, SER66, ARG69, LYS96, and SER98 in CD172a (Table 1). They should play a crucial role in the PTM-induced enhancement of CD47 affinity with CD172a, although PTM showed a stronger H-bond (with occupancy of 77%) between ASP51 on CD47 and ARG95 on CD172a very weak (Table 1). The five key residues on CD172a were located in the CD, DE and FG loop (Figure 1C–E). Additionally, three GLN1-involved H-bonds were weak, with survival ratios smaller than 0.25, and would vanish completely with the formation of a moderate H-bond (with a survival ratio of 0.52) between PCA1 on CD47 N-terminus and SER66 on the DE loop of CD172a, hinting that a reinforcement signal for CD47 binding with CD172 was triggered through converting a glutamine at N-terminus of CD47 to a pyroglutamic acid. The H-bond between GLU35 on CD47 and ARG69 on CD172a was strong, with high occupancy larger than 80% no matter whether PTM occurred at the CD47 terminal or did not, indicating that this residue pair should be crucial for binding of CD47 to CD172a.

Table 1. Key residues pairs in binding of CD172a to CD47 with or without PTM.

Number	Residue		Occupancy	
	CD47	CD172a	WT	GT
1	GLU106	LYS53	0.54 ± 0.04	0.48 ± 0.03
2	GLU104	LYS96	0.66 ± 0.04	0.52 ± 0.10
3	GLU104	GLN52	0.60 ± 0.03	0.22 ± 0.07
4	GLU100	ARG69	0.79 ± 0.06	0.69 ± 0.01
5	GLU97	LYS96	0.66 ± 0.09	0.30 ± 0.09
6	GLU97	LYS53	0.34 ± 0.01	0.28 ± 0.04
7	ASP51	ARG95	0.07 ± 0.06	0.77 ± 0.07
8	ASP46	SER98	0.53 ± 0.06	0.30 ± 0.13
9	GLU35	ARG69	0.82 ± 0.01	0.81 ± 0.02
10	PCA1 or GLN1	SER66	0.52 ± 0.04	0.21 ± 0.08
11	PCA1 or GLN1	GLU54	0	0.08 ± 0.03
12	PCA1 or GLN1	GLN52	0	0.15 ± 0.04

2.3. Mechanical Constraint on CYS15 of CD47, Instead of N-Terminal PTM, Might Improve Tensile Strength of CD172a/CD47 Complex

A stable transmembrane signal transduction through the CD47/CD172a axis required better mechanical strength for the CD47/CD172a complex, especially in the mechano-microenvironment. Herein, we performed "force-ramp" SMD simulation over 10ns thrice with pulling velocity of 5 Å/ns for each system to investigate the mechanical strength of the mechano-constrained complex with or without PTM. Two loading modes, mode 1 and 2, were applied in SMD simulations, no matter whether the bound CD47 was one with N-terminal PTM or not. In each loading mode, the Cα atom of CD172a C-terminal (ALA115) was selected as the steered one; and in mode 2, two Cα atoms of CD47, one on C-terminal (VAL115) and another on CYS15, were fixed, but in mode 1, only the C-terminal Cα-atom of CD47 was fixed (Figure 4C).

The force–time curve showed that the tension rose rapidly to its peak (the rupture force) of about 530 pN in mode 2 or about 420 pN in mode 1 (Figure 4A,D) in either PTM presence or absence, exhibiting a strong resistance to the pull-induced CD172a dissociation from CD47 in a manner independent of PTM and the loading modes. However, the present data revealed that the mechanical constraint on CYS15 of CD47 improved the tensile strength of the complex and did not improve the CD47 N-terminal PTM. It was believed that a disulfide bond could form between CYS15 in the extracellular domain and CYS245 in transmembrane domains of CD47 [17], meaning that the formation of the disulfide bond provided a mechanical intramolecular constraint on CD47, and made CD47 anchor tightly to the cellular membrane. The present data revealed that this disulfide bond enhanced resistance to the pull-induced CD172a dissociation from CD47 by modifying the CD47 orientation relative to the cellular membrane surface and the tensile direction along CD172a/CD47 axis under a mechano-microenvironment (Figure 4A,C,D).

The patterns of interfacial H-bonds indicated a diverse pull-induced evolution of the H-bond interaction on the binding site of WT or GT complex with one or two fixed atoms (Figure 4B). The synergistic interactions of the interfacial H-bonds contributed to resistance pull-induced dissociation of the complex. A similarity existed in the two systems for WT complex, and for GT complex (Figure 4B), although all the H-bond occupancy patterns were different from each other. However, the residue pairs, LYS96 on CD172a with GLU104 on CD47 and ARG69 on CD172a with both GLU100 and GLU35 on CD47, contributed three H-bonds with higher occupancies (Figure 4B), meaning that these residues provided a dominant resistance to the pull-induced dissociation of the WT or GT complex with or without constraint on CYS15 on CD47.

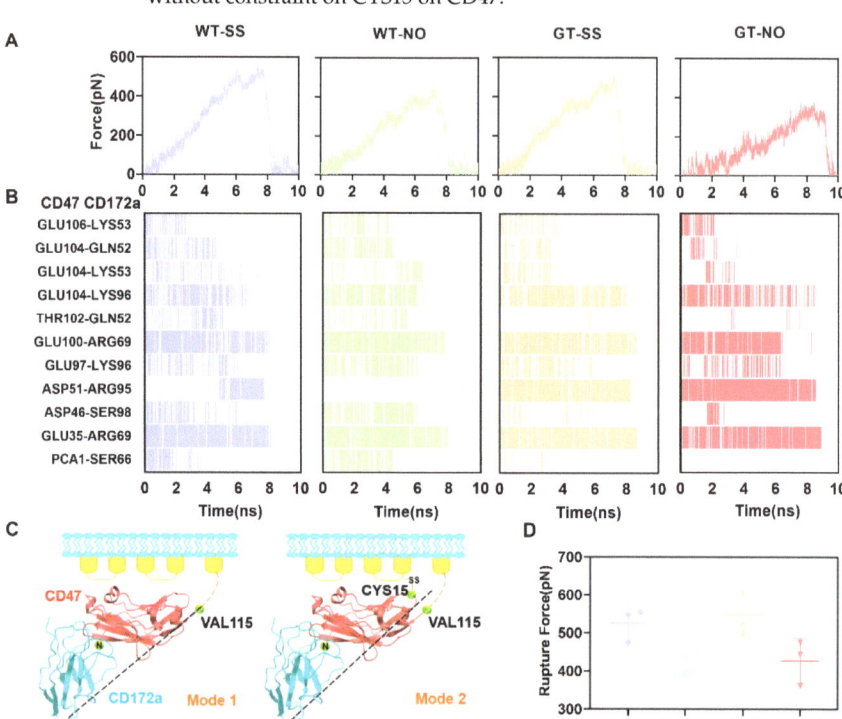

Figure 4. Pull-induced dissociation of CD172a from CD47. The data were from "force-ramp" SMD runs over thrice 10ns using pulling velocity of 5 Å/ns along direction from steered Cα atom on ALA115 on CD172a to either the Cα atom on VAL115 of CD47 or the center of the line between two fixed Cα atoms on VAL115 and CYS15 of CD47. (**A**) The representative time curves of loading force for four different systems, the WT or GT complex with one or two fixed Cα atom. The symbols, WT-SS and GT-SS, expressed the WT and GT complexes, in which the two Cα atoms on VAL115 and CYS15 of CD47 were fixed, respectively. The symbols, WT-NO and GLN-NO, were assigned to the WT and GT complexes, in which only the Cα atom on VAL115 of CD47 was fixed, respectively. (**B**) The occupancy patterns of interfacial H-bonds for a representative for each system. (**C**) Schematic diagram of the ligated CD47 under stretching. (**D**) The scatter plot of the rupture force for each system.

2.4. Catch Bond Mechanism Mediated the PTM-Enhanced CD172a-CD47 Interaction by Making the CD172a Binding Pocket Contract Further

To examine mechanical regulation of CD47 dissociation from CD172a under a physiological mechano-microenvironment, we performed "ramp-clamp" SMD simulation on each of the WT and GT molecular systems for 40 ns thrice at a tensile force of 25 pN just using loading mode 2, which was selected to model the physiological native disulfide bond from CYS245 paired with CYS15 in CD47 [17]. We found that the Cα-RMSD fluctuated in a range from 1 to 2 Å during the simulation duration (Figure 5A), while the distance from the steered atom to the fixed one stayed at a plateau with a slight roughness of height of about 4 Å (Supplementary Figure S1), suggesting a stable conformation of the stretched complex with an allosteric modulation. The tensile force of 25 pN caused an increase in H_{NB} (the mean interfacial H-bond number over 40 simulation time thrice) and a slight decrease in either the dissociation probability (f_D) or the binding energy (E_B), no matter whether the CD172a-bound CD47 was WT or GT (Figure 5B–D). In spite of that, the tensile change in either Buried-SASA or in Rgyr of the complex with or without PTM was negligible (Supplementary Figure S2). However, these data suggested a catch bond mechanism for the PTM-enhanced CD172a-CD47 interaction.

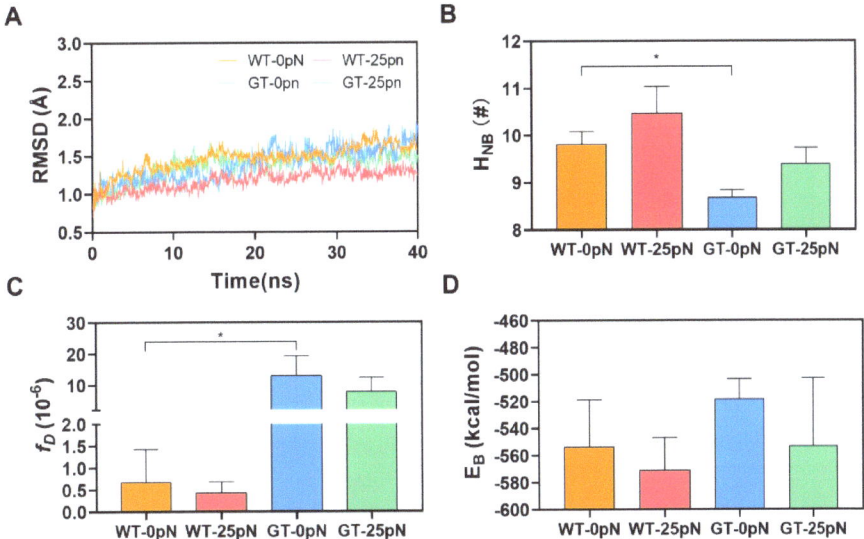

Figure 5. Interaction of CD47 with CD172a under tensile force of 25 pN. All data were sampled from "force-clamp" SMD runs of 40 ns thrice for WT-SS and GT-SS system at static state or under tensile force of 25 pN. (**A**) The representative time-courses of Cα-RMSD, (**B**) the mean hydrogen bond number (H_{NB}), (**C**) the dissociation probability (f_D), and (**D**) the mean binding energy of WT or GT CD47/CD172a complex at static state or under tensile force of 25 pN. The p-values of two-way analysis of variance (ANOVA) with Tukey's multiple comparisons were shown to indicate the statistical difference significance (* $p < 0.05$), or lack thereof. Data were shown with mean ± S.D of three runs.

To uncover the structural basis of the catch bond phenomenon in interaction of CD47 with CD172a, we measured the three abovementioned geometric characteristics (H, α and θ) (Figure 3A–C) of the complex from the "ramp-clamp" SMD runs at static and tensile force of 25 pN. In comparison with the case at static state, we found that the tensile force of 25 pN caused a slight decrease in either the distance H and the angle α (Figure 6A,B), leading to a force-induced contraction of the CD172a binding pocket. This force-induced change in H and α was pronounced more clearly in the GT rather than the WT complex (Figure 6A,B), and the angle θ remained almost constant for the WT complex but had a slight increase for the GT one (Figure 6C), indicating a PTM-mediated enhancement of mechanical stabilization of the complex structure. The present data suggested that a catch bond mechanism might mediate PTM-enhanced CD172a-CD47 interaction by making the CD172a binding pocket contract further.

Additionally, we found that all the thirteen interfacial H-bonds could be clustered to three types, type I, II, and III, based on their responses to the tensile force (Figure 6D). The H-bonds in the type I were force desensitized, and consisted of two members contributed by GLU104 on CD47 with LYS96 on CD172a and GLU35 on CD47 with ARG69 on CD172a. All six force-enhanced H-bonds were assigned to type II, and contributed by GLU104 on CD47 with GLN52 and LYS53 on CD173, ASP46, PCA1, THR102, and ASP51 on CD47 with their respective partners, such as SER98, SER66, ARG69, and ARG95 on CD172a. Meanwhile, five residue pairs, consisting of GLU106, GLU100, GLU97, LYS39, and LYS36 on CD47 with their respective partners, such as LYS53, ARG69, LYS96, ASP100, and GLU54 on CD172a, contributed the other five force-weakened H-bonds in the type III (Figure 6D).

However, it was the synergistic effect of the thirteen H-bonding interactions that induced the catch–bond phenomenon in CD47-CD172a interaction. Additionally, the residue ARG69 on CD172 with its partner on CD47 might be crucial to the structural stability of the complex because of their relevant H-bonds of high occupancies (Figure 6D).

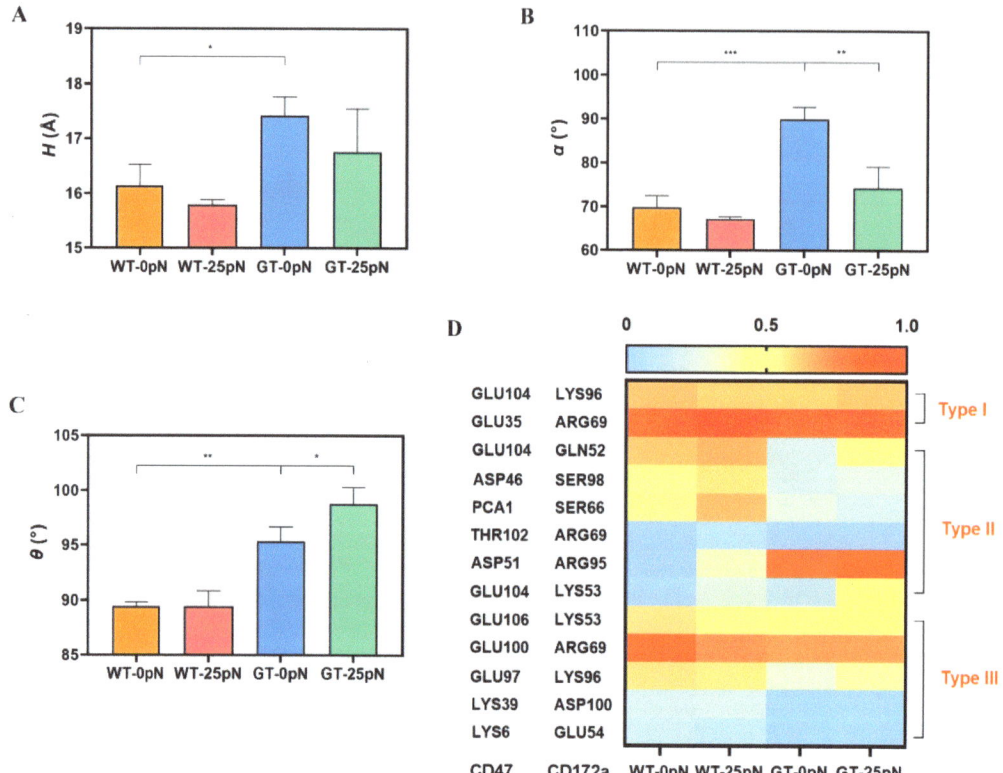

Figure 6. Tension-induced structural change in CD172a and the involved residue pairs at binding site. The plots of the distance H (A), the angle α (B), and θ (C) over thrice 40 ns "force-clamp" SMD runs for four systems, the WT-SS and GT SS systems at static state or under tensile force of 25 pN. (D) The heatmap of H-bond occupancies and their involved residue pairs under different conditions. The p-values of two-way analysis of variance (ANOVA) with Tukey's multiple comparisons were shown to indicate the statistical difference significance (* $p < 0.05$; ** $p < 0.01$; *** $p < 0.001$), or lack thereof. Data were shown with mean ± S.D of three runs.

3. Discussion

The post-translated modification (PTM) of CD47, or the conversion of glutamine (the 1st residue in CD47 N-terminus) to pyroglutamic acid via catalysis of glutaminyl cyclase on Golgi apparatus, was highly related to CD47 functions [17,22,23]. Meanwhile, the cells, which adhered to ECM or interacted with each other, would deform or move away from each other in response to mechanical stimulus, such as fluid shear stresses, squeezing, and tearing forces from surrounding tissues or cells [31]. Together with cell rolling or tethering, the cytoskeletal rearrangement, membrane expansion, and contraction of the deformed cells triggered transmembrane mechano-signaling, possibly leading to a remarkable tensile force on the CD47-CD172a linker between the cell and ECM or cell. Better mechanical strength and stability of the CD47-172a complex should be required for formation and stability of cell–cell cross-talking under diverse mechano-microenvironments in the presence and absence of PTM. With a series of free and steered MD simulations, we demonstrated that

PTM induced enhancement of CD47 binding to CD172a through contraction of the CD172a binding pocket either at static or under loading status. Support for this statement came from the fact the N-terminal PTM did not affect CD47 expression on cytomembrane but enhanced binding of CD172a in solution to CD47 on membrane, and the pharmacological inhibition or absence of the glutaminyl cyclase would inhibit CD47 signaling [22,23,37].

CYS15 in extracellular domain could form a disulfide bond with CYS245 on the transmembrane domain in CD47 [17]. This disulfide bond is sensitive to changes in redox potential, providing an additional membrane anchor point to likely constrain the orientation of CD47 relative to the cell membrane. It thereby prompted CD47–ligand binding and subsequent signal transduction [14,17,18]. To model the CYS15 anchor, we fixed the two Cα-atoms on CYS15 and VAL115 of CD47 for SMD runs. The present data exhibited that the constraint on CYS15 would induce enhancement of the mechanical strength of CD172a/CD47 complex with or without N-terminal PTM (Figure 4D), in better consistency with the statement for the disulfide bond (between CYS245 and CYS15)-induced improvement of CD47 adhesion function [14,18]. The residue GLN52 on CD172a with its partner CLU104 on CD47 might serve as anther pivot in responding to PTM, coming from the dramatic increase in the occupancy of H-bonding between this residue pair (Figure 4B; Table 1).

Conceivably, CD47 N-terminal PTM would trigger a cascade of intramolecular events for the CD47/CD172a complex. Through hydrogen interaction with SER66 on CD172a, the pyroglutamic acid on N-terminus of CD47 triggered remodeling of the interfacial H-bond occupancy pattern of the complex (Table 1) first, then caused an increase in the interfacial surface and a change in interfacial topology (Figures 2E and 3). It thereby enhanced binding of CD47 with CD172a. Being highly relevant to remodeling of interfacial H-bonding and binding site, the dynamics-driven and/or tensile force-induced contraction of the CD172a binding pocket with better flexibility was a crucial event in the cascade mentioned above (Figure 3). Additionally, we found that the stretched complex was stable under the given tensile force of 25 pN, and the force would enhance binding of CD172a to CD47 with or without PTM, but the complex with PTM had better mechanical stability in comparison with one without PTM (Figure 5C). This catch–bond phenomenon might be required for a stable cell–cell crosstalk, has been observed in various molecular systems, including P-selectin bound with PSGL-1 using the force clamp assay with AFM and a flow chamber [38], as well as Kindlin2 bound with β3 integrin and Mac-1 bound with GPIbα using molecular dynamics simulations [39,40]. So, the aforementioned data from SMD simulations stated that CD47 was mechano-sensitive, had a stable tension-induced allostery with an enhanced affinity to CD172a, and served not only linker molecules but also as a mechano-chemical signaling axis in cell–cell cross talking through binding with CD172a, while PTM raised the mechanical strength of the CD47-CD172a axis. The residue ARG69 on CD172a with its partner on CD47 might be crucial to the structural stability of the complex (Figure 6D), because of their relevant H-bonds of high occupancies. However, the effects of N-terminal PTM on CD47/CD172a complex dissociation were more significant than that of tensile force.

In conclusion, we determined from the MD simulation that both N-terminal PTM and tensile force did regulate binding of CD47 and CD172a. Our results suggested that the N-terminal PTM of CD47 mediated the dynamics-driven interfacial H-bonding enhancement of the complex, along with contracting of the binding pocket in the bound CD172a. An additional constraint on CYS15 on CD47 significantly improved the tensile strength of the complex with or without PTM, and a catch–bond phenomenon would also happen in complex dissociation under tensile force of 25 pN in a PTM-independent manner. Additionally, several binding site residues, such as GLN52, SER66, and ARG95 on CD172a, were qualified as candidates of the key residues in the PTM-enhanced binding of CD172a to CD47. This MD simulation prediction provided an explanation for the mechano-chemical regulation on the CD47 binding with CD172a at atom level, which might serve as the molecular basis for the PTM-induced function improvement of CD47. It would also be helpful for a deeper understanding CD47-relevant cellular immune response and cancer development and for developing new strategies of immunotherapy targeting this adhesion molecule interaction.

4. Materials and Methods

4.1. System Setup

Two molecular systems were set up using the CD47/CD172a complexes of wild-type (WT) and GLN-type (GT). A post-translational modification occurred on the N-terminal of CD47 in the WT complex and did not in the GT one. The crystal structure of WT CD47 (the 1st–115th residue)/CD172a (the 2nd–115th) complex was read from the Protein Data Bank (PDB ID: 2JJT) (Figure 1) [17], while the initial structural model of the GT complex was created by replacing residues 1–3 in the WT complex with residues 1–3 of CD47 bound with antibody B6H12 (PDB ID:5TZU) [41]. The initial model of the GT complex was energy minimized along the protocol, such that all atoms except either the third–fourth residues of CD47 at the first 1000 minimization steps or the first–fourth residues of CD47 at the followed 2000 minimization steps were fixed, and the last 5000 minimization steps were run for all the unconstrained atoms [42]. Each of the WT and GT complex structures was solvated in TIP3P water in a rectangular box with walls at least 15 Å away from any protein atom using VMD 1.9.2 [43]. The system was then neutralized with 150 mM NaCl to resemble the physiological environment.

4.2. Molecular Dynamics Simulation

Two software packages, VMD 1.9.2 and ChimeraX 1.4, were used for visualization and modeling [44], while MD simulations were performed using NAMD 2.13 with CHARMM36 force field [45]. Each system was subjected to an energy minimization first and then equilibrated thrice for 50 ns. A stable complex structure in each equilibrated system was chosen as the initial conformation for production simulations for each system. Free MD simulation was run thrice on each system over 40 ns.

The so-called "ramp-clamp" SMD simulations, a force-clamp MD simulation followed by a force-ramp one, were performed on each equilibrated system to examine the force-induced changes in complex conformation and function in the presence or absence of the post-translational modification of CD47. Two loading modes, mode 1 and mode 2, were applied in SMD simulations for studying mechanical constraint effects on CD47 function. In mode 1, just the Cα-atom of VAL115 on CD47 was fixed, and the steered atom, the C-terminal Cα atom of CD172a (ALA115), was pulled along a direction from the fixed atom to the pull atom. In mode 2, two Cα-atoms of CD47, one on VAL115 and another on CYS15, were fixed, and the Cα atom of CD172a C-terminal (ALA115) was the steered atom, but pulling took place along the direction from the center of the line between two fixed atoms to the steered atom. For the loading manner in mode 2, we fixed CYS15 in the extracellular domain of CD47 to form a disulfide bond with CYS245 in its transmembrane domain [17]. Model 2 was used to model the disulfide bond between CYS15 and CYS245, which lay in the extracellular and transmembrane domain of CD47, respectively [17]. The virtual spring connecting the dummy atom and the steered atom had a spring constant of 34.74 pN/nm. For each system at either loading mode 1 or 2, the force-camp MD simulation ran over 10 ns thrice with a time step of 2 fs and a constant velocity of 5 Å/ns, at which the pulling would be contributed to complex dissociation with secondary structure conservation.

Once tensile force reached 25 pN, the SMD simulation was transformed from the force-ramp mode to a force-clamp one, at which the complex was stretched with the given constant tensile force for the following 40 ns thrice. Herein, only loading mode 2 with two fixed CD47 Cα-atoms (one on VAL115 and another on CYS15) was applied to the force-clamp simulations, for simplicity. Each event of hydrogen bonding under stretching was recorded to examine the involved residues and their functions.

4.3. Data Analysis

VMD tools were used to analyze atom trajectories from MD simulations. The Cα-root mean square deviation (RMSD), the solvent accessible surface area (SASA, with a 1.4 Å probe radius) and the rotation radius (Rgyr) of complex were measured to estimate the structural stability, the complex interface geometry characteristics, and the tightness of complex structure, respectively. The number of hydrogen bonds (H-bonds) on the complex interface was calculated with a cutoff donor–acceptor distance of 3.5 Å and a cutoff donor–hydrogen–acceptor angle of 30°. The occupancy (or survival ratio) of a H-bond was evaluated based on the fraction of bond survival time in the simulation period. Additionally, the probability of ligand dissociation from receptor was calculated to approximately evaluate the receptor's affinity to its ligand, as described in our previous works [39]. The binding energy consisted of gas-phase energy difference between the complex and the separated receptor and ligand, the polar solvation free energy, and the nonpolar solvation-free energy.

4.4. Statistical Analysis

Statistical analysis was performed using GraphPad Prism version 9.0.0 (GraphPad Software, Inc., La Jolla, CA, USA). Unpaired two-tailed Student's t-tests were used to determine the statistical significance of the differences between the data obtained from the WT and GT systems (* $p < 0.05$; ** $p < 0.01$; *** $p < 0.001$). Two-way analysis of variance (ANOVA) with Tukey's multiple comparisons was employed to assess the statistical significance of the data obtained from the SMD simulations (* $p < 0.05$; ** $p < 0.01$; *** $p < 0.001$). All error bars represent the mean plus or minus standard deviation of the mean based on several independent experiments.

Supplementary Materials: The following supporting information can be downloaded at: https://www.mdpi.com/article/10.3390/molecules28104224/s1. Figure S1: Time-courses of distance between Cα atom of VAL115 in CD47 and Cα atom of ALA115 in CD172. All data were sampled from "force-clamp" SMD runs of 40 ns thrice for WT-SS and GT-SS system at static state or under tensile force of 25 pN. Figure S2: The plots of the Buried-SASA and the Rgyr over thrice 40 ns "force-clamp" SMD runs for four systems, the WT-SS and GT SS systems at static state or under tensile force of 25 pN.

Author Contributions: J.W. and Y.F. designed this research; Y.Z. overall and L.F. and P.G. partly performed research and analyzed data; Y.Z., P.G., Y.F. and J.W. wrote this paper. All authors have read and agreed to the published version of the manuscript.

Funding: This work was supported by National Natural Science Foundation of China (NSFC) Grant 12072117 (to J.W.) and 12072137 (to Y.F.).

Institutional Review Board Statement: Not applicable.

Informed Consent Statement: Not applicable.

Data Availability Statement: The data are available on request from the author.

Acknowledgments: The authors thank Xubin Xie for his help with the data treatment of this work.

Conflicts of Interest: The authors declare that they have no conflict of interest with the contents of this article.

Sample Availability: Samples of the compounds are not available from the authors.

References

1. Oldenborg, P.-A.; Zheleznyak, A.; Fang, Y.-F.; Lagenaur, C.F.; Gresham, H.D.; Lindberg, F.P. Role of CD47 as a Marker of Self on Red Blood Cells. *Science* **2000**, *288*, 2051–2054. [CrossRef] [PubMed]
2. Kim, H.; Bang, S.; Jee, S.; Paik, S.S.; Jang, K. Clinicopathological Significance of CD47 Expression in Hepatocellular Carcinoma. *J. Clin. Pathol.* **2021**, *74*, 111–115. [CrossRef] [PubMed]
3. Lamy, L.; Ticchioni, M.; Rouquette-Jazdanian, A.K.; Samson, M.; Deckert, M.; Greenberg, A.H.; Bernard, A. CD47 and the 19 KDa Interacting Protein-3 (BNIP3) in T Cell Apoptosis. *J. Biol. Chem.* **2003**, *278*, 23915–23921. [CrossRef] [PubMed]

4. Legrand, N.; Huntington, N.D.; Nagasawa, M.; Bakker, A.Q.; Schotte, R.; Strick-Marchand, H.; de Geus, S.J.; Pouw, S.M.; Böhne, M.; Voordouw, A.; et al. Functional CD47/Signal Regulatory Protein Alpha (SIRPα) Interaction Is Required for Optimal Human T- and Natural Killer- (NK) Cell Homeostasis in Vivo. *Proc. Natl. Acad. Sci. USA* **2011**, *108*, 13224–13229. [CrossRef]
5. Olsson, M.; Bruhns, P.; Frazier, W.A.; Ravetch, J.V.; Oldenborg, P.-A. Platelet Homeostasis Is Regulated by Platelet Expression of CD47 under Normal Conditions and in Passive Immune Thrombocytopenia. *Blood* **2005**, *105*, 3577–3582. [CrossRef]
6. Brittain, J.E.; Han, J.; Ataga, K.I.; Orringer, E.P.; Parise, L.V. Mechanism of CD47-Induced A4β1 Integrin Activation and Adhesion in Sickle Reticulocytes. *J. Biol. Chem.* **2004**, *279*, 42393–42402. [CrossRef]
7. Hu, T.; Liu, H.; Liang, Z.; Wang, F.; Zhou, C.; Zheng, X.; Zhang, Y.; Song, Y.; Hu, J.; He, X.; et al. Tumor-Intrinsic CD47 Signal Regulates Glycolysis and Promotes Colorectal Cancer Cell Growth and Metastasis. *Theranostics* **2020**, *10*, 4056–4072. [CrossRef]
8. Tseng, D.; Volkmer, J.-P.; Willingham, S.B.; Contreras-Trujillo, H.; Fathman, J.W.; Fernhoff, N.B.; Seita, J.; Inlay, M.A.; Weiskopf, K.; Miyanishi, M.; et al. Anti-CD47 Antibody–Mediated Phagocytosis of Cancer by Macrophages Primes an Effective Antitumor T-Cell Response. *Proc. Natl. Acad. Sci. USA* **2013**, *110*, 11103–11108. [CrossRef]
9. Matlung, H.L.; Szilagyi, K.; Barclay, N.A.; van den Berg, T.K. The CD47-SIRPα Signaling Axis as an Innate Immune Checkpoint in Cancer. *Immunol. Rev.* **2017**, *276*, 145–164. [CrossRef]
10. Anderson, N.R.; Minutolo, N.G.; Gill, S.; Klichinsky, M. Macrophage-Based Approaches for Cancer Immunotherapy. *Cancer Res.* **2021**, *81*, 1201–1208. [CrossRef]
11. Wang, Z.; Li, B.; Li, S.; Lin, W.; Wang, Z.; Wang, S.; Chen, W.; Shi, W.; Chen, T.; Zhou, H.; et al. Metabolic Control of CD47 Expression through LAT2-Mediated Amino Acid Uptake Promotes Tumor Immune Evasion. *Nat. Commun.* **2022**, *13*, 6308. [CrossRef] [PubMed]
12. Zhao, H.; Song, S.; Ma, J.; Yan, Z.; Xie, H.; Feng, Y.; Che, S. CD47 as a Promising Therapeutic Target in Oncology. *Front. Immunol.* **2022**, *13*, 757480. [CrossRef] [PubMed]
13. Feng, R.; Zhao, H.; Xu, J.; Shen, C. CD47: The next Checkpoint Target for Cancer Immunotherapy. *Crit. Rev. Oncol. Hematol.* **2020**, *152*, 103014. [CrossRef] [PubMed]
14. Rebres, R.A.; Vaz, L.E.; Green, J.M.; Brown, E.J. Normal Ligand Binding and Signaling by CD47 (Integrin-Associated Protein) Requires a Long Range Disulfide Bond between the Extracellular and Membrane-Spanning Domains. *J. Biol. Chem.* **2001**, *276*, 34607–34616. [CrossRef] [PubMed]
15. Lindberg, F.P.; Gresham, H.D.; Schwarz, E.; Brown, E.J. Molecular Cloning of Integrin-Associated Protein: An Immunoglobulin Family Member with Multiple Membrane-Spanning Domains Implicated in Alpha v Beta 3-Dependent Ligand Binding. *J. Cell Biol.* **1993**, *123*, 485–496. [CrossRef]
16. Reinhold, M.I.; Lindberg, F.P.; Plas, D.; Reynolds, S.; Peters, M.G.; Brown, E.J. In Vivo Expression of Alternatively Spliced Forms of Integrin-Associated Protein (CD47). *J. Cell Sci.* **1995**, *108*, 3419–3425. [CrossRef]
17. Hatherley, D.; Graham, S.C.; Turner, J.; Harlos, K.; Stuart, D.I.; Barclay, A.N. Paired Receptor Specificity Explained by Structures of Signal Regulatory Proteins Alone and Complexed with CD47. *Mol. Cell* **2008**, *31*, 266–277. [CrossRef]
18. Pan, Y.; Wang, F.; Liu, Y.; Jiang, J.; Yang, Y.-G.; Wang, H. Studying the Mechanism of CD47–SIRPα Interactions on Red Blood Cells by Single Molecule Force Spectroscopy. *Nanoscale* **2014**, *6*, 9951–9954. [CrossRef]
19. Lee, W.Y.; Weber, D.A.; Laur, O.; Severson, E.A.; McCall, I.; Jen, R.P.; Chin, A.C.; Wu, T.; Gernet, K.M.; Parkos, C.A. Novel Structural Determinants on SIRPα That Mediate Binding to CD47. *J. Immunol.* **2007**, *179*, 7741–7750. [CrossRef]
20. Kharitonenkov, A.; Chen, Z.; Sures, I.; Wang, H.; Schilling, J.; Ullrich, A. A Family of Proteins That Inhibit Signalling through Tyrosine Kinase Receptors. *Nature* **1997**, *386*, 181–186. [CrossRef]
21. Barclay, A.N.; Brown, M.H. The SIRP Family of Receptors and Immune Regulation. *Nat. Rev. Immunol.* **2006**, *6*, 457–464. [CrossRef] [PubMed]
22. Logtenberg, M.E.W.; Jansen, J.H.M.; Raaben, M.; Toebes, M.; Franke, K.; Brandsma, A.M.; Matlung, H.L.; Fauster, A.; Gomez-Eerland, R.; Bakker, N.A.M.; et al. Glutaminyl Cyclase Is an Enzymatic Modifier of the CD47- SIRPα Axis and a Target for Cancer Immunotherapy. *Nat. Med.* **2019**, *25*, 612–619. [CrossRef] [PubMed]
23. Wu, Z.; Weng, L.; Zhang, T.; Tian, H.; Fang, L.; Teng, H.; Zhang, W.; Gao, J.; Hao, Y.; Li, Y.; et al. Identification of Glutaminyl Cyclase Isoenzyme IsoQC as a Regulator of SIRPα-CD47 Axis. *Cell Res.* **2019**, *29*, 502–505. [CrossRef] [PubMed]
24. Gillman, A.L.; Jang, H.; Lee, J.; Ramachandran, S.; Kagan, B.L.; Nussinov, R.; Teran Arce, F. Activity and Architecture of Pyroglutamate-Modified Amyloid-β (Aβ $_{pE3-42}$) Pores. *J. Phys. Chem. B* **2014**, *118*, 7335–7344. [CrossRef]
25. Kuriyama, T.; Takenaka, K.; Kohno, K.; Yamauchi, T.; Daitoku, S.; Yoshimoto, G.; Kikushige, Y.; Kishimoto, J.; Abe, Y.; Harada, N.; et al. Engulfment of Hematopoietic Stem Cells Caused by Down-Regulation of CD47 Is Critical in the Pathogenesis of Hemophagocytic Lymphohistiocytosis. *Blood* **2012**, *120*, 4058–4067. [CrossRef]
26. Johnson, L.D.S.; Banerjee, S.; Kruglov, O.; Viller, N.N.; Horwitz, S.M.; Lesokhin, A.; Zain, J.; Querfeld, C.; Chen, R.; Okada, C.; et al. Targeting CD47 in Sézary Syndrome with SIRPαFc. *Blood Adv.* **2019**, *3*, 1145–1153. [CrossRef] [PubMed]
27. Liu, F.; Jiang, C.C.; Yan, X.G.; Tseng, H.-Y.; Wang, C.Y.; Zhang, Y.Y.; Yari, H.; La, T.; Farrelly, M.; Guo, S.T.; et al. BRAF/MEK Inhibitors Promote CD47 Expression That Is Reversible by ERK Inhibition in Melanoma. *Oncotarget* **2017**, *8*, 69477–69492. [CrossRef]
28. Suzuki, S.; Yokobori, T.; Tanaka, N.; Sakai, M.; Sano, A.; Inose, T.; Sohda, M.; Nakajima, M.; Miyazaki, T.; Kato, H.; et al. CD47 Expression Regulated by the MiR-133a Tumor Suppressor Is a Novel Prognostic Marker in Esophageal Squamous Cell Carcinoma. *Oncol. Rep.* **2012**, *28*, 465–472. [CrossRef]

29. Casey, S.C.; Baylot, V.; Felsher, D.W. The MYC Oncogene Is a Global Regulator of the Immune Response. *Blood* **2018**, *131*, 2007–2015. [CrossRef]
30. Burger, P.; Hilarius-Stokman, P.; de Korte, D.; van den Berg, T.K.; van Bruggen, R. CD47 Functions as a Molecular Switch for Erythrocyte Phagocytosis. *Blood* **2012**, *119*, 5512–5521. [CrossRef]
31. Zhu, C.; Chen, W.; Lou, J.; Rittase, W.; Li, K. Mechanosensing through Immunoreceptors. *Nat. Immunol.* **2019**, *20*, 1269–1278. [CrossRef] [PubMed]
32. Orazizadeh, M.; Lee, H.; Groenendijk, B.; Sadler, S.J.; Wright, M.O.; Lindberg, F.P.; Salter, D.M. CD47 Associates with Alpha 5 Integrin and Regulates Responses of Human Articular Chondrocytes to Mechanical Stimulation in an in Vitro Model. *Arthritis Res. Ther.* **2008**, *10*, R4. [CrossRef] [PubMed]
33. Rehfeldt, F.; Engler, A.; Eckhardt, A.; Ahmed, F.; Discher, D. Cell Responses to the Mechanochemical Microenvironment—Implications for Regenerative Medicine and Drug Delivery. *Adv. Drug Deliv. Rev.* **2007**, *59*, 1329–1339. [CrossRef] [PubMed]
34. Wu, A.-L.; Wang, J.; Zheleznyak, A.; Brown, E.J. Ubiquitin-Related Proteins Regulate Interaction of Vimentin Intermediate Filaments with the Plasma Membrane. *Mol. Cell* **1999**, *4*, 619–625. [CrossRef] [PubMed]
35. Azcutia, V.; Routledge, M.; Williams, M.R.; Newton, G.; Frazier, W.A.; Manica, A.; Croce, K.J.; Parkos, C.A.; Schmider, A.B.; Turman, M.V.; et al. CD47 Plays a Critical Role in T-Cell Recruitment by Regulation of LFA-1 and VLA-4 Integrin Adhesive Functions. *Mol. Biol. Cell* **2013**, *24*, 3358–3368. [CrossRef]
36. Ticchioni, M.; Raimondi, V.; Lamy, L.; Wijdenes, J.; Lindberg, F.P.; Brown, E.J.; Bernard, A. Integrin-associated Protein (CD47/IAP) Contributes to T Cell Arrest on Inflammatory Vascular Endothelium under Flow. *FASEB J.* **2001**, *15*, 341–350. [CrossRef]
37. Park, E.; Song, K.-H.; Kim, D.; Lee, M.; Van Manh, N.; Kim, H.; Hong, K.B.; Lee, J.; Song, J.-Y.; Kang, S. 2-Amino-1,3,4-Thiadiazoles as Glutaminyl Cyclases Inhibitors Increase Phagocytosis through Modification of CD47-SIRPα Checkpoint. *ACS Med. Chem. Lett.* **2022**, *13*, 1459–1467. [CrossRef]
38. Marshall, B.T.; Long, M.; Piper, J.W.; Yago, T.; McEver, R.P.; Zhu, C. Direct Observation of Catch Bonds Involving Cell-Adhesion Molecules. *Nature* **2003**, *423*, 190–193. [CrossRef]
39. Zhang, Y.; Lin, Z.; Fang, Y.; Wu, J. Prediction of Catch-Slip Bond Transition of Kindlin2/β3 Integrin via Steered Molecular Dynamics Simulation. *J. Chem. Inf. Model.* **2020**, *60*, 5132–5141. [CrossRef]
40. Jiang, X.; Sun, X.; Lin, J.; Ling, Y.; Fang, Y.; Wu, J. MD Simulations on a Well-Built Docking Model Reveal Fine Mechanical Stability and Force-Dependent Dissociation of Mac-1/GPIbα Complex. *Front. Mol. Biosci.* **2021**, *8*, 638396. [CrossRef]
41. Pietsch, E.C.; Dong, J.; Cardoso, R.; Zhang, X.; Chin, D.; Hawkins, R.; Dinh, T.; Zhou, M.; Strake, B.; Feng, P.-H.; et al. Anti-Leukemic Activity and Tolerability of Anti-Human CD47 Monoclonal Antibodies. *Blood Cancer J.* **2017**, *7*, e536. [CrossRef] [PubMed]
42. Liu, G.; Fang, Y.; Wu, J. A Mechanism for Localized Dynamics-Driven Affinity Regulation of the Binding of von Willebrand Factor to Platelet Glycoprotein Ibα. *J. Biol. Chem.* **2013**, *288*, 26658–26667. [CrossRef] [PubMed]
43. Humphrey, W.; Dalke, A.; Schulten, K. VMD: Visual Molecular Dynamics. *J. Mol. Graph.* **1996**, *14*, 33–38. [CrossRef] [PubMed]
44. Pettersen, E.F.; Goddard, T.D.; Huang, C.C.; Meng, E.C.; Couch, G.S.; Croll, T.I.; Morris, J.H.; Ferrin, T.E. UCSF CHIMERAX: Structure Visualization for Researchers, Educators, and Developers. *Protein Sci.* **2021**, *30*, 70–82. [CrossRef] [PubMed]
45. Phillips, J.C.; Hardy, D.J.; Maia, J.D.C.; Stone, J.E.; Ribeiro, J.V.; Bernardi, R.C.; Buch, R.; Fiorin, G.; Hénin, J.; Jiang, W.; et al. Scalable Molecular Dynamics on CPU and GPU Architectures with NAMD. *J. Chem. Phys.* **2020**, *153*, 044130. [CrossRef]

Disclaimer/Publisher's Note: The statements, opinions and data contained in all publications are solely those of the individual author(s) and contributor(s) and not of MDPI and/or the editor(s). MDPI and/or the editor(s) disclaim responsibility for any injury to people or property resulting from any ideas, methods, instructions or products referred to in the content.

Article

Molecular Dynamics Study on the Aggregation Behavior of Triton X Micelles with Different PEO Chain Lengths in Aqueous Solution

Jin Peng, Xiaoju Song, Xin Li, Yongkang Jiang, Guokui Liu, Yaoyao Wei * and Qiying Xia *

School of Chemistry and Chemical Engineering, Linyi University, Linyi 276000, China; linqingya1007@163.com (J.P.); 2015110104@lyu.edu.cn (X.S.); lx754362430@126.com (X.L.); justy838@163.com (Y.J.); liuguokui@lyu.edu.cn (G.L.)
* Correspondence: weiyaoyao@lyu.edu.cn (Y.W.); xiaqiying@lyu.edu.cn (Q.X.)

Citation: Peng, J.; Song, X.; Li, X.; Jiang, Y.; Liu, G.; Wei, Y.; Xia, Q. Molecular Dynamics Study on the Aggregation Behavior of Triton X Micelles with Different PEO Chain Lengths in Aqueous Solution. *Molecules* **2023**, *28*, 3557. https://doi.org/10.3390/molecules28083557

Academic Editors: Enrico Bodo and Benedito José Costa Cabral

Received: 11 March 2023
Revised: 2 April 2023
Accepted: 16 April 2023
Published: 18 April 2023

Copyright: © 2023 by the authors. Licensee MDPI, Basel, Switzerland. This article is an open access article distributed under the terms and conditions of the Creative Commons Attribution (CC BY) license (https://creativecommons.org/licenses/by/4.0/).

Abstract: The aggregation structure of Triton X (TX) amphiphilic molecules in aqueous solution plays an important role in determining the various properties and applications of surfactant solutions. In this paper, the properties of micelles formed by TX-5, TX-114, and TX-100 molecules with different poly(ethylene oxide) (PEO) chain lengths in TX series of nonionic surfactants were studied via molecular dynamics (MD) simulation. The structural characteristics of three micelles were analyzed at the molecular level, including the shape and size of micelles, the solvent accessible surface area, the radial distribution function, the micelle configuration, and the hydration numbers. With the increase of PEO chain length, the micelle size and solvent accessible surface area also increase. The distribution probability of the polar head oxygen atoms on the surface of the TX-100 micelle is higher than that in the TX-5 or TX-114 micelle. In particular, the tail quaternary carbon atoms in the hydrophobic region are mainly located at the micelle exterior. For TX-5, TX-114, and TX-100 micelles, the interactions between micelles and water molecules are also quite different. These structures and comparisons at the molecular level contribute to the further understanding of the aggregation and applications of TX series surfactants.

Keywords: Triton X; micelle; hydration number; solvent accessible surface area

1. Introduction

Micelles aggregated from surfactant molecules have been widely used owing to their ability to dissolve hydrophobic compounds effectively. People's interest in micelle solutions comes from their application potential as functionally molecular assemblies [1]. Different cationic, anionic, zwitterionic, and non-ionic surfactants have been widely studied. Compared with ionic surfactants, non-ionic surfactants are much less toxic and have more efficient surface active properties [2]. Among them, the series of Triton X (TX) non-ionic surfactants have been studied in depth and characterized [3]. Because of their unique molecular structure and amphiphilic nature, the micelles formed by TX surfactants have very flexible surface-active properties. The critical micelle concentration (CMC) of TX surfactant is very low, especially in aqueous solution. Different TX surfactants have been widely used in the fields of microbiology and biomedicine [4].

The TX-100 as one typical nonionic surfactant [5–7], has been extensively studied [8–10]. This surfactant consists of one hydrophilic chain of 9–10 ethylene oxide units linked to a benzene ring with an octyl chain [6]. Many researchers have explored the structural characteristics of TX-100 micelles in aqueous solution and their interaction with water by experimental techniques such as 2D NOESY NMR [11–13], pulsed field gradient NMR [8], solvent paramagnetic relaxation enhancement [7], fluorescence spectra [14,15], surface tension [15], light scattering [16], static and dynamic light scattering [15,17], turbidimetric method [18], small-angle X-ray scattering [13], quasi-elastic light scattering spectroscopy [19]. It is very important to study the hydration and the size of TX-100 micelles [20], because the effectiveness of micellar applications depends on their size and their effects on the properties of the solution [21]. Robson and Dennis [16], Paradies [13], Streletzky and Phillies [19], Phillies et al. [22] used different experimental methods to determine the hydration degree of TX-100 micelles. Other relevant researchers studied the size of the TX-100 micelle by measuring its radius. The aggregation characteristics of TX-100 micelles are also the key aspects that affect their potential applications [7]. Yuan et al. [12] revealed the dependence of the conformation of TX-100 micelles on the environment, and emphasized that the spatial arrangement of PEO chains in micelles is to a certain extent very compact. In addition, the self aggregation and supramolecular micelle structure of TX-100 surfactant molecules in aqueous solution determined by Denkova et al. [11] shows that TX-100 may aggregate in a double-layer or multilayer spherical conformation interlaced between PEO chains and octyl phenyl parts. In order to further explore the structural characteristics, Zhang et al. [7] proposed a more detailed aggregation mode of TX-100 micelles, pointing out that this micelle is more likely to be a multilayer staggered spherical micelle and octyl phenyl is partially dispersed in different layers.

With the development of the computer and algorithm, molecular dynamics (MD) simulation has been widely applied in scientific studies from a micro-perspective. MD simulation is a method of calculating and simulating the time evolution of a group of interacting atoms using Newtonian equations of motion. By solving classical Newton's equations of motion, the motion trajectory of the system can be obtained. This trajectory includes a large number of samplings of molecular configurations, as well as information of the position and velocity of the particles. Further analysis of the obtained configuration ensemble can provide macroscopic observable properties (density, surface tension, solubility, viscosity, and thermodynamic information) and microscopic properties (intuitive structure, conformational distribution, degree of structural fluctuation, position and intensity of non-bonded interactions, particle spatial distribution) of the system. The description (force field) of the interactions between molecules and atoms is the key for this method to obtain reliable results. This calculation method shows a high degree of relevance in the detailed characterization of self-assembly related systems, including complementarity with experimental data, optimization of experimental design, and prediction of related properties of chemical systems. This may be expensive or difficult for experimentation [23]. Moreover, MD simulations have been effectively applied to investigate the self-assembly and interfacial adsorption of surfactant systems.

In addition to many experimental studies, some literature studies [24–29] further discussed the aggregation characteristics of TX-100 micelles at the molecular level by MD simulation. Among these literature studies, the simulation results of several of these literature studies [24,25,28,29] show that the shape and behavior of micelles mainly depend on the aggregation number of aggregates. An important application of TX-100 micelles is that the hydrophobic part can dissolve insoluble organic compounds [30]. A detailed understanding of the pure micelle system is a prerequisite for the study of organic matter and micelle composite systems [31]. In order to better understand this solubilization mechanism, researchers [26,27] explored the basic structure of micelles under a specific aggregation number. Farafonov et al. [26] first observed that the surface of the micelles was highly irregular when they were finally balanced in aqueous solution. Generally, when the temperature was 298 K, the aggregation number of the TX-100 micelle measured by experiments [8,19,22] was about 100. Therefore, Pacheco-Blas and Vicente [27] studied the morphology of TX-100 micelles with an aggregation number of 100 with MD simulation. It was observed that the micelles appeared as a quasi-sphere. The hydrophobic parts of the surfactants preferentially gather in the core of the micelles; however, there is some probability that these hydrophobic groups may be located at the outer region of the micelle.

Micelle structure in surfactant solutions is an important factor affecting the properties and application fields of self-assembly [11]. In this paper, we used all-atom MD simulations to study the self-aggregation properties of TX-5, TX-114, and TX-100 surfactant systems. By analyzing micelle shape and size, the solvent accessible surface area (SASA) of the micelle, the micelle morphology, as well as the interactions between micelles and water molecules, the differences and connections of these three micelles could be compared.

2. Results and Discussion

Figure 1 shows the time evolutions of radius of gyration (R_g), the solvent accessible surface area (SASA), and the energy of the system during a simulation run to determine the equilibriums of the systems. As can be seen from Figure 1a,b, the R_g and SASA of three micelles tend to be stable after 150 ns. The total energies of the three systems hardly change during the whole process. These changes of R_g, SASA, and total energies indicated that our simulated systems all reach equilibrium after 150 ns, and the last 30 ns trajectory of each system was used for result analyses in this paper.

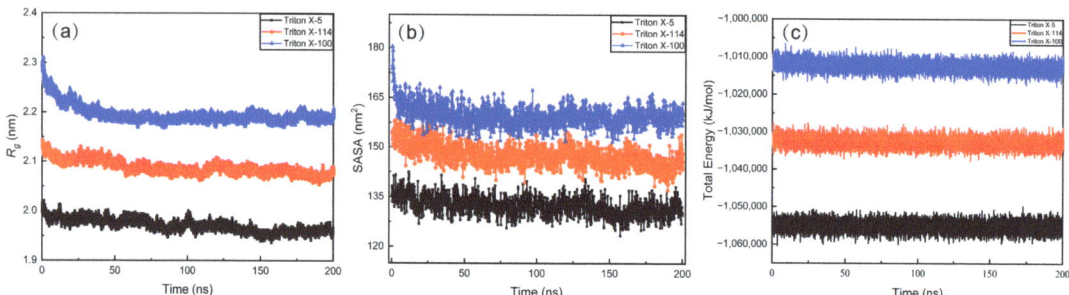

Figure 1. (a) The radius of gyration of TX-5, TX-114, and TX-100 micelles with time evolution, (b) the solvent accessible surface area of TX-5, TX-114, and TX-100 micelles with time evolution, and (c) the total energy of TX-5, TX-114, and TX-100 micelles with time evolution.

2.1. Micelle Shape and Size

2.1.1. Micelle Shape

The ratio of the moments of inertia I_{max}/I_{min} is usually used to characterize the shape of the micelle, where I_{max} and I_{min} are the largest and the smallest moment of inertia along the x, y, or z axis, respectively. When I_{max}/I_{min} is equal to 1, it is a perfect spherical micelle [32]. According to the results obtained in Table 1, the ellipsoidal degree of TX-100 micelle is significantly greater than that of TX-5 and TX-114 micelles. The TX-114 micelle is slightly more spherical than the TX-5 micelle. In addition, eccentricity (e) is also an important indicator to judge the shape of the micelle, which is defined as $e = 1 - I_{min}/I_{avg}$ [32]. The I_{avg} is the average moment of inertia of micelle. When e is equal to 0, the micelle shows a perfect spherical shape. When micelle shape is stable, the eccentricity will converge near a certain value and small fluctuations will appear [33]. In order to further verify the accuracy of the calculation results, we calculated the e values of the three micelles. The e of the TX-100 micelle is about 0.12, which is close to the value (0.08) of the TX-100 micelle obtained by Pacheco-Blas et al. [27]. The results show that the ellipsoidal degree of the TX-100 micelle is the largest, and the spheroidal degree of the TX-114 micelle is slightly larger than the TX-5 micelle.

2.1.2. Micelle Size

The size of the micelle is one of the important characteristics of micelle structure, in which R_g is a standard to characterize the size of the micelle [34]. It can be seen from Table 1 that R_g increases with the increase of the PEO chain length. In addition, the average micelle radius (R_s) is also used to judge the size of the micelle, and its definition [35] is as follows:

$$R_s = \sqrt{\frac{5}{3}} R_g$$

According to the results obtained in Table 1, the change trend of R_s is the same as that of R_g. The radius value of the TX-100 micelle ($\approx$2.83 nm) obtained in this paper is consistent with the radius value of the TX-100 micelle ($\approx$2.87 nm) simulated by Pacheco-Blas et al. [27] and the experimental result (3.1 nm) obtained by Brown et al. [8]. Our calculated results are therefore considered to be reliable.

Table 1. Structural characteristics of micelles Triton X-5, Triton X-114, Triton X-100.

	TX-5	TX-114	TX-100
I_{max}/I_{min}	1.15 ± 0.03	1.12 ± 0.04	1.26 ± 0.03
e	0.07 ± 0.02	0.06 ± 0.02	0.12 ± 0.02
R_g (nm)	1.96 ± 0.01	2.07 ± 0.01	2.19 ± 0.01
R_s (nm)	2.53 ± 0.01	2.67 ± 0.01	2.83 ± 0.01
SASA(total) (nm^2)	132.24 ± 2.77	147.52 ± 3.25	159.54 ± 2.97
SASA(hydrophilic) (nm^2)	30.62 ± 2.33	32.72 ± 2.27	35.81 ± 1.96
SASA(hydrophobic) (nm^2)	101.66 ± 2.19	114.77 ± 2.29	123.71 ± 2.54

2.1.3. Solvent Accessible Surface Area

Solvent accessible surface area (SASA) is the surface area of the TX micelle in contact with solvent, which can well reflect the properties of the micelle surface. To calculate the SASA of the micelle, the double cubic lattice method [36] was applied. With this method, all water molecules are first removed from the system [35], and then one spherical probe molecule with a radius of 1.4 Å is used to simulate the rolling of water molecules on the surface of the micelle [37]. Table 1 lists the total SASA and the hydrophilic and hydrophobic SASA values of the three micelles. We find that all SASA values increase with the increase of PEO chain length, which is consistent with the increasing trend of micelle size. The hydrophobic surface area of each micelle is much larger than the hydrophilic surface area, and the proportion of the hydrophobic area increases with the rise of the polar head chain. This is on account of the polar head of the research system being polymerized by PEO, and each PEO contains a hydrophobic C_2H_4 group and a hydrophilic O group. Therefore, the hydrophobic part accounts for a large proportion. In addition, some hydrophobic tails are distributed outside of the micelle. These hydrophobic groups will also increase the proportion of the hydrophobic surface area of the micelle.

2.2. Micelle Structure

The spatial distribution of atoms in a micelle can characterize the structural properties of the micelle [38]. The main difference between the non-ionic surfactant molecules of the TX family lies in the number of ethylene oxide units or the chain length [29]. In order to analyze the structures of the three micelles TX-5, TX-114, and TX-100 studied in this paper, we calculated the probability distributions of O atoms at different positions in each micelle system and the carbon atoms at the end of the hydrophobic tail chain relative to the micelle center of mass (COM). All the results are shown in Figure 2.

In general, the nonpolar alkyl chain of amphiphilic surfactant molecules usually aggregates in the hydrophobic region of the micelle, while the polar chain is exposed in the hydrophilic region. Sodium dodecyl sulfate (SDS) is a commonly used ionic surfactant and has been widely studied. Shelley et al. explored the structure of the SDS micelle in aqueous solution by MD simulation. It was found that most of the terminal C atoms of the SDS hydrophobic chain were distributed in the interior of the micelle, and only a small part were distributed on the surface of the micelle [39]. Gao et al. [40], Palazzesi et al. [38], MacKerell et al. [41], and Liu et al. [35] also found a similar phenomenon. It should be noted that these studies are ionic surfactants with small tail chains and polar heads. For TX micelle, there is a relatively large benzene ring near the end of the hydrophobic tail chain and long polar PEO chain head, which is very different from the SDS ionic surfactant. Pacheco-Blas et al. [27] used MD simulation to study the behavior of the non-ionic surfactant Triton X-100 in extracting metal ion Cd^{2+} in an aqueous environment. In the paper, it was found that the hydrophobic chain had a relatively high distribution in the outside of the micelle. In addition, Zhang et al. [7] considered that the TX-100 micelle was more likely to be a multilayer spherical micelle. Molecules from different layers were staggered, and octyl phenyl was dispersed in different layers. In this study, we found that the probability distribution of the tail C of the TX-100 micelle with respect to micelle COM is relatively wide, ranging from the micelle core to the outside of the micelle. In particular, the probability distribution of the terminal C in the outer region of the micelle is higher. The TX-5 and TX-114 micelles also show a similar phenomenon. This may be due to the large benzene rings in the hydrophobic region of the micelles, which are spatially exclusive and widely distributed. Although the tail C has a high probability in the outer region of the micelle, its distribution is also located on the left side of the distribution of the outermost polar O atoms. This means that most tail C atoms are included by the external polar O atoms, to avoid contact with environmental water as much as possible. The distributions of both hydrophobic terminal C and polar head O atoms are multimodal, indicating the existence of the micelle multilayer structure.

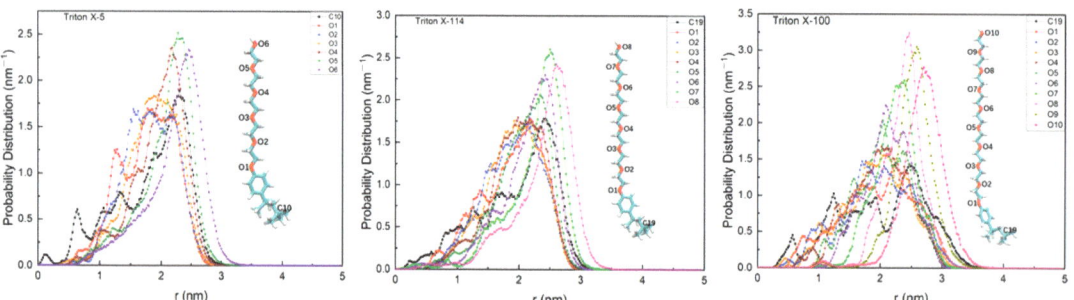

Figure 2. Probability distribution of O atoms at different position and the carbon at the end of the hydrophobic tail chain in TX-5, TX-114, and TX-100 micelle systems with respect to micelle COM.

In order to show the differences among the three micelle structures intuitively, we highlighted the C atom at the end of each micelle and the outermost O atom on the polar head to observe their internal structures more clearly. It can be seen from Figure 3 that the micelle shape is consistent with that obtained by the ratio of the moment of inertia and the eccentricity. In agreement with the multimodal probability distributions of terminal C atoms from the micelle center to the micelle surface, some terminal C atoms of the hydrophobic chains are distributed widely in the whole micelle.

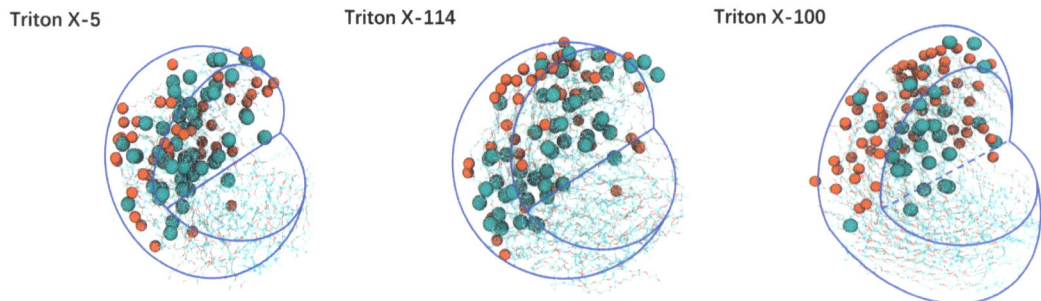

Figure 3. Representative structures of TX-5, TX-114 and TX-100 micelles. Note: the dark green ball in the figure symbolizes C atom, the red ball symbolizes O atom, and the blue auxiliary line symbolizes the outline.

2.3. Interactions between Micelle and Water

2.3.1. Hydration Numbers

The interactions between the micelle and water are characterized by calculating the hydration numbers of C and O atoms at different positions in TX-5, TX-114, and TX-100 surfactant molecule. For convenient analysis, the selected C and O atoms are grouped. Among them, a C atom with a similar chemical environment was selected and averaged for calculation, as shown in Figure 4. The hydration numbers were obtained by calculating the radial distribution function (RDF) integral of water molecules around the selected C atom or O atom in the range of 0.35 nm [40,42]. It can be seen from Figure 4 that the three systems have a similar distribution. From the end of the hydrophobic chain to the position of the polar head, the hydration numbers of the selected atoms show increasing trends. The hydration numbers of C atoms 1–4 on the hydrophobic chain are very low. The hydration numbers of C atoms in the benzene ring increase slightly. For C atoms and O atoms in the PEO chain, the hydration numbers present a zigzag trend. The hydration number of the O atom in each ethylene oxide (EO) unit is greater than that of the C atom. Moreover, the hydration numbers of the O atom closer to the position of the polar head is larger. The hydration numbers of the outermost O atoms of the polar head are significantly larger than those of other atoms (>3) with the decreased sequence of TX-100 > TX-114 ≈ TX-5. The hydrations numbers of H atoms on the OH groups also show this sequence. In addition, the hydration numbers of the C atoms at the same position in the three micelles are similar.

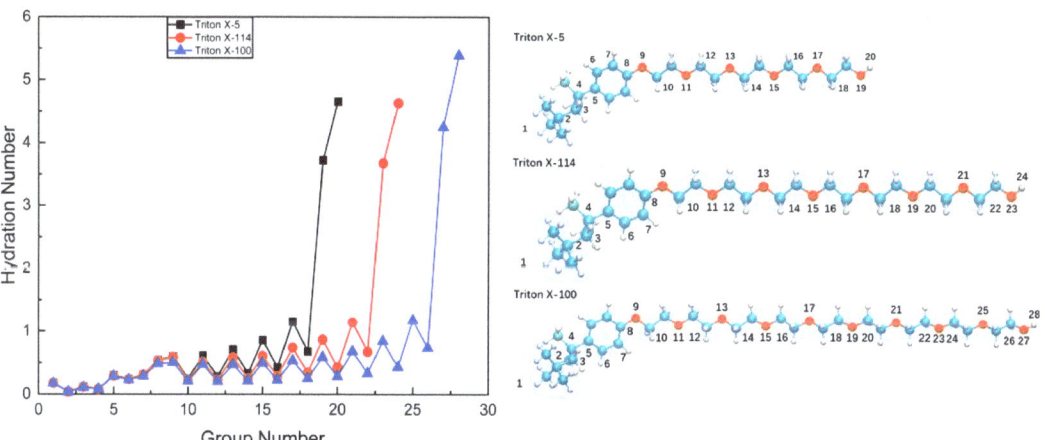

Figure 4. Hydration numbers of C and O atoms at different positions in TX-5, TX-114, and TX-100 surfactant molecules. Note: the Triton X molecular diagram on the right describes the atoms corresponding to each number. Dark green sphere represents C atoms, red sphere represents O atoms, and gray-white sphere represents H atoms.

2.3.2. Number of Hydrogen Bonds

In order to further characterize the interactions between the three micelles and water, we calculated the average numbers of hydrogen bonds between the O atoms at different positions and the water molecules. The geometric criterion for the existence of hydrogen bonds is that the distance between donor and acceptor pairs is within 3.5 Å and the angle between hydroxyl and hydrogen atoms is less than 120° [43]. As shown in Figure 5, the numbers of hydrogen bonds of the three systems show a significant upward trend from inner O atom to micelle surface O atom. Among them, the numbers of hydrogen bonds formed between the O atom close to the hydrophobic moiety and the water molecules are the lowest. In previous discussions, the hydration numbers of the outmost polar head O atoms of the three systems were the largest, ca. 4. However, the number of hydrogen bonds formed between these O atoms and water molecules is ca. 0.3, which means that not all the hydrated water molecules can form hydrogen bonds.

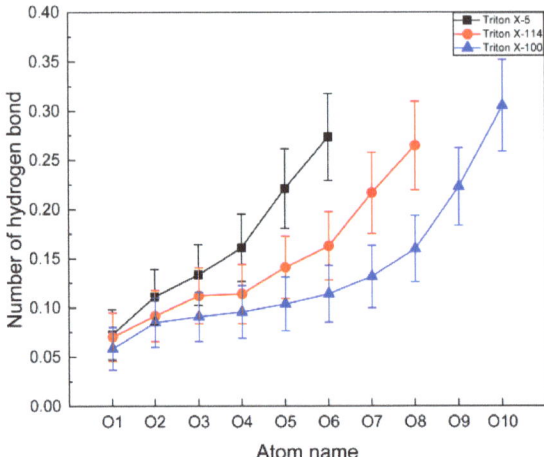

Figure 5. The average numbers of hydrogen bonds formed between O atoms at different positions and water molecules in a surfactant molecule in TX-5, TX-114, and TX-100 micelles. The O atom label is same as the schematic diagram of Figure 2.

2.3.3. Time Correlation Function of the Hydrogen Bond

One method to analyze the dynamic properties of the hydrogen bond is to calculate the intermittent hydrogen bond time correlation function ($C_{HB}(t)$), which is defined as follows:

$$C_{HB}(t) = \frac{\langle h(t)h(0) \rangle}{\langle h \rangle}$$

This function is considered to be the criterion for judging the formation or fracture of hydrogen bonds, and $C_{HB}(t)$ allows the hydrogen bonds to break and reform within the time interval t [44]. The faster $C_{HB}(t)$ decays, the more unstable is the hydrogen bond [45].

In this paper, the $C_{HB}(t)$ changes of the hydrogen bonds between O atoms at different positions and water molecules and between the outermost hydroxyl group and O atoms of water molecules in the three systems were calculated by this function. As shown in Figure 6, TX-5, TX-114, and TX-100 systems all show three fast decay curves for O atoms connected to hydrophobic chain, the outermost O and H atoms of the polar head. Among these three curves, the outermost H atoms decay fastest. The hydration numbers of the outermost H atoms are the largest and they have a spatial advantage, so that the water molecules that form hydrogen bonds with them exchange faster. This may result in the faster decay curve of the hydrogen bond. The outmost O atoms also behave the same. For internal O atoms, they may behave due to the environmental impact of being connected to the benzene ring of the hydrophobic chain and the spatial limits.

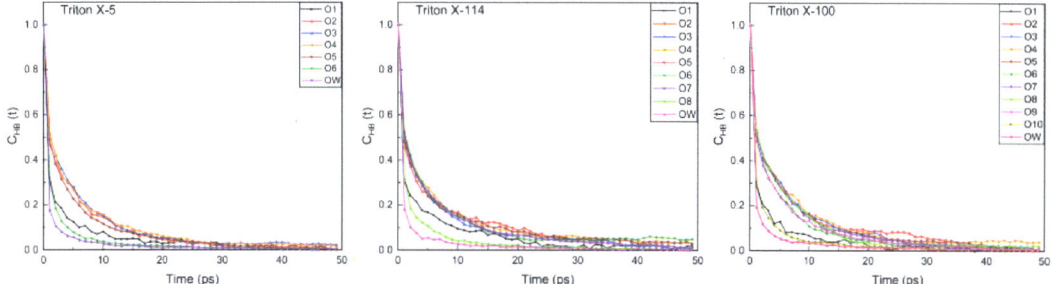

Figure 6. Time correlation function between O atoms at different positions and water molecules, and between the outermost hydroxyl group and O atoms in water molecules in TX-5, TX-114, and TX-100 micelle systems. The O atom label is the same as the schematic diagram of Figure 2. Note: OW represents the O atom in a water molecule.

3. Materials and Methods

In this paper, we chose the aggregation number of TX-5, TX-114, and TX-100 micelles as 100 according to the experimental values [46,47]. The molecular structures of TX-5, TX-114, and TX-100 were optimized by Gaussian 09 software [48] at the 6-31g(d) level. Then, the force field parameters of all studied molecules were obtained through the automated force field topology builder (ATB) [49,50]. With Packmol software [51], 100 pre-aggregated surfactant molecules for the three micelles were constructed. Then, the TX-5, TX-114, and TX-100 systems were treated with Gromacs 2019 software [52]. Pre-constructed micelle was placed in the center of a $10 \times 10 \times 10$ nm^3 cube box for each system by inbuilt tools. With the gmx solvate tool, water molecules were randomly added to fill the box. This inbuilt tool can ensure that water molecules are added in the box at locations larger than the volume of water molecules. Subsequently, the tension of the system can be released through pre-equilibrium processes. The simulated systems can achieve equilibrium and obtain a stable conformation. The single point charge (SPC) model was selected to describe water molecules. Specific compositions of all the studied systems are shown in Table 2. After minimizing the energy of the system, a reasonable initial structure was obtained for each system. With the V-rescale method [53], the 2 ns NVT equilibrium was performed to reach the required 298 K. The constant τ_T of 0.1 ps was used. Then, the NPT equilibrium with 2 ns was carried out to stabilize the system at 1 bar with a Berendsen pressure controller [54]. In this step, a constant τ_p of 2.0 ps was applied. After balancing the temperature and pressure of the system, the 200 ns NVT simulations were carried out to obtain the MD production trajectories. The same temperature controlling method as the NVT equilibrium step was used in the MD production step. In the process of simulation, periodic boundary conditions were used in all directions of the simulation system. The LINCS algorithm [55] was used to limit hydrogen-included bonds, and the particle-mesh Ewald (PME) [56] summation method was used for electrostatic interactions. The 1.4 nm cutoff value was used for van der Waals interactions and short-range electrostatic interactions. The integration time step was 2 fs. For every 10 ps interval, the MD sampling configurations were stored.

Table 2. Specific composition of micelle system Triton X-5, Triton X-114, Triton X-100.

Molecule	TX-5 (n = 5)	TX-114 (n = 7)	TX-100 (n = 9)
N_{agg}	100	100	100
H_2O	30,830	30,264	29,751

n represents oxyethylene groups.

4. Conclusions

Through MD simulations, we studied the structural characteristics of TX-5, TX-114, and TX-100 micelles at the same aggregation number and for the related properties of each micelle in aqueous solution. With the increase of the PEO chain length, the degree of micelle sphericity was reduced while the micelle radius increased. It is worth noting that the tail C atoms of each micelle were distributed widely with multimodal distributions extending from the micelle COM to the micelle surface. Large parts of the hydrophobic tail of the C atoms were located in the outer region of the micelle. Some polar O atoms close to the hydrophobic chains also showed multimodal distributions. These distributions indicate the existence of a multilayer structure of the TX micelles. The distributions of the tail chains and the polar heads of these three micelles are significantly different from those of traditional cationic and anionic surfactant micelles. For TX-5, TX-114, and TX-100 micelles, the outermost O and H atoms have the largest hydration numbers and hydrogen bond numbers, but the decay of the hydrogen bonds is the fastest. The outermost OH groups of TX-100 show stronger interactions with the surrounding water molecules than those of TX-5 and TX-114. More importantly, the representative structures of TX-5, TX-114, and TX-100 micelles can serve as the structural basis for understanding the properties of the relevant micelles. This study provides a reference for further expanding the exploration and application of TX-5, TX-114, and TX-100 micelles in related fields and it is of great significance for the exploration of other non-ionic surfactant micelle systems.

Author Contributions: Conceptualization, J.P., G.L., Y.W. and Q.X.; methodology, J.P. and X.S.; software, J.P. and X.L.; validation, G.L., Y.W. and Q.X.; formal analysis, J.P., X.S., X.L. and Y.J.; investigation, J.P., X.S., X.L., Y.J., G.L., Y.W. and Q.X.; data curation, G.L., Y.W. and Q.X.; writing—original draft preparation, J.P. and X.S.; writing—review and editing, G.L., Y.W. and Q.X.; supervision, G.L., Y.W. and Q.X.; funding acquisition, G.L. and Y.W. All authors have read and agreed to the published version of the manuscript.

Funding: This research was funded by the Natural Science Foundation of Shandong Province, grant number ZR2021QB153 and ZR2022QB043.

Institutional Review Board Statement: Not applicable.

Informed Consent Statement: Not applicable.

Data Availability Statement: The data presented will be made available on request by the corresponding authors.

Conflicts of Interest: The authors declare no conflict of interest.

Sample Availability: Not applicable.

References

1. Barzykin, A.V.; Tachiya, M. Interpretation of passive permeability measurements on lipid-bilayer vesicles. Effect of fluctuations. *BBA Biomembr.* **1997**, *1330*, 121–126. [CrossRef] [PubMed]
2. Azum, N.; Rub, M.A.; Azim, Y.; Asiri, A.M. Micellar and spectroscopic studies of amphiphilic drug with nonionic surfactant in the presence of ionic liquids. *J. Mol. Liq.* **2020**, *315*, 113732. [CrossRef]
3. Jaiswal, S.; Mondal, R.; Paul, D.; Mukherjee, S. Investigating the micellization of the triton-X surfactants: A non-invasive fluorometric and calorimetric approach. *Chem. Phys. Lett.* **2016**, *646*, 18–24. [CrossRef]
4. Abu-Ghunmi, L.; Badawi, M.; Fayyad, M. Fate of Triton X-100 Applications on Water and Soil Environments: A Review. *J. Surfactants. Deterg.* **2014**, *17*, 833–838. [CrossRef]
5. Molina-Bolívar, J.A.; Aguiar, J.; Ruiz, C.C. Growth and Hydration Of Triton X-100 Micelles In Monovalent Alkali Salts: A Light Scattering Study. *J. Phys. Chem. B* **2002**, *106*, 870–877. [CrossRef]
6. Parra, J.G.; Iza, P.; Dominguez, H.; Schott, E.; Zarate, X. Effect of Triton X-100 surfactant on the interfacial activity of ionic surfactants SDS, CTAB and SDBS at the air/water interface: A study using molecular dynamic simulations. *Colloid. Surface. A* **2020**, *603*, 125284. [CrossRef]
7. Zhang, L.; Chai, X.; Sun, P.; Yuan, B.; Jiang, B.; Zhang, X.; Liu, M. The Study of the Aggregated Pattern of TX100 Micelle by Using Solvent Paramagnetic Relaxation Enhancements. *Molecules* **2019**, *24*, 1649. [CrossRef]
8. Brown, W.; Rymden, R.; Van Stam, J.; Almgren, M.; Svensk, G. Static and dynamic properties of nonionic amphiphile micelles: Triton X-100 in aqueous solution. *J. Phys. Chem.* **1989**, *93*, 2512–2519. [CrossRef]

9. Li, M.; Rharbi, Y.; Huang, X.; Winnik, M.A. Small variations in the composition and properties of Triton X-100. *J. Colloid Interface Sci.* **2000**, *230*, 135–139. [CrossRef]
10. Thakkar, K.; Patel, V.; Ray, D.; Pal, H.; Aswal, V.K.; Bahadur, P. Interaction of imidazolium based ionic liquids with Triton X-100 micelles: Investigating the role of the counter ion and chain length. *Rsc. Adv.* **2016**, *6*, 36314–36326. [CrossRef]
11. Denkova, P.S.; Lokeren, L.V.; Verbruggen, I.; Willem, R. Self-Aggregation and Supramolecular Structure Investigations of Triton X-100 and SDP2S by NOESY and Diffusion Ordered NMR Spectroscopy. *J. Phys. Chem. B* **2008**, *112*, 10935–10941. [CrossRef] [PubMed]
12. Yuan, H.-Z.; Cheng, G.-Z.; Zhao, S.; Miao, X.-J.; Yu, J.-Y.; Shen, L.-F.; Du, Y.-R. Conformational Dependence of Triton X-100 on Environment Studied by 2D NOESY and 1H NMR Relaxation. *Langmuir* **2000**, *16*, 3030–3035. [CrossRef]
13. Paradies, H.H. Shape and size of a nonionic surfactant micelle. Triton X-100 in aqueous solution. *J. Phys. Chem.* **1980**, *84*, 599–607. [CrossRef]
14. Kumbhakar, M.; Goel, T.; Mukherjee, T.; Pal, H. Role of Micellar Size and Hydration on Solvation Dynamics: A Temperature Dependent Study in Triton-X-100 and Brij-35 Micelles. *J. Phys. Chem. B* **2004**, *108*, 19246–19254. [CrossRef]
15. Ruiz, C.C.; Molina-Bolívar, J.A.; Aguiar, J.; MacIsaac, G.; Moroze, S.; Palepu, R. Thermodynamic and Structural Studies of Triton X-100 Micelles in Ethylene Glycol-Water Mixed Solvents. *Langmuir* **2001**, *17*, 6831–6840. [CrossRef]
16. Robson, R.J.; Dennis, E.A. The size, shape, and hydration of nonionic surfactant micelles. Triton X-100. *J. Phys. Chem.* **1977**, *81*, 1075–1078. [CrossRef]
17. Kumbhakar, M.; Nath, S.; Mukherjee, T.; Pal, H. Solvation dynamics in triton-X-100 and triton-X-165 micelles: Effect of micellar size and hydration. *J. Chem. Phys.* **2004**, *121*, 6026–6033. [CrossRef]
18. Fedyaeva, O.A.; Poshelyuzhnaya, E.G. Dimensions and Orientation of Triton X-100 Micelles in Aqueous Solutions, According to Turbidimetric Data. *Russ. J. Phys. Chem. A* **2019**, *93*, 2559–2561. [CrossRef]
19. Streletzky, K.; Phillies, G.D. Temperature dependence of Triton X-100 micelle size and hydration. *Langmuir* **1995**, *11*, 42–47. [CrossRef]
20. Mandal, A.; Ray, S.; Biswas, A.; Moulik, S. Physicochemical studies on the characterization of Triton X 100 micelles in an aqueous environment and in the presence of additives. *J. Phys. Chem.* **1980**, *84*, 856–859. [CrossRef]
21. Phillies, G.D.; Yambert, J.E. Solvent and solute effects on hydration and aggregation numbers of Triton X-100 micelles. *Langmuir* **1996**, *12*, 3431–3436. [CrossRef]
22. Phillies, G.D.; Stott, J.; Ren, S. Probe diffusion in the presence of nonionic amphiphiles: Triton X 100. *J. Phys. Chem.* **1993**, *97*, 11563–11568. [CrossRef]
23. Leach, A.R. *Molecular Modelling: Principles and Applications*; Pearson Education: London, UK, 2001.
24. De Nicola, A.; Kawakatsu, T.; Rosano, C.; Celino, M.; Rocco, M.; Milano, G. Self-Assembly of Triton X-100 in Water Solutions: A Multiscale Simulation Study Linking Mesoscale to Atomistic Models. *J. Chem. Theory. Comput.* **2015**, *11*, 4959–4971. [CrossRef] [PubMed]
25. Murakami, W.; De Nicola, A.; Oya, Y.; Takimoto, J.-I.; Celino, M.; Kawakatsu, T.; Milano, G. Theoretical and Computational Study of the Sphere-to-Rod Transition of Triton X-100 Micellar Nanoscale Aggregates in Aqueous Solution: Implications for Membrane Protein Purification and Membrane Solubilization. *ACS Appl. Nano Mater.* **2021**, *4*, 4552–4561. [CrossRef]
26. Farafonov, V.; Lebed, A.; Khimenko, N.; Mchedlov-Petrossyan, N. Molecular Dynamics Study of an Acid-Base Indicator Dye in Triton X-100 Non-Ionic Micelles. *Vopr. Khimii I Khimicheskoi Tekhnologii* **2020**, *1*, 97–103. [CrossRef]
27. Pacheco-Blas, M.D.A.; Vicente, L. Molecular dynamics study of the behaviour of surfactant Triton X-100 in the extraction process of Cd^{2+}. *Chem. Phys. Lett.* **2020**, *739*, 136920. [CrossRef]
28. Ritter, E.; Yordanova, D.; Gerlach, T.; Smirnova, I.; Jakobtorweihen, S. Molecular dynamics simulations of various micelles to predict micelle water partition equilibria with COSMOmic: Influence of micelle size and structure. *Fluid Phase Equilib.* **2016**, *422*, 43–55. [CrossRef]
29. Yordanova, D.; Smirnova, I.; Jakobtorweihen, S. Molecular Modeling of Triton X Micelles: Force Field Parameters, Self-Assembly, and Partition Equilibria. *J. Chem. Theory Comput.* **2015**, *11*, 2329–2340. [CrossRef]
30. Rauf, M.A.; Hisaindee, S.; Graham, J.P.; Al-Zamly, A. Effect of various solvents on the absorption spectra of dithizone and DFT calculations. *J. Mol. Liq.* **2015**, *211*, 332–337. [CrossRef]
31. Faramarzi, S.; Bonnett, B.; Scaggs, C.A.; Hoffmaster, A.; Grodi, D.; Harvey, E.; Mertz, B. Molecular Dynamics Simulations as a Tool for Accurate Determination of Surfactant Micelle Properties. *Langmuir* **2017**, *33*, 9934–9943. [CrossRef]
32. Wei, Y.; Wang, H.; Liu, G.; Wang, Z.; Yuan, S. A molecular dynamics study on two promising green surfactant micelles of choline dodecyl sulfate and laurate. *RSC Adv.* **2016**, *6*, 84090–84097. [CrossRef]
33. Allen, D.T.; Saaka, Y.; Lawrence, M.J.; Lorenz, C.D. Atomistic description of the solubilisation of testosterone propionate in a sodium dodecyl sulfate micelle. *J. Phys. Chem. B* **2014**, *118*, 13192–13201. [CrossRef] [PubMed]
34. Wei, Y.; Liu, G.; Wang, Z.; Yuan, S. Molecular dynamics study on the aggregation behaviour of different positional isomers of sodium dodecyl benzenesulphonate. *RSC Adv.* **2016**, *6*, 49708–49716. [CrossRef]
35. Liu, G.; Zhang, H.; Liu, G.; Yuan, S.; Liu, C. Tetraalkylammonium interactions with dodecyl sulfate micelles: A molecular dynamics study. *Phys. Chem. Chem. Phys.* **2016**, *18*, 878–885. [CrossRef] [PubMed]

36. Eisenhaber, F.; Lijnzaad, P.; Argos, P.; Sander, C.; Scharf, M. The double cubic lattice method: Efficient approaches to numerical integration of surface area and volume and to dot surface contouring of molecular assemblies. *J. Comput. Chem.* **1995**, *16*, 273–284. [CrossRef]
37. Tang, X.; Koenig, P.H.; Larson, R.G. Molecular dynamics simulations of sodium dodecyl sulfate micelles in water-the effect of the force field. *J. Phys. Chem. B* **2014**, *118*, 3864–3880. [CrossRef]
38. Palazzesi, F.; Calvaresi, M.; Zerbetto, F. A molecular dynamics investigation of structure and dynamics of SDS and SDBS micelles. *Soft Matter.* **2011**, *7*, 9148–9156. [CrossRef]
39. Shelley, J.; Watanabe, K.; Klein, M.L. Simulation of a sodium dodecylsulfate micelle in aqueous solution. *Int. J. Quantum Chem.* **1990**, *38*, 103–117. [CrossRef]
40. Gao, J.; Ge, W.; Hu, G.; Li, J. From Homogeneous Dispersion to Micelles A Molecular Dynamics Simulation on the Compromise of the Hydrophilic and Hydrophobic Effects of Sodium Dodecyl Sulfate in Aqueous Solution. *Langmuir* **2005**, *21*, 5223–5229. [CrossRef]
41. MacKerell, A.D. Molecular Dynamics Simulation Analysis of a Sodium Dodecyl Sulfate Micelle in Aqueous Solution: Decreased Fluidity of the Micelle Hydrocarbon Interior. *J. Phys. Chem.* **1995**, *99*, 1846–1855. [CrossRef]
42. Chun, B.J.; Choi, J.I.; Jang, S.S. Molecular dynamics simulation study of sodium dodecyl sulfate micelle: Water penetration and sodium dodecyl sulfate dissociation. *Colloid. Surface. A* **2015**, *474*, 36–43. [CrossRef]
43. Liu, G.; Li, R.; Wei, Y.; Gao, F.; Wang, H.; Yuan, S.; Liu, C. Molecular dynamics simulations on tetraalkylammonium interactions with dodecyl sulfate micelles at the air/water interface. *J. Mol. Liq.* **2016**, *222*, 1085–1090. [CrossRef]
44. Pal, S.; Bagchi, B.; Balasubramanian, S. Hydration Layer of a Cationic Micelle, C_{10}TAB: Structure, Rigidity, Slow Reorientation, Hydrogen Bond Lifetime, and Solvation Dynamics. *J. Phys. Chem. B* **2005**, *109*, 12879–12890. [CrossRef]
45. Song, X.; Zhang, X.; Peng, J.; Li, Y.; Leng, X.; Liu, G.; Xia, Q.; Wei, Y. Molecular dynamics study on the aggregation behaviours of Platonic micelle in different NaCl solutions. *J. Mol. Liq.* **2022**, *353*, 118828. [CrossRef]
46. Pal, A.; Pillania, A. Modulations in surface and aggregation properties of non-ionic surfactant Triton X-45 on addition of ionic liquids in aqueous media. *J. Mol. Liq.* **2017**, *233*, 243–250. [CrossRef]
47. Qiao, L.; Easteal, A.J. Mass transport in Triton X series nonionic surfactant solutions: A new approach to solute-solvent interactions. *Colloid Polym. Sci.* **1996**, *274*, 974–980. [CrossRef]
48. Frisch, M.J.; Trucks, G.W.; Schlegel, H.B.; Scuseria, G.E.; Robb, M.A.; Cheeseman, J.R.; Scalmani, G.; Barone, V.; Mennucci, B.; Petersson, G.A.; et al. *Gaussian 09, Revision. B.01*; Gaussian, Inc.: Wallingford, CT, USA, 2010.
49. Koziara, K.B.; Stroet, M.; Malde, A.K.; Mark, A.E. Testing and validation of the Automated Topology Builder (ATB) version 2.0: Prediction of hydration free enthalpies. *J. Comput. Aided Mol. Des.* **2014**, *28*, 221–233. [CrossRef] [PubMed]
50. Malde, A.K.; Zuo, L.; Breeze, M.; Stroet, M.; Poger, D.; Nair, P.C.; Oostenbrink, C.; Mark, A.E. An Automated Force Field Topology Builder (ATB) and Repository: Version 1.0. *J. Chem. Theory Comput.* **2011**, *7*, 4026–4037. [CrossRef]
51. Martínez, L.; Andrade, R.; Birgin, E.G.; Martínez, J.M. PACKMOL: A package for building initial configurations for molecular dynamics simulations. *J. Comput. Chem.* **2009**, *30*, 2157–2164. [CrossRef]
52. Lindahl, E.; Abraham, M.; Hess, B.; van der Spoel, D. GROMACS 2019.3 Source Code (2019.3). Zenodo. 2019. Available online: https://doi.org/10.5281/zenodo.3243833 (accessed on 10 March 2023).
53. Bussi, G.; Donadio, D.; Parrinello, M. Canonical sampling through velocity rescaling. *J. Chem. Phys.* **2007**, *126*, 014101. [CrossRef]
54. Berendsen, H.J.C.; Postma, J.P.M.V.; Van Gunsteren, W.F.; DiNola, A.; Haak, J.R. Molecular dynamics with coupling to an external bath. *J. Chem. Phys.* **1984**, *81*, 3684–3690. [CrossRef]
55. Hess, B.; Bekker, H.; Berendsen, H.J.C.; Fraaije, J.G.E.M. LINCS: A linear constraint solver for molecular simulations. *J. Comput. Chem.* **1997**, *18*, 1463–1472. [CrossRef]
56. Essmann, U.; Perera, L.; Berkowitz, M.L.; Darden, T.; Lee, H.; Pedersen, L.G. A smooth particle mesh Ewald method. *J. Chem. Phys.* **1995**, *103*, 8577–8593. [CrossRef]

Disclaimer/Publisher's Note: The statements, opinions and data contained in all publications are solely those of the individual author(s) and contributor(s) and not of MDPI and/or the editor(s). MDPI and/or the editor(s) disclaim responsibility for any injury to people or property resulting from any ideas, methods, instructions or products referred to in the content.

MDPI AG
Grosspeteranlage 5
4052 Basel
Switzerland
Tel.: +41 61 683 77 34

Molecules Editorial Office
E-mail: molecules@mdpi.com
www.mdpi.com/journal/molecules

Disclaimer/Publisher's Note: The statements, opinions and data contained in all publications are solely those of the individual author(s) and contributor(s) and not of MDPI and/or the editor(s). MDPI and/or the editor(s) disclaim responsibility for any injury to people or property resulting from any ideas, methods, instructions or products referred to in the content.

www.ingramcontent.com/pod-product-compliance
Lightning Source LLC
LaVergne TN
LVHW072346090526
838202LV00019B/2491